Building Energy
Management Systems

Other titles from E & FN Spon

Air Conditioning
A practical introduction
2nd Edition
D. V. Chadderton

Building Services Engineering
2nd Edition
D. V. Chadderton

Combustion Engineering and Gas Utilization
3rd Edition
British Gas

Heating and Water Services Design in Buildings
K. Moss

HVAC Control Systems
C. Underwood

Illustrated Encyclopedia of Building Services
D. Kut

Naturally Ventilated Buildings
Building for the senses, the economy and society
T. D. J. Clements-Croome

Site Management of Building Services Contractors
J. Wild

Spon's Mechanical and Electrical Services Price Book
Davis Langdon and Everest

Ventilation of Buildings
H. B. Awbi

For more information about these and other titles published by E & FN
Spon, please contact:

The Marketing Department, E & FN Spon, 11 New Fetter Lane, London,
EC4P 4EE. Tel: 0171 583 9855; Fax: 0171 842 2303; or visit our web site
at **www.efnspon.com**

Building Energy Management Systems

Applications to low-energy HVAC and
natural ventilation control

2nd Edition

G. J. LEVERMORE

Department of Building Engineering, UMIST, UK

E & FN SPON

ALERE FLAMMAM

Taylor & Francis Group

First published 1992
by E & FN Spon

Second edition published 2000
by E & FN Spon
11 New Fetter Lane, London EC4P 4EE

Simultaneously published in the USA and Canada
by E & FN Spon
29 West 35th Street, New York, NY 10001

E & FN Spon is an imprint of the Taylor & Francis Group

© 1992, 2000 G. J. Levermore

The right of G. J. Levermore to be identified as the Author of this Work has been asserted by him in accordance with the Copyright, Designs and Patents Act 1988.

Typeset in Times by Graphicraft Limited, Hong Kong
Printed and bound in Great Britain by St Edmundsbury Press,
Bury St Edmunds, Suffolk

The publisher makes no representation, express or implied, with regard to the accuracy of the information contained in this book and cannot accept any legal responsibility or liability for any errors or omissions that may be made.

British Library Cataloguing in Publication Data
A catalogue record for this book is available from the British Library

Library of Congress Cataloging in Publication Data
Levermore, G. J.
    Building energy management systems : applications to low energy
HVAC and natural ventilation control / G. J. Levermore. — 2nd ed.
        p.   cm.
    1. Buildings—Energy conservation.   I. Title.
TJ163.5.B84L46   2000
696—dc21   99–39663
                                                               CIP

ISBN 0–419–26140–0 (hbk)
ISBN 0–419–22590–0 (pbk)

Contents

Foreword

This long-awaited second edition of *Building Energy Management Systems* is an excellent in-depth review of the fundamentals associated with environmental control in modern buildings. Levermore's detailed analysis of control and energy balance equations provides an informed basis for all engineering students, supplemented with practical examples to aide explanation and understanding. The book provides an excellent overview of BEMS, personal computers and transducer hardware along with how these are configured and communicate. Constant and variable flow air and water systems are analysed, and this second edition contains a new description of several low-energy HVAC developments like displacement ventilation and chilled ceilings. A number of analysis tools and numerical examples are given which provide for better understanding of these systems and how they operate and respond to varying conditions.

As well as to students who are studying HVAC, this book will be useful to practising engineers who wish to improve their understanding of additional design and analysis techniques, and new HVAC developments and control. It will be a valuable addition to the bookshelf of those interested in the design of state-of-the-art systems and buildings.

Don Colliver
University of Kentucky

Preface

Much of what stands in the preface to the first edition still stands, but advances have been made, especially in PCs and communications, which are reflected in substantial revisions to three chapters and the addition of four completely new chapters. The new CIBSE Guide A was published as this book was being typeset so only brief reference is made to the new notation and modified thermal models in it.

Chapter 13 on network analysis was added as more air and water systems are using variable-speed drives to regulate fans and pumps and the systems contain VAV dampers and two-port valves which can be difficult to design and control. A number of analysis tools and examples are given to aid greater understanding of these systems.

Natural ventilation is increasingly popular, and Chapter 11 deals with its control and the use of night ventilation and building fabric to reduce daytime peak temperatures. There is also a detailed description of wind- and stack-driven ventilation and the relevant equations. Case studies are also discussed in detail.

Displacement ventilation, chilled ceilings, chilled beams and mixed-mode air conditioning are discussed in Chapter 12 as well as the control of lighting. As shading and blinds become more essential, the importance of artificial lighting control is stressed. Conventional air-conditioning systems using fan coils and VAV boxes are still very much in use, and Chapter 10 gives an in-depth analysis from first principles.

Sick building syndrome, however nebulus and difficult to define, haunted building services in the late 1980s and early 1990s. Chapter 1 has been revised to include it in Section 1.6 on occupant feedback. Occupant feedback is a vital part of the commissioning and management process for new and

existing buildings. As a profession we must heed the complaints and some-
· times compliments of the occupants regarding the building and its facilities.
A questionnaire is discussed that allows a fingerprint and score to be derived
for the occupants' satisfaction with the building and its services. Hopefully
this will allow a more user-friendly analysis of the questionnaire.

Much of the work is based on my research and lectures since I have been
at UMIST. My thanks are therefore to my colleagues, researchers and students who have helped me directly or indirectly to revise and rewrite this
book. Especial thanks are due to Ivan Khoo, Steve Fotios, Li Mei, Virginia
Cooper and Andy Wright; Don McIntyre for his useful comments on the
four new chapters; Jonathan Dewsbury for his comments on the natural
ventilation chapter; Tony Sung and Ken Letherman; Eric Keeble for comments related to the weather data; and Peter Tunnel, Peter Day and Malcolm
Clapp for practical comments. However, any errors are mine alone.

Finally I would like to thank Carolyn for her help and for bearing with my
long periods at the PC. I thank Tom for helping with the figures and Alison
for being patient in waiting to use the PC. I also thank my mother.

Geoff Levermore

Preface to the first edition

This book is intended as both a student text on the control of a building services plant and a practitioner's guide to the basics of practical control. With the advent of the microprocessor the building energy management system takes a prominent position in the book, hence its title.

As an energy manager I found that there was little understanding of controls, even though they were crucial to energy efficiency and comfort in buildings. Then, as a lecturer, I discovered the difficulty of teaching control, with most of the textbooks relating to electronic or process control where the reaction times are much faster than those of a building. I also found in teaching short courses and in some consultancy work that the potential of BEMSs was not being realized, partly due to a lack of training and understanding but also due to insufficient time and money allowed on the design, installation and commissioning of plant and controls. Indeed I often ask consultants, clients and students to tell me of any building that they know which works properly after commissioning, but the response is rare and little. This book, I hope, will help alleviate these problems.

As control is a large subject it was decided to devote this book to the application of control to heating and energy efficiency to appeal to energy managers and consultants as well as to students. At the heart of the control of heating and energy efficiency are the building and plant responses. These contribute to the difficulties of using standard control texts and theory as the building fabric can take hours, even days, to heat whereas the plant can respond in minutes, perhaps seconds. So Chapter 7, on building heat loss and heating, has been included. In Chapter 9 on optimizer control this response is treated in a simplified manner as a first order differential equation in order to give an appreciation of the physical processes occurring. To treat the response in more detail with numerical solutions would have overly complicated the treatment.

The book starts by detailing the development of BEMSs, the advantages and disadvantages from case studies and the future prospects. Some of these case studies are ETSU Demonstration Schemes, which are now a little dated but they illustrate cardinal points which later studies do not cover in such depth. Chapter 2 is devoted to the outstation, its structure, operation and configuration. Of necessity there is some electronics here, such as the operational amplifier, but no previous knowledge of electronics is assumed. Chapter 3 covers the central station, its elements and software, which is the heart of a BEMS.

There is a tendency amongst some traditional engineers that unless they can measure the temperature on a gauge or with a thermometer then somehow any other reading is suspect. Perhaps this has evolved from the extensive use of strap on sensors used in retrofit BEMs installations. Chapter 4 addresses the problem of strap-on sensors and also emphasizes that all measurements contain inaccuracies, but that these can be limited.

Chapters 5 and 6 develop basic control into three-term PID control. Although there is more theory here than in earlier chapters, the explanation of the use of BEMSs for analysing temperature curves will be of particular use to practitioners who have to examine their plant and buildings' operation. The control strategies adopted in the book are based as far as possible on those in commercial BEMSs, although there are a number of alterations to simplify the explanation.

Compensator control of heating systems is covered in Chapter 8, with sufficient explanation for the reader to appreciate the implications of the control settings and the size of the heating plant on the internal temperature attained. Chapter 9 deals with the other important heating control, the optimizer.

A major use of BEMS is for monitoring plant and buildings' energy consumption and the subsequent targeting of the consumptions. This is dealt with in Chapter 10. If this chapter alone is read and acted upon then the book will have produced a reasonable payback.

In producing this book I was helped by a number of people, including many of my students who provided their feedback on my lectures mainly by their exam marks. In the actual production of the book I would particularly like to thank the late Mark Coppens, Mrs Pat Cross and Shashi Patel for translating my rough sketches into good graphics. Also I would like to record my appreciation of conversations and discussions with Chris Chapman and Bill Freshwater of Trend Control Systems Ltd and Malcolm Clapp and Sultan Siddiqui of Satchwell Control Systems Ltd. Colleagues, Professor Ken Letherman and Tony Sung kindly commented on a number of the chapters.

I would finally like to express my thanks to my wife Carolyn for her help and useful advice in the writing of this book, and also to my parents, and Alison and Tom for helping me in many various ways to write this book.

There will undoubtedly be some mistakes that have crept into the book, for which I take responsibility. Feedback on these and any other points in the book will be gratefully appreciated.

Geoff Levermore

1
An introduction to BEMSs

Building energy management systems (BEMSs) are now an established part of modern buildings, although their potential has yet to be fully realized in practice. However, as computer and communication technology advance inexorably, so the contribution of BEMSs will increase. With the power of microprocessors doubling every 18 months (sometimes known as Moore's law [1]), the potential is great. Although BEMSs advanced with the developments in microprocessors, the current rapid advances of communications and networks will shape the increased use of BEMSs.

The exact nature of BEMS, BMS, EMS and BAS is also becoming less easy to define as even small components, such as switches and radiator valves, can have small chips attached to them so that they can be linked into a communication and control bus. Figure 1.1 shows the different levels of a

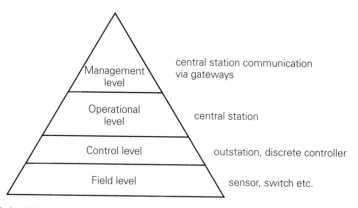

Fig. 1.1 The levels of control in a BEMS.

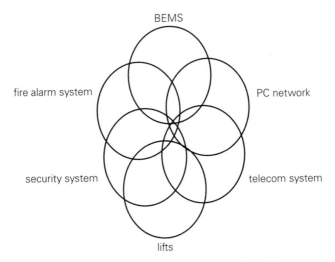

Fig. 1.2   The potential overlap of microprocessor-based systems.

BEMS from the field device, which is becoming more intelligent, to the head-end management PC.

1.1   Intelligent buildings

Intelligence is a much misused word but here it is used to imply that a microprocessor is incorporated in the intelligent device. However, it is also applied to buildings, where again there is no simple definition. One definition is a building that provides a productive and cost-effective environment through optimization of its four basic elements: structure, services and management and the interrelationships between them [2, 3]. Figure 1.2 gives an indication of the interaction of the microprocessor-controlled systems that can exist in an intelligent building, such as the PC network with its server, or the security system with its PC head end. The more these system circles overlap, i.e. the more common the sharing of bus systems and intercommunications, the more intelligent the building becomes.

1.2   The development of BEMSs

Building energy management systems have developed alongside, and been a result of the microelectronics and computing revolution of recent years. This is because BEMSs are simply microcomputer systems used for controlling and monitoring building services plant. BEMSs have also benefited from the knowledge and technology in the application of computer control to manufacturing and the process industry.

The earliest ancestor of the BEMS was the hard-wired, central system. It first appeared in the 1960s and was employed in large buildings [4]. The system was basically an extension of the conventional control wires to a central console, with dials, indicator lights and a chart recorder, which enabled an operator at the console to monitor distant plant and see temperatures displayed. No computers or microelectronics were involved, and it relied on the operator to change control settings and times [5].

These hard-wired systems were then improved with the telephone technology of the day to enable individual items of plant to be switched, via **data-gathering panels** local to the plant, into a central multicore trunk cable running around the building from the central console. This switching, or **multiplexing**, saved on cabling by using the same trunk cable for a number of data-gathering panels.

With the rapid advances of microelectronics, and hundreds of transistor devices being integrated onto one **large-scale integrated** (LSI) chip of silicon, area about 5 mm^2, the first computer-based monitoring and control systems emerged. These early BEMSs were **centralized energy management systems** and they first appeared in the 1970s [6], having been developed in the United States. The **central station** was based on a minicomputer, which contained all the computing power or 'intelligence' in the system, with 'dumb' or unintelligent **outstations**, which were boxes or cabinets for relays and connections to sensors and actuators, similar to the earlier data-gathering panels. The term 'intelligent' is used because the central station (the minicomputer) had the ability to calculate and make decisions using the data it received from the outstations.

The systems were very expensive, so they were only viable for large sites such as prestige air-conditioned headquarters [7]. Although they related initially to the control and monitoring of HVAC plant and were therefore energy management systems, they were also capable of controlling the lighting and the lifts and monitoring the security and fire alarms. (In the United Kingdom, regulatory decisions meant that security and fire alarms were rarely linked to BEMSs.) In fact, the systems were considered as building management systems to help in the management of large and complex sites, without specifically saving energy. These early building energy management sytems were actually in existance before the energy crisis of 1973/74.

Although these early BEMSs were capable of monitoring and controlling fire and security systems, they rarely did, as dedicated systems were used for the potentially life-saving fire detection devices and also the security devices. There are still problems in integrating all systems such as fire alarm systems and security systems into BEMSs today, mostly due to the different and disparate disciplines and standards, rather than the technology, and also the different manufacturing companies involved.

Since about 1980 the rapid development of LSI and VLSI (very large scale integration) led to thousands of devices being put on a chip (for the

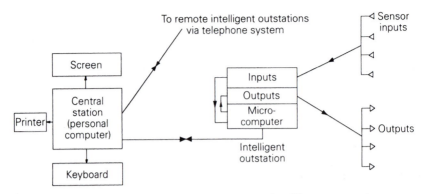

Fig. 1.3 A microprocessor-based BEMS with intelligent outstations.

Pentium chip it is now millions of transistors). Hence microcomputers, or **personal computers** (PCs), became as powerful, perhaps more powerful than the previous minicomputer. And the outstations, small microcomputers themselves, or more correctly, they contain microprocessor chips (see later), have gained considerably in processing power, giving them 'intelligence' (Fig. 1.3). This enables them to operate on their own, or to become stand-alone outstations, dependent only on the central station for a small proportion of their operating time. These outstations have considerably more control functions than the older, dumb outstations, which tended to have more of a monitoring role. Indeed, each intelligent outstation can control a small building on its own, and it is economic to install these intelligent outstations in small and medium-sized buildings.

The cost of a BEMS system can best be judged using cost per *point* (a point is an input or an output, e.g. a temperature input to the outstation or a control signal output to a valve) or cost per floor area. However, the buyer must beware of the implications of marginal costing [8] and a loss-leader pricing policy to gain entry to an organization in the hope that the system will later be expanded, when the vendor will hope to recoup its profit [9].

Table 1.1 shows the cost of a BEMS at 1998 prices for the basic shell and core elements of a building (i.e. the central plant and distribution duct and pipework without the final units, such as fan coil units or variable air volume (VAV) boxes, that in a speculative office building will be put in later during the fit-out). Note that the cost of the BEMS central station, or head-end supervisor, the outstations and the software engineering is only 13% of the total cost. Table 1.2 considers the additional cost for a category A fit-out with 'intelligent' outstations at each fan coil unit subsequently fitted [10]. Alternative costs are included for simple fan coil unit control if the outstations are to be omitted.

Airside control of the fan coil, using dampers (Chapter 10), will save £70 per unit compared with waterside control using valves. Simply using a room

Table 1.1   Shell and core for a speculative office [10]

	Cost per item (£)	Cost (£)	GIA (£ m⁻²)	NLA (£ m⁻²)	Per cent
Head-end supervisor	2 800	2 800	0.28	0.37	2
DDC outstations (1 × 32 point and 2 × 72 point outstation)	15 000	15 000	1.5	2	8
Motor control panels (average of 80 points per panel)	12 500	25 000	2.5	3.33	14
90 Loose controls (sensors and valves)		14 400	1.44	1.92	8
Power wiring (SWA cable, clipped to dedicated tray)		57 600	5.76	7.68	32
Control wiring (dedicated containment)	160	32 000	3.2	4.27	18
Allowance for network wiring	2 700	2 700	0.27	0.36	1
Allowance for commissioning		4 500	0.45	0.6	2
Allowance for software engineering		5 000	0.5	0.67	3
Allowance for design, organization costs and site management		21 000	2.1	2.8	12
Total		180 000	18	24	100
Cost per point		**1022.7**	**0.1**	**0.14**	

GIA = gross internal floor area
NLA = net lettable floor area

Table 1.2   Costs of BEMS for fitted-out building (with fan coil units)

Category A fit-out	Cost per item (£)	Cost (£)	GIA (£ m⁻²)	NLA (£ m⁻²)	Percentage of total
220 Fan coil unit intelligent controllers	115	25 300	2.53	3.37	32
Fan coil controls (440 four-port valves, 220 flying-lead return air temperature sensors)		25 000	2.5	3.33	31
Allowance for control wiring (screened beldan type, dedicated containment)		3 000	0.3	0.4	4
Network wiring (twin-twisted screened beldan type, dedicated containment)		15 400	1.54	2.05	19
Allowance for commissioning		4 500	0.45	0.6	5
Allowance for software engineering		800	0.08	0.11	1
Allowance for design, organization costs and site management		6 000	0.6	0.8	8
Total		80 000	8	10.66	100

GIA = gross internal floor area
NLA = net lettable floor area

temperature sensor and a standard non-communicating ('unintelligent') controller would cost £90 per unit. With speed control of the fan as well, the price would go up to £130 per unit. Table 1.3 shows the indicative costs of BEMS for other HVAC systems and building types [10].

Table 1.3 Indicative costs of BEMS for HVAC systems and building types

System or building	Cost per square metre of gross internal floor area (£)
Four-pipe fan coil, office, cat. A fit-out	20–30
Variable air volume, office, cat. A fit-out	15–25
Chilled ceiling, office, cat. A fit-out	20–25
Variable refrigerant volume, office, cat. A fit-out	12–18
Supermarket	3–5
Department store	15–20
Leisure centre	10–15
Swimming-pool complex	25–35
Four- or five-star hotel	18–25

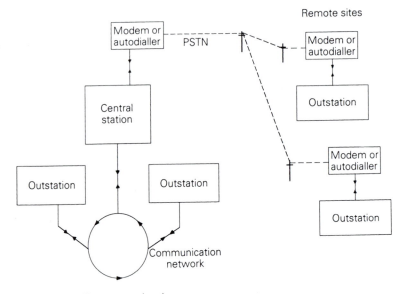

Fig. 1.4 BEMS communications.

The central station can communicate with many outstations (Fig. 1.4), when it needs to, either on a local communications network or to remote outstations over telephone lines, the public switched telephone network (PSTN), using modems and autodiallers (Chapter 3). BEMS communications can also use ISDN lines (integrated services digital network; higher-quality telephone lines capable of transmitting data and video signals as well as the voice), LANs (local area networks) and the Internet.

Some conventional controls, which are **serially interfaceable** (Chapter 3), can be connected to computers to show setpoints and allow them to be adjusted. However, these controls do not have the full capabilities of most BEMSs.

As microprocessors have become more powerful and less expensive, out-stations too are becoming very much smaller and cheaper, so they can now control individual items of plant. The plant is becoming intelligent. BEMSs manufacturers are supplying these small outstations to original equipment manufacturers (OEMs), so that their plant can be commissioned at the factory and simply connected into the BEMS communication system when delivered.

Communications networks and bus systems have been developed to allow small building services equipment to communicate, even down to light switches and electric socket outlets. This is dealt with in more detail in Chapter 3.

1.3   Some benefits of BEMSs

There are cost benefits from using a BEMS, and the International Energy Agency gives rough estimates of the savings from the different functions it performs [11]. Most savings are about 5% with paybacks of 2–3 years. It is difficult to assess the savings exactly; indeed, most of the savings may be because the BEMS allows poor plant performance to be identified [12]. However, there are some general benefits from BEMSs.

1.3.1   Monitoring

One of the main benefits of BEMSs has been the constant monitoring of the plant, and the ability to recall the monitored data at a later time. This has enabled engineers and technicians to achieve a better understanding of their buildings and plant and has often led to plant improvements and energy saving as a result. Energy efficiency can be checked as a BEMS can monitor and log data from fuel and electricity meters.

1.3.2   Communication

An added bonus of the BEMS is that information can be communicated over the telephone system from remote sites to the central station. This saves the considerable time and effort involved in travelling to and from sites to check the plant, unless there is a local plant operator.

A swimming-pool complex provides a good example of monitoring and communication. Without a BEMS, the complex had to run its ventilation fans 24 hours a day to stop condensation (there was no pool cover). It was decided to install two-speed fan motors to run the fans at half-speed at night to save energy when the evaporation of water would be reduced. The heating was also reduced at night, again to save energy. However, when this work was done, there appeared to be no saving of fuel at night. A technician visited the site to check the heating control, which seemed to be working satisfactorily. To further check it, he placed a data logger and thermocouple in the supply air duct. A few days later he returned to see that the data showed no evidence that the heating controller was reducing the heating at night.

He made two more site visits to meet technicians from the controls company, who asserted that the controller was indeed working satisfactorily when they tested it. Eventually a senior technician from the company came, after a copy of the temperature graph had been sent to him. Again he was sure the controller was satisfactory, but a check of the heater battery valve, in a rather inaccessible place, showed that the valve stem had broken. Although the controller was working, it was not moving the control valve, so there was no control of the heating.

If the swimming pool had had a BEMS, the local authority technician could have checked the heating system from his central station in the civic centre, and having realized it was not working, he may even have been able to identify the fault. If the controls company had had a BEMS, it could have dialled up the pool and checked it. Many staff hours of visits and checking would have been saved.

1.3.3   Staff savings and maintenance

There are also staff savings to be made with BEMSs. The caretaker or plant operator can often be replaced by a communicating BEMS outstation, or one operator can cover more buildings. Primarily because of this remote monitoring facility, most energy management, maintenance and facilities management companies use BEMSs.

When outstations are communicating regularly with the central station at the organization's headquarters, alarms are automatically sent from the outstations if problems arise (e.g. a low temperature or a switched off boiler), and the situation is restored to normality often before the client is aware of any plant malfunction. BEMSs can also allow the plant condition to be monitored (condition monitoring), which provides better maintenance.

1.3.4   Commissioning

Once the BEMS itself has been commissioned, it is very useful for fine-tuning the controls as the building and plant are monitored in use. This is especially true in low-energy, mixed-mode buildings (Chapter 12), where thermal storage in the building fabric and night ventilation are complex processes to control.

It is also possible to commission the mechanical and electrical services from the BEMS if there are communicating flow measuring devices and flow regulating or balancing valves and dampers. This would be more possible with VAV and fan coil systems (Chapter 10), which often have their own intelligent controls or outstations. It is interesting that the benefits mentioned here are primarily management related as opposed to better control *per se*.

1.3.5 Potential problems with BEMSs

There are many benefits from a BEMS when properly applied and used, but it would be naive to assume that people do not experience problems. In a study of 33 sites with BEMSs in four countries, more than 50% of the BEMS users noticed problems in the design, commissioning or operation phase [12], although on average 70% of users were satisfied with their systems.

One can put this into context by remembering that BEMSs are simply computers monitoring and controlling plant and the problems with BEMSs are often those associated with computers themselves. Like computers, BEMSs have large user manuals to explain the many functions and operations that they can perform. Therefore, although the basics can be mastered fairly quickly, some study is required to master the whole system. Also, many user manuals are not particularly well written (or user friendly), so users have to rely on manufacturers' training courses. This training is not to be decried, as future operators may well cause problems and upset plant operation by experimenting in order to learn. However, training courses are expensive and not all operators are sent on them. Experienced operators can change jobs or leave the organization, so impairing the benefits from the BEMS. The BEMS programs (software) and equipment will often be updated as technology advances, then operators will often require further study or training.

Another common problem is that inexperienced users can program the BEMS to generate a considerable amount of data, wishing to use as much of the data logging function and alarm reporting facilities as possible. Consequently, the BEMS spews out data and alarms as if it were suffering from data diarrhoea, and after the initial enthusiasm of the operator wanes, the BEMS becomes a hindrance rather than a help. The manipulation of energy data for monitoring and targeting buildings is also a problem with BEMSs. A survey of 50 energy managers showed that 82% had BEMSs but only 2% could use them for targeting. Clearly BEMSs are not being used to their full potential.

Prospective purchasers should be aware that BEMSs from different manufacturers are not compatible with each other. For instance, one cannot easily link one manufacturer's outstation to another's central station or use another's sensors. This problem stems more from the incompatibility of computer systems than from the BEMS industry itself. Open protocols, such as ASHRAE's BACnet, Batibus, LonWorks and EIBus, will slowly allow more interoperability between systems (Chapter 3). This should avoid users getting locked into one manufacturer's system.

Table 1.4   Analysis of fuel savings due to BEMS

Source of saving	Percentage fuel saved
Increased heating system efficiency	6.6
Improved starting of heating at optimum start time in morning	5.1
Reduced inside temperature	4.6
Public holiday schedule	4.1
Stopping of heating at optimum time at end of occupancy	3.1
Heating pump overrun to use residual boiler heat	1.6
Use of gains from lighting	1.0
Heating turned off on warm days in spring	1.0
Total savings	27.1

1.4   Case studies

Most case studies relate to the early days of BEMSs, the early 1980s, when they were a lot more expensive than conventional analogue controls [13–16]. BEMSs are now accepted as a necessity in larger buildings, hence there is not such a pressing need for detailed case studies. However, it is interesting that a detailed study [17] of the savings of a BEMS in an office block in the 1980s (Table 1.4) shows that a number of standard malfunctions and inefficiencies were identified by the monitoring process. Only a small number of items, such as optimum start of the heating, could be directly related to the BEMS.

A number of the savings in Table 1.4 were due in part to the 'before' monitoring and subsequent remedial action. For instance, the heating system efficiency was improved when monitoring revealed that the high/low flame burner nozzles on one boiler were in the wrong positions, and they were changed over. Some floors were found to be getting warmer than others, so the heating system was rebalanced. This allowed the overall building temperature to be reduced.

Based on saving 27.1% of the fuel, the simple payback was estimated at 5.8 years. Allowing for a possible saving in maintenance cost, Birtles and John estimated a lower payback of 4.8 years.

1.5   Commissioning

Proper commissioning is an important consideration for BEMSs and the mechanical and electrical systems they control. Often sufficient time is not allowed in the haste to get the new building occupied. Delays in general construction work can often mean that commissioning time is squeezed as there is a fixed deadline for the completion of the building. The result is that many BEMS and HVAC systems have not performed to their full potential [18–20].

In one 14 500 m^2 commercial building with a VAV system, 'commissioned' in a fraction of the time required, it was found that 30% of its sensors gave dubious readings, based on a sample of 400 sensors from a total of 2000. The static pressure sensor was also positioned too close to the supply fan, hindering good BEMS control. These cases happen when commissioning is not sufficiently thorough, perhaps because of time pressure, a shortage of trained commissioning staff, or a lack of recognized procedures [21, 22]. Being tail-end Charlie on a project means that the BEMS and HVAC installer is often reliant upon the many other subcontractors finishing their work and their contractual wrangles.

General figures for commissioning times for BEMSs themselves are between 8 min per point to 1 h per point. Typical times might be 0.5 h for a sensor, 0.25 h for a thermostat or pressure switch, 0.5 h to 1 h for a valve or damper actuator, and 1 h for an inverter drive. In practice these times may be shortened due to the time pressures mentioned above. Temperature sensors may sometimes be situated in the wrong places; but instead of investigating, the technician may simply apply an offset to give a 'more correct' reading.

1.5.1   Specification and commissioning codes

A main contributor to a successful BEMS and HVAC system is a good specification which should emanate from a consultant's design rather than from a critical evaluation of a vendor's products [23]. A number of countries have specifications for BEMSs, some of which are reviewed in the literature [24]. There is also the NIST Guide Specification [25] and the United Kingdom has developed a standard specification [26] and a commissioning code specifically for BEMSs. The UK specification does not specify a BEMS in terms of hardware, which might be unduly restrictive and possibly biased, but on performance requirements. A guide to the specification runs to 57 pages and the specification itself is 58 pages long. There are 12 clauses on commissioning, one of which relates to the desirability of commissioning as much as possible off-site. One clause requires a complete record of parameter values and switches set during the commissioning process to be handed over on completion. There is also a section on performance tests that may need to be carried out after practical completion. Complementing the specification is a recent 86-page commissioning code divided into the following sections and appendices [27]:

- Introduction
- Site safety
- Planning the project

- Precommissioning
- BEMS commissioning
- Specific plant commissioning
- Completion
- Documentation
- Definitions

A good aspect of the code is the way it allocates responsibility for the various actions to be taken. It emphasizes careful planning and good project management of BEMS commissioning, so that the contract programme can be met. Mason also emphasises this point, suggesting a systems engineering approach [28].

The section on planning states that the consultant should give detailed control requirements at the main plant design stage, and also a comprehensive performance specification supplemented with flow diagrams. To aid commissioning of air and water systems, relevant items of plant controlled by the BEMS should be fitted with manual–off–auto switches; the manual position is used when commissioning the plant.

The code also states that it is the responsibility of the consultant, perhaps in discussion with the BEMS contractor, to specify the correct location of the sensors. This is a crucial area and can be the source of poor control, especially relating to VAV fans and heating and cooling coils. It explains the importance of having a specification which stresses the requirement for witnessing completion tests, something many specifications omit. Also on-site commissioning time can be reduced by off-site commissioning of software and graphics, using simulators. Later sections consider the development of simulators or, more correctly, emulators. A sample questionnaire of 19 questions is provided for the BEMS contractor to send to the client; it covers operating times, setpoints and other information which is often sadly lacking until the last minute.

The section on precommissioning covers the off-site checking of software and control panels and the on-site checking of wiring, sensors and actuators connected to the panels. Once this is satisfactorily completed, the system is suitable for commissioning. The code states that the consultant should ask to see and check the logic diagrams and software configuration, and the graphics in hard copy. Unfortunately, there is currently no common standard in the configuration and logic diagrams used by various manufacturers.

The checking of control panels is another main element of precommissioning included in the code. Among a number of checklist points, it advises that screen and ground connections should be carefully checked at the factory, and also the grounding of the wiring on-site as electromagnetic interference can be a problem, especially with starters and motors in a mechanical equipment room.

Subcontractors often wire the panels and outstations. Thus the code advises that there should be a completion certificate before they are handed over to the BEMS contractor. If a panel is to be left on-site for some time before commissioning, it is advised that care should be taken to protect it from dirt and moisture, two elements that can ruin microelectronic equipment. This should be heeded as all too frequently one sees BEMS cards exposed to the dust on-site.

Another important aspect covered in the precommissioning section is checking the sensors and their accuracy. An error in a mixed air temperature sensor of +2.8°C can produce a 60% increase in cooling energy consumption and a −2.78°C error can produce a heating waste of over 30% [29]. The accuracy quoted in the code is within ±0.9°F (±0.5°C) for air and chilled water sensors, ±2°C for water sensors and ±5% RH for humidity sensors. A suitable checklist for the precommissioning is shown in Fig. 1.5.

Once the precommissioning has been satisfactorily completed, and when the building services plant and BEMS controls have been commissioned, the BEMS itself can be commissioned. Section 5 of the code deals with BEMS commissioning and the next section covers commissioning checks for a number of specific items of plant, such as fans, dampers and air handling units.

The code reflects current good practice but gradually the BEMS will play a more useful role in the overall commissioning of a building's HVAC plant. At present this is unpopular because of the probable contractual difficulties in defining responsibilities. This should change as more contractors and commissioning engineers realize the value of the BEMS. With intelligent serially interfaceable flow sensors, where necessary, and intelligent actuators on balancing valves and dampers, commissioning the HVAC system could be done from the BEMS central station. This would save commissioning time and effort, although it might increase capital costs. A bonus is that the consultant can witness the commissioning on a BEMS in their own office, and the building can easily be recommissioned when it is altered.

1.5.2 Configuration strategy

The process of establishing the control strategy of the BEMS starts from the consultant who writes the specification, often with reference to plant diagrams. The manufacturer then interprets the specification and configures the BEMS strategy in the lines of a computer program. With different computer languages and different manufacturers' software routines, the consultant will probably be unable to check the written strategy. The manufacturer may well have had difficulty understanding the consultant's written specification in the first place. This is a recipe for disaster, so a number of manufacturers are now employing block diagrams to help consultants and users to understand the BEMS configuration. This will help in checking the control strategy

Contract No. ——————————————
Location ——————————————
Client ——————————————

		Signed .... Date ....
Data commissioned		
Data Cleared		
Snags (if none state none)		
Volt free contact		Outstation totally commissioned and tested to central
Digital output		
Digital input		
Sensor — Check reading		
Sensor — Reading (field)		
Actuator — Stroke		
Actuator — Start		
Actuator — Span		
Identified		
Wired		
Installed		
Field Device point ref		Signed .... Date ....

Fig. 1.5   A checklist for commissioning a sensor. (Adapted from [28])

before installing the BEMS in the building. So far there is no standard set of blocks, but standardization cannot be far off, perhaps as a consequence of BACnet or as a separate ASHRAE standard [30]. Figure 1.6 shows part of the block diagram for a VAV box [31].

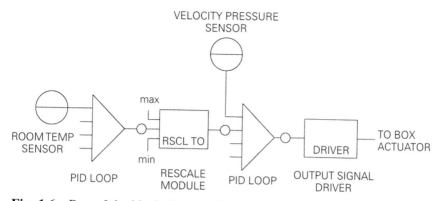

Fig. 1.6   Part of the block diagram for VAV box control. (Adapted from [33])

A further step to speeding up the configuration process is for the consultant to send the CAD files of the HVAC design drawings and control details to the manufacturer. The manufacturer can then input the consultant's files into their computer, which will then add on the standard control elements from the manufacturer's computer library for the required control strategy. From these files the manufacturer's computer can also produce wiring diagrams for the BEMS outstations and commissioning checksheets. This procedure of electronic design from consultant's drawing to commissioning sheets is a thing for the near future. But one UK manufacturer has already written an expert design program with a library of common control configurations. This means it is possible to take a point file and quickly produce the control strategy, wiring diagrams, commissioning checksheets and a description of operation for the operator's manual (Fig. 1.7). This has cut the manufacturer's engineering time for a system to one-sixth [32].

As to the electronic exchange from the consultant to the manufacturer, standard file formats and standard data exchange methods are still being agreed. One prototype is based on the STEP standard (ISO standard for the exchange of product model data). This uses NIAM (Nijssen information analysis method) diagrams [33], which graphically express the relationships between objects and ideas that were conventionally expressed in natural language. Figure 1.8 shows the NIAM diagram of a fan in a VAV system. For further details see the work by Wix and Storer [34].

1.5.3   Emulators

Once the BEMS strategy has been configured, can we be certain that it will work? For instance, what should be the static pressure setpoint in a conventionally controlled VAV system? And is two-thirds of the way down the

Engineering
stage

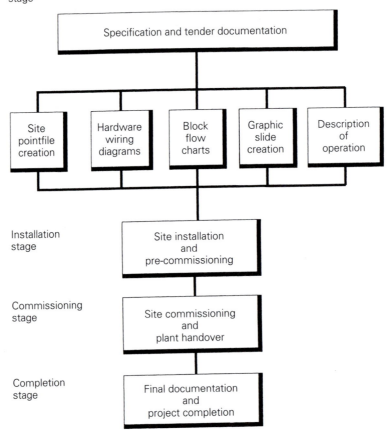

Installation
stage

Commissioning
stage

Completion
stage

Fig. 1.7   The BEMS engineering process. (Adapted from [32])

index run a good position for the sensor? Does the polling of VAV boxes for controlling the fan starve some VAV boxes? And for temperate climates, when should the fan control be stopped and the chilled water reset be started?

These are difficult questions but emulators are rapidly developing that will help the designer to answer them and to train operators before the BEMSs have been installed [35, 36]. An emulator is a computer system to which the BEMS is attached; its software emulates the building shell and the plant, sending sensor input signals to the BEMS and responding to output signals from the BEMS. The computer system is called an emulator rather than a simulator; this is because it runs in real time whereas a simulation runs as fast as the computer allows.

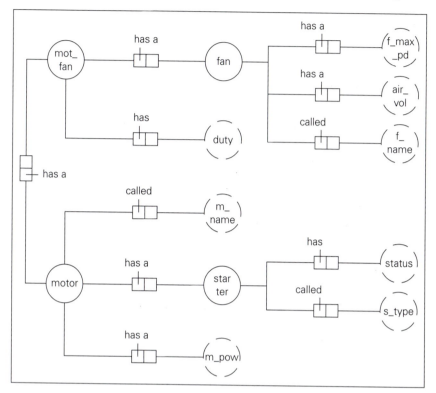

Fig. 1.8 NIAM diagram for a fan: f = fan, pd = pressure difference, m = motor; full circles are actual objects; dashed circles are attributes. (Adapted from [34])

At present, BEMSs are often precommissioned with variable resistances replacing sensors to check whether lamps react to the resultant output signals. This is useful, but full emulation will allow one to do so much more, such as answering the VAV questions above, checking optimum start/stop programs, free-cooling strategies, boiler and chiller sequencing, etc.

1.6 Occupant feedback

One aid to commissioning HVAC systems and BEMSs is an occupant survey with a self-assessed questionnaire; it also reveals how the building itself is perceived by the occupants. This is becoming increasingly important since the advent of sick building syndrome, concerns over the influence of environmental conditions on staff productivity, and reports of occupant dissatisfaction about lack of control over their environment.

Table 1.5   A seven-point scale of likes and dislikes

	Do you like the .... **dislike like**							How important is this in the design of your ideal office? **unimportant important**						
Noise level	−3	−2	−1	0	1	2	3	−3	−2	−1	0	1	2	3
Comments:														

A questionnaire has been developed which is especially suitable for commissioning as it produces an easily understood fingerprint and score without the need for a lot of statistics. A novel feature of the questionnaire is the double Likert scale for liking and importance of 22 or 24 factors relating to the interior environment and the organization. A seven-point scale for like and dislike is used for the questionnaire (Table 1.5).

A like/dislike scale rather than a scale of satisfaction and dissatisfaction is used to reduce the fog index and ease understanding, which is important for a self-administered questionnaire. Alongside the seven-point like/dislike scale is another seven-point scale for the respondent to indicate how important they consider the factor should be in designing their ideal office. A score and fingerprint can then be calculated to simplify the results and enable its use in management feedback and commissioning. I will concentrate on the results from this section. The organization factors are in two questions asking about colleagues and the management. The initial inclusion of these questions was justified in that colleagues are often considered the most liked and the most important factor. Here are the factors:

- Noise level
- Electric lighting
- Daylight
- Glare level in the room
- Office temperature
- Ventilation
- Draught level
- Freshness of your room
- Humidity
- Smell in the building
- Colours of the room
- Attractiveness of the room
- The management
- Your office in general
- Control you have over your local environment
  (later replaced by three factors)
- Working space you have in the room

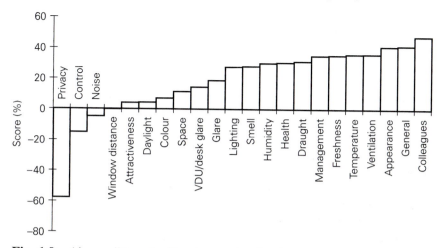

Fig. 1.9 Air-conditioned office in prestige headquarters: fingerprint of ranked liking scores (overall score = +19%).

- Your privacy in the room
- Your immediate colleagues
- Outward appearance of your building
- Glare level around your desk or VDU
- Your state of health when in the building
- Your distance away from the window

In later questionnaires 'control you have over your local environment' was replaced by these three factors:

- The control you have over the ventilation
- The control you have over the lighting
- The control you have over the heating

A fingerprint and score can be derived from the responses [37]. The score is expressed as a percentage and most occupants' scores range between +20% and −30%. A positive score indicates a degree of liking and a negative score a degree of disliking. Most overall scores from respondents in a building have ranged between +19% and −34%. Figures 1.9 and 1.10 give examples of fingerprints.

Other sections of the questionnaire deal with general occupant details: age, sex, hours at work on a PC, etc. Further sections of the questionnaire can be used for establishing whether it is too hot or too cold in winter and summer, assessing the building sickness score and dealing with the stress of the occupant.

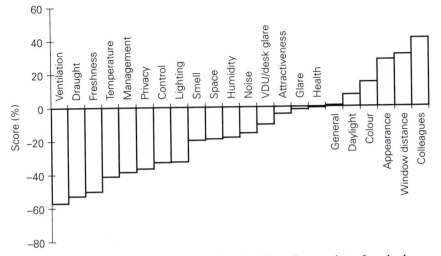

Fig. 1.10   Old, small, naturally ventilated office: fingerprint of ranked liking scores (overall score $= -14\%$).

1.6.1   The liking score

A liking score can be derived from an occupant's liking and importance vote for each factor. Each importance vote is transformed (by adding 4 to all values to make a positive value) then multiplied by the corresponding liking vote. This yields a score between +21 and −21. The overall liking score for an individual (ILS) is the average of all the 22 or 24 multiplications normalized through dividing by the maximum possible score and expressing as a percentage [37]. The fingerprint shows the 22 or 24 factors' normalized scores for the group of individuals, as explained in the literature [37]. The group overall liking score (OLS) is the average of the ILSs for the group.

1.6.2   Distribution

For each respondent there are over 400 possible ILS values. The distribution of ILSs for eight buildings (395 respondents) is close to a normal cumulative distribution (Fig. 1.11). The individual scores for these eight buildings are given in Table 1.6. Other independent surveys have been conducted on buildings 1, 2 and 3, and generally they agree with the findings of the fingerprints and scores [38–40].

For building 4, an academic office and lecture block, some students were surveyed in addition to the staff, whose score is shown in Table 1.5. Two groups of students (40 students in total) were resurveyed to test the robustness of the scores. One group were resurveyed a week apart and the other

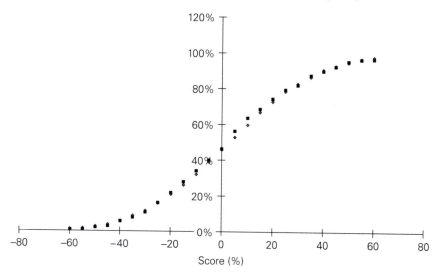

Fig. 1.11 Actual scores (∎) from 395 responses compared to a normal cumulative distribution (♦).

Table 1.6 Individual scores for the buildings

Building	Type	Score (%)
1	Full AC	+19
2	AC atrium, openable windows	+17
3	AC	+2
4	Shallow-plan NV vent	+1
5	Part AC, part NV	−11
6	Full AC	−12
7	Shallow-plan NV	−14
8	Deep-plan NV with atria	−15

group two hours apart. The resurvey scores were not significantly different from their original scores ($p < 0.001$ two-tailed t-test [41]).

1.6.3 Fingerprint

In addition to the liking score, the individual factor scores (FLSs) can be shown as a fingerprint, as in Fig. 1.10 for building 1. The factors have been ranked for ease of viewing. Building 1 has one of the best scores. This is not surprising as it is a prestige headquarters building. An independent survey confirms a number of the findings here [38]. Privacy is the greatest dislike of this open-plan office followed by control and noise.

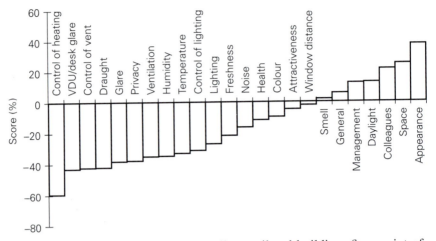

Fig. 1.12   Modern, deep-plan, naturally ventilated building: fingerprint of ranked liking scores (overall score = −15%).

In contrast, Fig. 1.11 shows a poor fingerprint (OLS = −14%). Although −14% is poor, a score of −34% has been achieved in a building with no windows. The building with OLS = −14% is building 7, an old, small office with high ceilings and large windows and an inadequate heating system. The ventilation and associated factors, draught and freshness, cluster at the end of the scale, along with temperature. It is noteworthy here that the management is in a low position, although immediate colleagues are liked almost to the level in building 1.

Figure 1.12 shows an equally poor fingerprint (OLS = −15%); this is for building 8, a modern, deep-plan, naturally ventilated building with atria. This fingerprint has the added control factors. Two of these control factors are greatly disliked. This is not surprising as the control strategies for day and night ventilation, and also the heating, were still being tuned on the BEMS when it was surveyed. However, glare on the PC screens was the second greatest dislike. In contrast to building 7, the occupants here liked the management.

Figure 1.13 shows another modern building, building 2; this also has atria but it has underfloor VAV, controlled by a BEMS, and openable windows. Building 2 is much better liked (OLS = +17%). This building has also been assessed in the PROBE study [39].

1.6.4   Importance vote

The importance vote, or score, gives an indication of the importance of the 22 or 24 factors in the design of an occupant's ideal office. Over 90% of

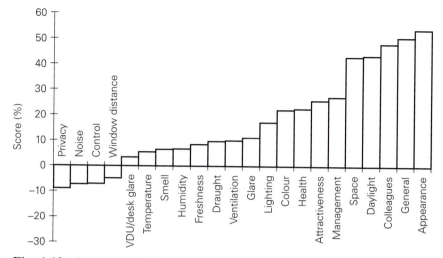

Fig. 1.13 Deep-plan office with atrium and underfloor air conditioning: fingerprint of ranked scores (overall score = +17%).

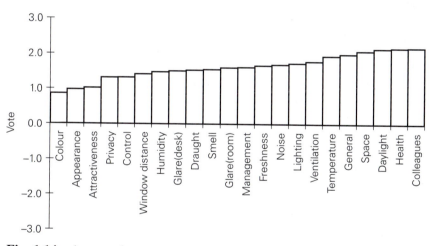

Fig. 1.14 Average importance votes, ranked.

votes by occupants to all factors were greater than or equal to 0, i.e. a large majority of respondents did not consider that the factors were unimportant [42].

Figure 1.14 shows the average importance votes, in rank order, for the first seven buildings. This has colleagues as the most important factor, about which building services engineers can do little. Daylight is also high ranking, although distance from a window is ranked 6/22. Temperature is ranked

6/22 and ventilation 7/22. Control is rather low in importance, although the control that occupants have is often disliked. This was explored further with the addition of control factors in two later surveys, one of which was building 8. Here the control of ventilation was placed 9 out of 24 factors (9/24) with control of lighting and heating close behind.

1.6.5   Industrial psychology

The questionnaire can be used as an aid to commissioning the BEMS controls and the HVAC system as well. It also gives an insight into the perception and reactions of the occupants. An appreciation of industrial psychology is required to understand people's perceptions and reactions to the built environment, its plant and its control. J. A. C. Brown traces the origin of industrial psychology to the work of Elton Mayo at the Hawthorne works in Chicago between 1924 and 1927 [43].

Production at the works, which manufactured telephone equipment, was low and there was dissatisfaction and grumbling among the 30 000 employees. The employer was considered progressive by the standards of the 1920s, with pension schemes, sickness benefit schemes, recreational and other facilities. Mayo's work started by assessing the environmental conditions of the workplace, on the assumption that people produce maximum output in perfect environmental conditions and using perfect methods of working.

Mayo examined the influence of increasing the lighting levels on production. To do this he took two groups of employees. For the control group, the lighting level was unchanged; and for the trial group it was raised. Production in the trial group increased, but unexpectedly it also increased in the control group. This was quite puzzling as the lighting had remained constant in the control group.

The trial group's lighting level was then reduced while the control group's lighting remained constant. Again the production increased and again it was for both groups. Clearly other factors were at work than just the lighting, and it would not make sense to reduce the lighting levels to increase production in the long term. Mayo did many years of research into these factors, and his conclusions are extensive (for a full discussion the interested reader is referred to Brown's book [43]). The main conclusions of interest are that the production levels for both groups increased as the employees felt important, more than mere cogs in a machine. Work is a group activity, and the Hawthorne experiments show that social factors can influence production.

The lesson for designers of buildings, plant and controls is that the human factor is vitally important; feedback and consultation can add to the success of a building. This is especially true with increasing demands for occupants' control over their own environment.

Another useful conclusion from the Hawthorne experiments is that a complaint was not necessarily an objective recital of fact. It was commonly a symptom manifesting itself as an individual's status position. So a complaint about a BEMS may not actually be about the BEMS itself, but about the organization and people using it.

Mayo's work was conducted many years ago and it has received extensive discussion and some criticism. But more recent work on energy efficiency supports the idea that human factors are important in operating buildings properly [44].

The installation of energy efficiency measures was studied in nine identical children's homes in a local authority [45]; it showed that energy efficiency measures did save fuel. Staff in the children's homes were briefly informed of the work before it was carried out and its purpose was explained to them. Although the measures were only to save fuel, it was found that all the homes had also saved considerable amounts of electricity, analogous to production increasing at the Hawthorne works.

1.7 Future developments

1.7.1 Artificial intelligence and *Expert systems*

Artificial intelligence (AI) is the study and process of enabling computers to mimic human learning and decision making [46]. AI systems cover **natural language processors** that eliminate the computer user having to learn a computer language, **robotics** that enables a machine to learn (machine learning) and to have **automated vision**, and finally the expert system.

An expert system is a computer program that learns, deduces, diagnoses and advises [46]. A more formal definition is that it is the embodiment in a computer of a knowledge-based component from an expert that can offer intelligent advice or take an intelligent decision, and in addition it can justify its own line of reasoning. The style adopted to attain these characteristics is **rule-based** programming [47]. The rules are based on logic, and a standard computing statement such as IF . . . THEN is a common element of expert systems [48], as are yes/no answers to questions. Standard probability theory and Bayes' theorem also contribute to expert systems [49]. If hard yes/no answers, or definite probabilities, are not easy or possible to give then **fuzzy logic** can be used for approximate reasoning [50, 51]. Three of the best-known proven expert systems are Prospector, Dendral and Mycin.

The earliest expert system (work started on it in 1964) and a significant advance in expert systems was Dendral, which identifies molecular compounds from their mass spectrograms. It performs better than the best human experts and has made some worthwhile contributions [48]. Prospector was developed in 1978 and is designed to help geologists in their search for ore deposits and to investigate the mineral potential of large areas of land. Tests against known sites of exploration revealed a 7% agreement [49].

Development on Mycin started in 1974 at Stanford University. The program helps doctors diagnose and treat bacterial infection. Although it is successful, it takes 20–30 min per consultation and is mostly used for teaching [49]. Mycin, like other expert systems, has two main elements: the **knowledge base** and the **inference engine**. The knowledge base contains the expert's knowledge such as the answers to questions and the consequences of the IF . . . THEN statements, whereas the inference engine uses the facts and rules to logically answer questions posed by the user. Further work on Mycin stripped out the medical knowledge to leave the inference engine and a system for adding new rules to create EMYCIN (Essential Mycin), effectively an **expert system shell**. Expert system shells are programs or software that allow a user to set up their own expert systems in any area, or **domain**, of expertise. This avoids the need to write the expert system program in LISP or PROLOG, common languages for such programs.

An example of an expert system, or a **knowledge-based system** (KBS), applied to a BEMS is BREXBAS, developed as a prototype by the Building Research Establishment (BRE) to demonstrate the use of expert systems in analysing data from a BEMS for fault diagnosis [52]. After some initial tests BREXBAS was then redesigned to monitor and interpret information from an online BEMS at a building in Epsom [53]. The KBS supplied advice on the performance of one of two heating plants and the underfloor heating in the atrium. Various flow temperatures and plant item statuses were monitored as well as the space temperature of the building. The knowledge of the heating system was obtained and structured for the expert shell by a programmer discussing the heating with the designer. Three main tasks were intended to be performed by BREXBAS:

- Ascertain the current space temperature.
- Evaluate the current conditions to determine acceptability and diagnose the cause of any problems.
- Provide advice and/or potential remedies to the user.

Figure 1.15 shows the general structure of the KBS on its own computer, linked to the BEMS central station.

1.7.2   Fuzzy logic

Fuzzy logic relates to how we describe objects and communicate in every day life [54, 55]. For instance, we do not say we are hot just as the temperature goes above 25°C; there may be a fuzzy region around 25°C when we start to feel hot. This type of logic can be useful in control (Chapter 5).

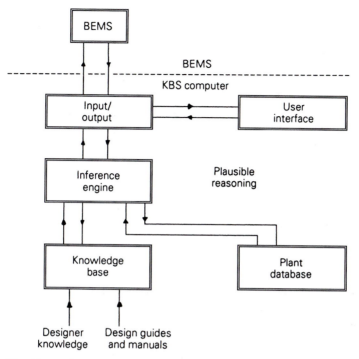

Fig. 1.15   General structure of BREXBAS.

1.7.3   Artificial neural networks

Artificial neural networks (ANNs) combine a number of simple transfer functions in parallel which can be trained to fit almost any mathematical function or series [56]. This can be useful for modelling non-linear heater batteries [57], e.g. for better control and automatic fault diagnosis and detection. Also the building response and its energy consumption can be modelled for better targeting of its energy consumption [58–60] than simple degree-day relationships (Chapter 14).

1.7.4   Simulation

Building designers will increasingly use dynamic simulation programs, with hourly based weather data, to examine how their building designs perform. Indeed, it is a recommendation that for successful natural ventilation a building design should be simulated through a near-extreme summer period (April to September), and the dry resultant temperature should not exceed 25°C for more than 5% of the occupied time in the year [61, 62].

1.7.5   Communications

The IT and communications revolution will allow even more HVAC components, down to valves and switches, to be interfaced to a communications bus and consequently to a BEMS. However, access of occupants to the Internet and to intranets [63–65] and connection of the BEMS to these networks will open up the possibility of more occupant control (Chapter 3). The era of the BEMS is just maturing.

References

[1]   Kaku, M. (1998) *Visions: how science will revolutionize the twenty-first century*, Oxford University Press, Oxford.

[2]   IBI (1998) Intelligent Building Definition, Intelligent Building Institute.

[3]   So, A. T. P., Chan, W. L. and Tse, W. L. (1996) Building automation on the information superhighway. *ASHRAE Trans.*, **102**(4196).

[4]   Haines, R. W. (1987) *Control Systems for Heating, Ventilating and Air Conditioning*, 4th edn, Van Nostrand Reinhold, New York.

[5]   Clapp, M. D. and Wortham, R. H. (1989) Developments in building-management systems. *GEC Review*, **5**(1).

[6]   Gardner, P. R. and Ward, L. (1987) *Energy Management Systems in Buildings*, Energy Publications, Newmarket, UK.

[7]   ETSU (1986) *Energy Management Systems*, Energy Efficiency Office of the Department of Energy, Harwell, UK.

[8]   Bull, R. J. (1972) *Accounting in Business*, Butterworth, London.

[9]   Levermore, G. J. (1989) Presenting a case for a building energy management system. BEMS Centre, Bracknell, UK.

[10]   Anon (1998) Cost model building energy management systems. *Building Services Journal*, **20**(5).

[11]   Hyvarinen, J. (1991) Cost benefit assessment methods for BEMS. International Energy Agency Annex 16, Building and energy management systems: user guidance, June 1991, Air Infiltration and Ventilation Centre, Coventry, UK.

[12]   Brendel, T. and Schneider, A. (1991) User experiences in BEMS applications. International Energy Agency Annex 16, Building and energy management systems: user guidance, October 1991, Air Infiltration and Ventilation Centre, Coventry, UK.

[13]   ETSU (1987) Central monitoring and control of heating schemes: a demonstration at Hereford and Worcester Council. ETSU Final Report ED/136/62, Harwell, UK.

[14]   ETSU (1987) Automatic remote monitoring and control of coal-fired boiler plant in schools: a demonstration in Staffordshire County Council schools. ETSU Final Report ED/111/64, Harwell, UK.

[15] ETSU (1986) Monitoring of a microprocessor based energy management system: a demonstration project at various tenanted office buildings. ETSU Final Report ED/105/127, Harwell, UK.

[16] ETSU (1984) The evaluation of an energy management system for the central control of a number of remote school heating installations in Hereford and Worcester. EBU Interim Report I/47/84/62, Harwell, UK.

[17] Birtles, A. B. and John, R. W. (1984) Study of the performance of an energy management system. *BSERT*, **4**(5).

[18] Elovitz, K. M. (1992) Commissioning building mechanical systems. *ASHRAE Trans.*, **98**, Pt 2.

[19] Bulbeck, A. (1993) A code of malpractice. *Building Services Journal*, April.

[20] Herzig, D. J. and Wajcs, F. F. (1993) Lessons learned from monitored office building data. *ASHRAE Trans. Symposia*, **99**, Pt 1.

[21] Pike, P. (1992) Getting it up and running. *Energy in Buildings*, April.

[22] Dexter, A. L., Haves, P. and Jorgensen, D. R. (1993) Development of techniques to assist in the commissioning of HVAC control systems. Paper presented at the CIBSE National Conference, Manchester.

[23] Bynum, H. D. (1991) Plan and specification documentation of a direct digital control/building management system. *ASHRAE Trans.*, **97**, Pt 1.

[24] Teekaram, A. J. H. and Grey, R. W. (1991) Specifications and standards for BEMS. International Energy Agency Annex 16, Building and energy management systems: user guidance, September 1991, BSRIA, Bracknell, UK.

[25] NISTIR (1991) NISTIR 4606 Guide specification for direct digital control based building information systems.

[26] BSRIA (1990) Standard specification for BEMS, version 3.1. BSRIA Application Handbook AH1/90, Volume 2, BSRIA, Bracknell, UK.

[27] Pike, P. and Pennycook, K. (1992) Commissioning of BEMS – a code of practice. BSRIA Application Handbook AH2/92, BSRIA, Bracknell, UK.

[28] Mason, P. (1993) How to make a BEMS work on the day. *BSERT*, **14**(4).

[29] Kao, J. Y. and Pierce, E. T. (1983) Sensor errors and their effects on building energy consumption. *ASHRAE Journal*, Dec.

[30] Newman, H. M. *Direct Digital Control of Building Systems*, John Wiley, Chichester.

[31] Levermore, G. J. (1994) Commissioning building energy management systems. *ASHRAE Journal*, Sept.

[32] Clapp, M. D. and Blackmun, G. (1992) Automatic engineering of building management systems. *GEC Review*, **8**(1).

[33] Griffin, C. R. (1993) The exchange of engineering information using complex data structures. PhD thesis, South Bank University, London.

[34] Wix, J. D. and Storer, G. (1993) STEP – a convergence between research and standards development. Paper presented at the CIBSE National Conference, Manchester.

[35] Haves, P., Dexter, A. L., Jorgensen, D. R., Ling, K. V. and Geng, G. (1991) Use of a building emulator to evaluate techniques for improved commissioning and control of HVAC systems. *ASHRAE Trans.*, **97**, Pt 1.

[36] Kelly, G. E. Park, C. and Barnett, J. P. (1991) *ASHRAE Trans.*, **97**, Pt 1.

[37] Levermore, G. J. (1994) Occupants' assessments on interior environments. *BSERT*, **15**(2).

[38] Dickson, D. and Collins, P. (1992) Healthy building: an energy efficient air conditioned office with good indoor quality. In *Proceedings of the 13th AIVC Conference*, Nice, France.

[39] Anon (1995) Probe 1: Tanfield House, building services. *CIBSE Journal*, **17**(9).

[40] Anon (1995) Probe 2: Aldermanbury Square, building services. *CIBSE Journal*, **17**(12).

[41] Levermore, G. J. and Meyers, D. (1996) Occupant questionnaire on interior environmental conditions: initial results. *BSERT*, **17**(1).

[42] Leventis, M. and Levermore, G. J. (1996) Occupant feedback – important factors for occupants in office design. Paper presented at the CIBSE/ASHRAE Conference, Harrogate, UK.

[43] Brown, J. A. C. (1977) *The Social Psychology of Industry*, Penguin, London.

[44] Watson, B. M. (1982) *Employee Participation in Energy-Saving Programmes*. Energy Publications, Cambridge, and Grafton Consultants.

[45] Levermore, G. J. (1985) Monitoring and targeting. Paper presented at the Construction Industry Conference, Energy Experience, London.

[46] Colantonio, E. S. (1989) *Microcomputers and Applications*, D. C. Heath, Lexington MA.

[47] Naylor, C. (1983) *Build Your Own Expert System*, Sigma Technical Press, Wilmslow, UK.

[48] Hartnell, T. (1985) *Exploring Expert Systems on Your Microcomputer*, Interface Publications, London.

[49] Sell, P. S. (1985) *Expert Systems – A Practical Introduction*, Macmillan, London.

[50] Zimmermann, H.-J. (1984) *Fuzzy Set Theory and Its Applications*, Kluwer, Dordrecht.

[51] Driankov, D., Hellendorm, H. and Reinfrank, M. (1996) *An Introduction to Fuzzy Control*, 2nd edn, Springer Verlag, Berlin.

[52] Shaw, M. (1988) Applying expert systems to environmental and management control problems. Paper presented at the Intelligent Buildings Conference.

[53] Shaw, M. and Willis, S. (1991) Intelligent building services control and management systems. Paper presented at the CIBSE National Conference, Canterbury, UK.

[54] Jang, J. S. R. and Gulley, N. (1995) *MATLAB Fuzzy Logic Toolbox User's Guide*, Mathworks Inc., Natick MA.

[55] Wang, L. X. (1997) *A Course in Fuzzy Systems and Control*, Prentice Hall, Englewood Cliffs NJ.

[56] Demuth, H. and Beale, M. (1992) *Neural Network Toolbox for Use with MATLAB*, Mathworks Inc., Natick MA.

[57] Hepworth, S. J., Dexter, A. L. and Willis, S. T. P. (1994) Neural network control of a nonlinear heater battery. *BSERT*, **15**(3).

[58] Anstett, M. and Kreider, J. F. (1993) Application of neural networking models to predict energy use. *ASHRAE Trans.*, **99**(1).

[59] Haberl, J. S., Smith, L. K., Cooney, K. P. and Stern, F. D. (1988) An expert system for building energy consumption analysis: applications at a university campus. *ASHRAE Trans.*, **94**(1).

[60] Kreida, J. F., Cooney, K., Graves, L., Meadows, K., Stem, F. and Weilert, L. (1990) An expert system for commercial buildings: HVAC and energy audits – a progress report. *ASHRAE Trans.*, **96**(1).

[61] CIBSE (1999) CIBSE Guide Design Data, Chartered Institution of Building Services Engineers, London.

[62] CIBSE (1999) CIBSE Guide Weather and Solar Data, Chartered Institution of Building Services Engineers, London.

[63] Tanenbaum, A. S. (1996) *Computer Networks*, 3rd edn, Prentice Hall, Englewood Cliffs NJ.

[64] Hodson, P. (1997) *Local area networks, including internetworking and interconnection with WANs*, 3rd edn, Letts, London.

[65] Derfler, F. J. and Freed, L. (1998) *How Networks Work*, 4th edn, Que Corp., Indianapolis IN.

2
The outstation

The benefits and problems of BEMSs have been discussed in Chapter 1, but in this chapter the nature and details of the BEMS will be examined. First the **outstation** will be considered. This is the BEMS unit closest to the sensors and actuators and is often situated in the plant room. Other terms for the outstation are **field processing unit** (FPU), **data gathering panel** (DGP), **distributed processing unit** (DPU), **field interface device** (FID) and **substation** [1, 2]. The term FPU implies limited local processing power, whereas the DPU is more like the modern outstation considered in this book.

2.1 Elements of an outstation

Figure 2.1 shows a typical BEMS, with its central station, often a personal computer (PC), and a number of outstations. The BEMS outstation is defined here as the unit with inputs and outputs that controls the plant, but it does not have a large keyboard and screen of a central station, although it may have a small display, similar to a calculator display, and a keypad, a small keyboard or set of pressure pads often with cursor arrows for limited access.

Figure 2.2 shows typical BEMS architectures. Inputs come in from sensors and switches; the outstation can then use them for a **control loop** or a **function**. The output from the control loop goes to the output section of the outstation to control the action of a valve, perhaps, whereas a function relates to time schedules, optimizers and **logic** expressions, and again it sends signals to the outstation's output unit. The rest of this chapter considers its components and operation in more detail.

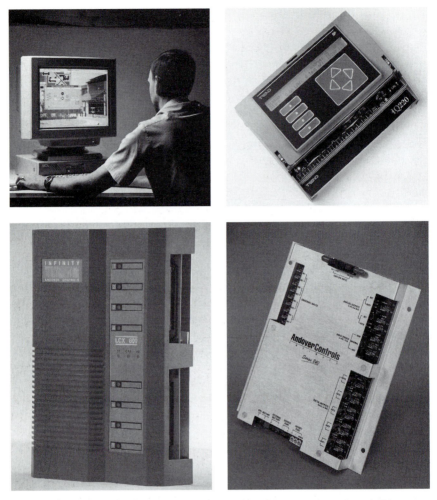

Fig. 2.1   A central station (top left) and a variety of outstations. (Photos courtesy Trend Control Systems Ltd and Andover Controls Ltd)

2.2   Intelligent outstations

DGP and, to a lesser extent, FPU describe older outstations that took input/output signals and performed initial processing before transmitting them to the central station. The outstations considered here are more powerful than that; they have their own microprocessor, hence they are commonly described as intelligent.

Intelligent outstations range from a single, main circuit board type (Fig. 2.3) to larger outstations with many printed circuit boards (PCBs), or cards, plugged into a **rack** or a **backplane**, a metal chassis with electrical connections to the PCBs (Fig. 2.4). Modular outstations are similar to the rack-mounted PCB outstations, except their electronics are housed in individual plastic or metal **modules** mounted next to each other on rails or a baseplate with suitable connections.

In the multiple-board and modular outstations, each PCB or module has a certain function: for inputs, for outputs, for communication, for the microprocessor, and so on. With these types of outstation the user has the advantage of being able to add more PCBs and modules, perhaps to expand the outstation by adding more inputs and outputs. If there is a fault, an individual unit can be replaced rather than the whole outstation.

The single-board outstations are limited to the number of inputs and outputs provided; they cannot be expanded. But some of them are very small and very cheap, and they may be dedicated to particular functions and equipment, e.g. boiler control.

The single-board versions are often contained in integral metal boxes which are housed in electrical cabinets or control panels with the associated relays and connections, whereas the multiple-board versions are already contained within their own electrical cabinets. But whatever the type of outstation, all its electronics and circuitry have to be protected from dust and damp in plant rooms, to stop electrical breakdown and tracking across the PCBs. To establish how much protection is given by the casings and cabinets, an **ingress protection** system designates the level of protection using the letters IP followed by two digits [3]. The first digit, 0 to 6, indicates the degree of protection from dust: 0 = not protected and 6 = dust-tight. The second digit, 0 to 8, indicates protection from water penetration: 0 = not protected and 8 = protected against submersion.

Metal cabinets offer protection against electromagnetic interference and noise; this can impair the performance of microprocessor equipment which operates with small voltages signals [4, 5]. This is particularly important when the outstation is located in a plant room where there are discharge lamps and motors for pumps and burners which can generate significant electromagnetic radiation.

Some outstations have small keyboards or keypads and small display panels of one or two lines for limited access to the outstation's data by staff in the plant room. Access to the outstation's data and programs can be restricted by use of a **password**. Password protection means that the heating temperature setpoint, for instance, cannot be increased by an unauthorized person. Differing security levels, or **operator access levels** (often signified by two-digit numbers attached to the password) can be set in the outstation

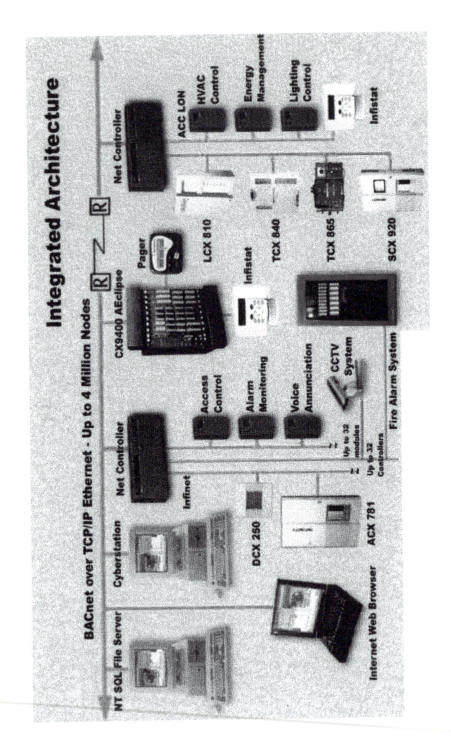

Integrated Architecture

BACnet over TCP/IP Ethernet - Up to 4 Million Nodes

NT SQL File Server

Cyberstation

Net Controller

Infinet

DCX 250

ACX 781

Internet Web Browser

CX9400 AEclipse

Pager

Infistat

Access Control

Alarm Monitoring

Voice Annunciation

CCTV System

Up to 32 modules

Up to 32 Controllers

Fire Alarm System

Net Controller

ACC LON

HVAC Control

Energy Management

Lighting Control

Infistat

LCX 810

Infistat

TCX 840

TCX 865

SCX 920

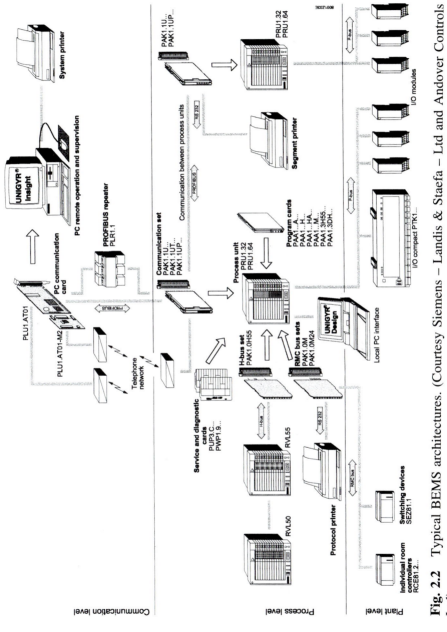

Fig. 2.2 Typical BEMS architectures. (Courtesy Siemens – Landis & Staefa – Ltd and Andover Controls Ltd)

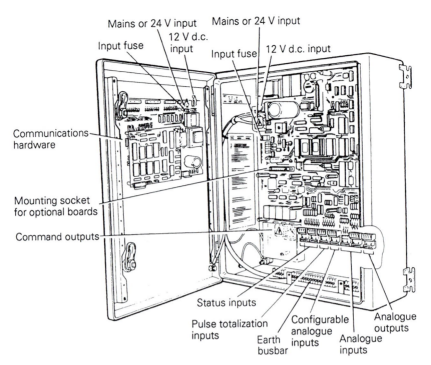

Mains or 24 V input   Mains or 24 V input

12 V d.c. input

Input fuse   Input fuse   12 V d.c. input

Communications hardware

Mounting socket for optional boards

Command outputs

Status inputs

Pulse totalization inputs

Configurable analogue inputs

Earth busbar

Analogue inputs

Analogue outputs

Fig. 2.3 Single-board outstation.

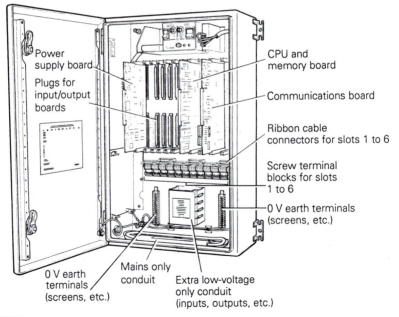

Power supply board

Plugs for input/output boards

CPU and memory board

Communications board

Ribbon cable connectors for slots 1 to 6

Screw terminal blocks for slots 1 to 6

0 V earth terminals (screens, etc.)

0 V earth terminals (screens, etc.)

Mains only conduit

Extra low-voltage only conduit (inputs, outputs, etc.)

Fig. 2.4 Multiple-board outstation.

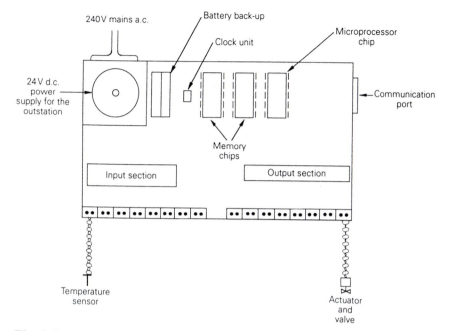

Fig. 2.5 Outstation structure.

for various passwords so that low-level information, such as the outside air temperature, can be easily accessed whereas only restricted, higher-level passwords allow one into the control parameters, such as temperature setpoints. Numerical passwords, or **personal identification numbers** (PINs) are used on some BEMSs.

2.2.1 Outstation structure

Figure 2.5 is a simplified diagram of a single-board outstation with its cover off, revealing the basics of the printed circuit board. The outstation is powered by an internal 24 V d.c. power supply which transforms and rectifies the 240 V a.c. supply. The outstation's microelectronics only operate on low-voltage d.c., and large a.c. voltages would damage them.

During a mains power cut, or if the mains has to be switched off for maintenance, the outstation is powered by the battery back-up. Without this back-up, all the control programs resident in the outstation would be lost. Although the outstation could be set to 'fail safe' (i.e. leave the heating on or switch off all plant for safety), it is extremely important that the battery back-up is reliable, has adequate life expectanncy, and has enough capacity to keep the outstation running for sufficient time until mains power is restored.

Figure 2.5 shows the input from one temperature sensor, although seven more sensors or input devices could be connected to this outstation. Similarly, seven more outputs could be employed than just the one shown (controlling the valve actuator). The temperature sensor cable is connected via terminals on the outstation to an input section, or **input unit** (a set of electronic chips and circuitry on the PCB, not identifiable as a single discrete component). The cable to the valve comes from terminals connected to the **output unit** of the outstation (a set of electronic chips and circuitry).

The **communications port** in Fig. 2.5 is the electrical access point for communications. A port is a connection point in an electronic circuit where a signal may be input or output. Inside the outstation would be an additional small circuit board (not shown in Fig. 2.5) to interface with a modem or a LAN. This board is often known as a **communications node controller** (CNC). For communication purposes, each outstation would have its individual identification number and name. If the outstation did not have a keypad and display, there would be an additional communications port for connecting to a portable PC for local programming and reconfiguration.

The outstation is simply a microcomputer that controls the plant, except it does not have a large keyboard as an input, or a monitor and printer as outputs. Instead it has sensors as inputs and actuators or relays as outputs. But just like a microcomputer, the outstation has a microprocessor chip and memory chips, as shown in Fig. 2.5.

2.3 Microprocessor

The microprocessor, also called the central processing unit (CPU) or the microprocessor unit (MPU), is the principal component of both the microcomputer, used as the BEMS central station, and the outstation. Produced by large-scale integration (LSI) on a small chip of silicon (about 1.5 mm^2), the microprocessor is the brains of a microcomputer. Later microprocessors are VLSI chips and correspondingly more powerful. The actual microprocessor chip is often contained in a dual-in-line (DIL) package with between 16 and 64 pins, like legs holding up the package.

A simplified schematic of the microprocessor (Fig. 2.6) shows several vital components: the **control unit**, a small store or memory consisting of **registers**, and the **arithmetic and logic unit** (ALU). The Pentium chip is more complicated than this simple microprocessor. Cache memory units are ultrafast memory for recently accessed or frequently accessed data. A floating-point unit (FPU) deals with very high or low decimal fractions to a high accuracy. The Pentium chip has two ALUs and some other extra units. For more details consult R. White's *How Computers Work* [15].

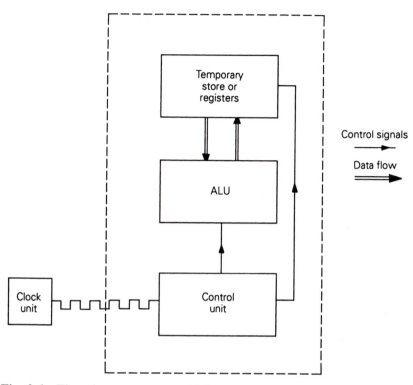

Fig. 2.6 The microprocessor or CPU.

The arithmetic and logic unit is the operational unit. It performs calculations such as addition and multiplication, as well as logical decision processes for selecting, sorting and comparing data (see later).

The control unit ensures the ALU performs the correct operations. It coordinates all the microcomputer's functions and interprets the instructions in the program to perform the control functions necessary to run the program. For instance, it will enable the latest temperature reading from a sensor to be accessed; the ALU will then compare it with the required temperature when instructed by the program.

The temporary store is a small memory composed of a number of registers; it will hold temperature values and the immediately required program instructions for the ALU and control unit to work on.

Also shown in Fig. 2.6 is the **clock unit** which sends out timing pulses from a quartz crystal. This enables the control unit to time and synchronize the operations of the microprocessor. Microprocessors operate at high speeds and the Intel 8080 microprocessor, a widely used LSI CPU chip, requires a

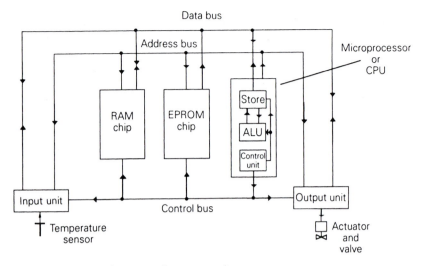

Fig. 2.7   Bus architecture of an outstation.

20 MHz crystal oscillator for its clock circuit. A frequency of 20 MHz means 20 million cycles or pulses a second! The clock unit speed is one of the main factors that determine the **processor cycle time** (or machine cycle time); this is the basic speed of executing an average program instruction and is about 250 ns (250×10^{-9} s).

The outstation itself will also contain a clock unit, a chip separate from the microprocessor but driven by it (Fig. 2.5); this is a **real-time clock** which keeps the time of day and the date, allowing the outstation to control plant items on a time program or schedule. This means that the outstation has its own clock for timing the switching of plant – to switch it off at night and during weekends and holidays.

Figure 2.7 shows how the microprocessor chip is connected to the memory unit and the input and output units; all these units have their own micro chips. Connection between the microprocessor and the other chips is by sets of parallel wires called buses (short for busbars). The **data bus** is for the transfer of data between chips, perhaps for transferring the sensor temperature from the input unit to the microprocessor temporary store, ready for the ALU to compare it with the required setpoint.

The **address bus** is for locating the required data in a memory or a register, or for locating a program instruction. The address is like a telephone number or a postal address. Every piece of data or program instruction stored in the microprocessor outstation must have an address. There could be many sensors connected to the input unit of the outstation and each will have its own definite address and its own storage area in the memory set aside for sensor readings.

Once the address of a piece of data or an instruction has been located on the address bus by the microprocessor, the microprocessor control unit sends a signal on the **control bus** to that address. This signal requests the data or instruction at that address to be sent on the data bus to the new address where it is required. For instance, the control bus could carry a signal for a sensor temperature reading to be sent on the data bus to the microprocessor temporary store.

Systems that use three buses to connect the chips are known as **three-bus architectures**. The three-bus architecture is a very common architecture for the early, small 8-bit and 16-bit microprocessor chips. These and similar chips are suitable for BEMS outstations; the power of a Pentium is just not required for the processing of input/output signals. In the single-board outstation considered here, the buses are 'printed' conductors on the single PCB. In outstations using many PCBs, the buses are contained within the backplane onto which the PCBs are slotted. This way an outstation can be easily expanded.

2.4 Memory

Much of the data from the sensors and the program instructions is stored in the memory chips. The microprocessor only has a small temporary store; but because it is so small, it operates very fast. The separate memory chips are larger and operate more slowly. Figure 2.7 shows the memory chips connected to the bus structure. One chip is a **read-only memory** (ROM) and the other is a **random access memory** (RAM), more accurately a read and write memory. Figure 2.8 shows the principle of a memory chip – its addresses and contents are similar to pigeonholes stuffed with letters. A complete set of memory details is called a **memory map**. A typical 8-bit microprocessor (a bit is a binary digit, a 1 or a 0) uses a 16-bit address bus, allowing $2^{16} = 65\,536$ address locations to be handled.

The ROM chip can only send data or instructions, it cannot receive and store them from other chips in the outstation. In other words, it can only be 'read' from not 'written' to. The ROM chip therefore contains the manufacturer's program and data which the user cannot alter. The program and data are 'burnt' into the ROM during manufacture. Microprocessors, and computers in general, can only deal with binary signals, where 1s and 0s correspond to high and low voltages. This means the memories simply have to store 1s and 0s. The ROM in Fig. 2.9 is an array or matrix of 'switches' that ar permanently set open or closed. The high and low voltages depend on the type of chip structure; many chips are CMOS (complementary metal–oxide–semiconductor) [4], and typical values are 5 V for a high voltage and 0.5 V for a low voltage.

These switches are based on **bipolar transistors** or **field-effect transistors** (FETs) [4]. The absence of a device (here a diode) will result in a

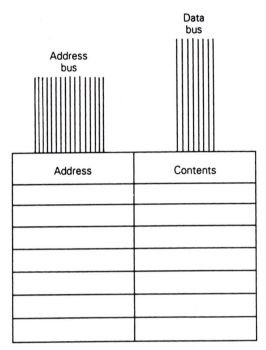

Fig. 2.8   Principle of a memory chip.

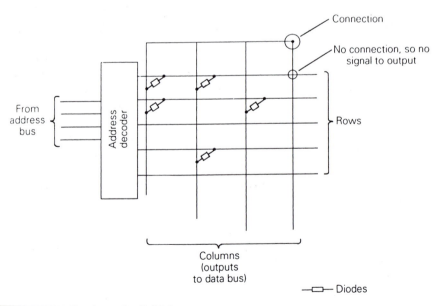

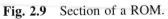

Fig. 2.9   Section of a ROM.

0 being stored, and the presence of a device will result in a 1 being stored. According to the signal on the address bus from the microprocessor, the **address decoder** locates the appropriate row for reading; here only four are shown, each with its 4-bit data word [6]. The decoder is like a set of switches for a particular line. When the switch for a particular line is closed, the signals from that line appear on the data bus. Buffers (not shown in the diagram), controlled by the control bus, store the data before it is sent along the data bus. For instance, if the decoder selects the top line of the memory matrix in Fig. 2.9 then the output on the data bus, going from left to right, is 1100. If the memory stores individually accessed bits of data instead of **bytes** then it is necessary to have a decoder for the column of the memory matrix for addressing individual bit memory locations. Figure 2.9 shows the storage for a ROM; for a RAM there are extra connections to the data bus and to extra buffers for writing data. More details of the memory operation are given in the literature [6, 11].

Standard control functions for the outstation, such as a time schedule, on/off control, optimizer loops and PID loops (explained in later chapters), are written by the manufacturer and stored in ROM chips. The RAM stores the times for switching plant on, the setpoints and the control functions, etc., required from the ROM library; all are aspects the user can set and alter as required. Selecting the control functions required in the outstation, i.e. the control strategy, is more complex than simply deciding setpoints and times, and it is often known as **configuring** the outstation. Many operators do not get involved with configuration; instead they buy preconfigured outstations. Some manufacturers do not allow easy access to the configuration. Configuration is dealt with in more detail later on.

The control functions, or programs, written in ROM chips are considered as **firmware**, whereas the user's data of setpoints and times are part of the user software because they can be easily altered – they are soft. Most familiar programs that run on personal computers are considered as software, and they are often known as **application software**.

The outstation and its chips, and in general any parts of the computer system which can be touched, these are called hardware. Similarly, an outstation's input channels (for sensors, etc.) and output channels (for relays, valves, etc.) are called **hard points**, **real points** or often just **points**. The number of points is a measure of the outstation's size. **Soft points**, also called **pseudo points** or **nodes**, relate to calculations and setpoints for controls.

Although ROMs are unalterable once manufactured, a number of manufacturers actually use erasable and reprogrammable ROMs (EPROMS) so that alterations, such as improvements in later outstation models, can be made using the same memory chips. The EPROMs are erased with low-intensity ultraviolet light and they can be programmed with special equipment.

Some of the smaller and cheaper outstations are totally preprogrammed, with the user only able to put in a rather restricted 'program' of setpoints, times, etc., in a small RAM memory. The larger outstations have greater memories and they tend to have a greater flexibility for configuring control strategies by selecting from a menu of various control functions in the EPROM (optimum start, compensation, PID, etc.). The user's particular configuration for the items of plant, with the relevant time and temperature settings, will be stored in RAM. RAM can be written to by the user so that times and temperatures can be easily changed as and when required.

The 1s and 0s in a ROM matrix are indicated by the presence or absence of simple transistor elements. But a RAM matrix consists of changeable switching elements that need a constant d.c. supply of electricity to maintain their state – the memory contents are volatile. In other words, the program and data stored in a RAM chip are lost if mains power to the outstation is cut. To avoid this catastrophe, outstations have back-up batteries (Fig. 2.5) in case of failure in the mains supply. However, these batteries can be of variable quality and they do not last many years, so they need to be regularly checked. It is also wise to heed the computer maxim: back-up every program. The outstation configuration (the control strategy program) can be copied onto the larger memory of the central station, and the outstation reprogrammed from the central station copy if it is lost at the outstation. Some outstations have **software clocks** instead of real-time clocks, and software clocks can stop if there is no battery back-up.

To appreciate why RAM is volatile memory, consider the **dynamic RAM**, or DRAM. Based on a charged capacitor, its storage elements need regular or dynamic recharging. A **static RAM**, or SRAM, uses flip-flop memory elements that stay in an on or off state as long as there is power to the chip. SRAM is substantially faster than DRAM, but it is also more expensive. There is also **synchronous dynamic RAM** (SDRAM); this synchronizes with the microprocessor clock speed on the data bus for high-speed operation. To find out about these and other types of RAM, consult R. White's *How Computers Work* [15]. The central station is often a personal computer, so it is very similar to an outstation except it also has a screen, keyboard and printer. Another significant difference is that the central station also has a considerably larger memory, not only RAM and ROM but disk storage too, and this is much larger.

Each binary signal (1 or 0, on or off) is called a bit (binary digit); a **byte** is a group of 8 bits which is treated as a unit and stored at a storage location. Many outstations are built around 8-bit microprocessor chips, which work with data and program instructions in 8-bit lengths. Nowadays PCs more commonly use 32-bit and 64-bit chips. Outstations are also following this trend; 16-bit and 32-bit outstations are now relatively common. A machine that can deal with more bits in a unit, or has a longer **word length**, is a more powerful machine and has more capabilities.

Often the RAM chips are put on single-in-line memory modules (SIMMs) for plugging into PCs in units of 256 Kbyte, 1 Mbyte and 2 Mbyte. To keep costs down, most outstations do not have extensive memories, although they vary considerably between manufacturers. Consequently, there is a limit to the data that an outstation can store, just as there is a limit to the number of inputs and outputs and programming that it can handle.

Example 2.1
Consider an outstation with 8 sensors whose temperatures are logged and stored every 15 min. How much data can be stored in 1 Kbyte of RAM?

Solution
If one temperature reading takes up 1 byte of memory then 32 bytes of data are used every hour, and

$$32 \times 24 = 768 \text{ bytes}$$

are used every day. So 1 Kbyte could hold just over one day's data. If the log (the logged data) held hourly readings then one day's data would be 192 bytes. However, if a control was being set up and tuned then its sensor could well be monitoring every 15 s, and 1 Kbyte would then store

$$(1024 \times 15)/3600 = 4.267 \text{ h}$$

of temperature readings, and for just one sensor. Once the memory had become full, unless the data was downloaded to the central station's larger memory, the initial readings would have been overwritten by the later readings. Care must therefore be exercised over data stored in an outstation.

2.5 Operation

Figure 2.10 shows an immersion sensor reading the temperature of water in a hot water cylinder being heated by primary hot water. The primary hot water is controlled by a valve. A very simplified version of how the outstation operates is that the sensor continuously sends back an electrical signal to the input unit of the outstation. The input unit samples this signal and converts it to suitable binary voltage signals for the CPU to work on. The CPU has two basic operations: **fetching** and **executing**.

Executing the relevant program instruction, via its **instruction decoder**, the CPU sends a signal to the input unit to fetch the appropriate converted temperature reading. In fact, the input unit may have signalled to the CPU, perhaps using an interrupt, to indicate that the unit has data ready for processing. Receiving the signal from the input unit that it has data ready, the CPU locates the address in the small input **buffer**, where the incoming signal is temporarily stored, and sends a control signal on the control bus to fetch the binary temperature, on the data bus, to temporary storage in the CPU memory.

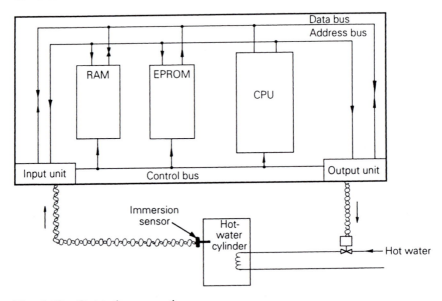

Fig. 2.10   Outstation operation.

The next step of the control program is read by the CPU from the RAM. A program counter in the CPU locates the relevant program step and loads it into the CPU's **instruction register**. This instructs the CPU to take from the EPROM a small control element program; the CPU then runs this program to compare the temperature reading with a temperature setpoint in the RAM, chosen by the user. This is all done by fetch and execute operations, each taking only approximately 250 ns to perform.

If the CPU finds that the sensor temperature is now above the setpoint, the EPROM program will instruct the CPU to send a binary signal to the correct address of the output unit to switch off the valve. The output unit will have to convert the binary signal to a suitable signal to operate the valve. This is an example of the outstation being used for simple on/off control.

In the middle of all this activity, the CPU may well have executed a program instruction to store the sensor temperature in the RAM at a particular address to form part of a temperature log. More details on microprocessor operations for control can be found in the literature [7–9, 11].

This whole procedure will take less than a second for the microelectronics to perform. In fact, the CPU will probably be able to service each input and output point in a matter of milliseconds, although this depends on the outstation program and the number of inputs and outputs available at the outstation. The outstation program has a **sequence table**, which is the order in which inputs, outputs and soft points, or control instructions, are serviced by the CPU. In the example of Fig. 2.10 the sequence table would be as follows:

Sequence order	Item
1	Immersion temperature sensor
2	Control program
3	Control valve output

The other inputs and outputs would then be serviced after these. An item can be serviced quicker by entering it in the sequence table more than once.

The outstation in Fig. 2.5 can accommodate a maximum of 150 sequence steps in its table, and the average **sequence cycle time**, the time to execute all the steps, is up to 5 s. This is effectively the **program cycle time**, and it means that each item is served by the CPU in 5/150 ≈ 33 ms.

Unlike the example in Fig. 2.5, some outstations may have all their program functions (control, alarm handling, reporting, logging, etc.) in one program, sequentially processed. This would produce a program cycle time nearing 20 s [10]. To reduce time, the program functions are split into smaller programs that can be run independently, effectively in parallel at the same time. This is achieved using a **multi-tasking operating system** where the CPU divides its time between the programs or tasks. As the CPU is so fast, each program is effectively as quick as its own program cycle time will allow.

In a multiple-board outstation each card would be served in turn by the CPU card. As these are often bigger outstations than single-board versions, the sequence cycle time will be slightly longer.

The outstation is continuously monitoring the hot water cylinder temperature and continuously executing the instructions in the control program. This is an example of **real-time** or **online computing**, where the microprocessor is interacting and responding to events as they happen. When a program and data are loaded into a PC and run, this is an example of **batch processing**. The software in most PC central stations is a combination of both these types, with an **interrupt service routine** (ISR) to interrupt the main PC operating program, perhaps to raise an alarm. After the interrupt, the PC returns to its operating program.

Although the outstation is performing real-time computing, not every temperature signal from the sensor would be stored for long; only those required for logs at definite times, e.g. every 15 min or as defined by the user, would be stored in the RAM memory.

2.6   Direct digital control

Figure 2.11(a) shows the signal flow diagram for the outstation control. The microprocessor is in complete control, and as it operates with digital signals, it is termed **direct digital control** (DDC). Some manufacturers' stand-alone

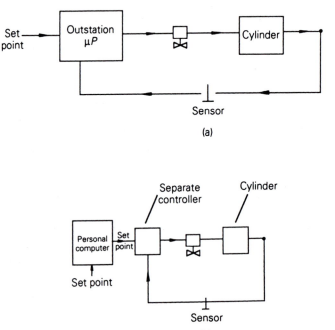

Fig. 2.11   Control: (a) direct digital control, (b) supervisory control.

controls, such as optimizers and compensators, can be interfaced to personal computers (Fig. 2.11(b)). This is not strictly DDC but **supervisory control**, where the PC can adjust the setpoint and in some cases monitor a sensor reading. But the control cannot be reconfigured or interrogated like an outstation, so it has a rather limited role.

A common feature of these two examples is that the sensor feeds back the temperature to the controller – the sensor is in a **feedback loop**. Hence control programs in outstations are known as control loops or sometimes just loops. In the sequence table of the outstation, the control loops should be listed sequentially, along with their inputs and outputs, so that the input signal produces an output signal as soon as possible to keep the signals in phase. The time between an input signal being processed and an output signal being generated is called the **control loop time**, or loop time. This time should be as short as possible for good control.

2.7   Input unit

The input unit, or section, takes signals from sensors, relays, meters, etc., and conditions them so that the microprocessor sees them as digital signals of the correct voltage. For instance, the temperature sensor in Fig. 2.10 is continuously sending back an electrical signal (a small voltage or current) to the input section of the outstation. This is an **analogue**, or continuous, signal; it needs to be converted into bits and bytes to form a **digital signal** that can be processed by the CPU. The input section therefore contains an analogue-to-digital converter (ADC).

The outstation can also receive on/off signals from **volt-free** contacts, which are relays, or spare contacts on equipments relays or contactors, with no voltage apart from the small voltage signal from the outstation to determine whether the relay is open or closed. The volt-free contacts are opened and closed by coils exerting a magnetic influence on them; the coils are in the equipment circuit, so they are energized when the equipment is switched on. Such contacts are useful for monitoring on/off status of plant or events and counting the units consumed on utility energy meters. As these signals are already in digital form, they do not need to be converted.

With volt-free contacts, the outstation is isolated and protected from any large voltages that are being switched. Other methods of isolation are often used to protect outststations, both at the inputs and the outputs, where large voltage equipment is being monitored and controlled. One method is **opto-isolation**, where a current through a light-emitting diode converts the signal into light which is picked up by a closely coupled **phototransistor** or **photodiode**.

2.7.1 Operational amplifiers

A major electronics component of outstations themselves and also of ADCs is the operational amplifier, or op-amp. There are two basic types of amplifier: (1) **small-signal amplifiers**, or voltage amplifiers, for amplifying small voltages or currents, and (2) **power amplifiers**, with power transistors and **thyristors**, for controlling equipment on the output side of an outstation, such as motors and valve actuators [12, 13]. The op-amp is a typical small-scale amplifier which has a high voltage gain (ratio of input voltage to output voltge), often much greater than 1000. The symbol for an op-amp is a triangle (Fig. 2.12 (a)), although a commonly used op-amp, the 741, is an integrated circuit in an 8-pin DIL package, of which only seven pins are required and only three pins are shown in the figure. The 741 is described in more detail in the literature [4, 14].

The word 'operational' in the name of the amplifier arises from its use to perform mathematical operations such as addition, subtraction, multiplication and division. The operational amplifier, which uses only analogue signals, is the building block of an analogue computer for solving mathematical equations. In fact, this type of computer was the predecessor of the digital computer. Op-amps are a main component of non-digital controls, often known as analogue controls; and these controls were the forerunners of

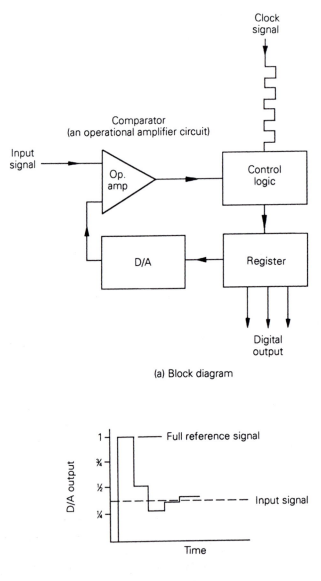

(a) Block diagram

(b) Conversion process

Fig. 2.12   A successive approximation ADC.

digital BEMSs. As op-amps can multiply, integrate and differentiate, it is hardly surprising that a standard control is a proportional plus integral plus differential (PID) controller.

Although an operational amplifier performs its operations with a high degree of accuracy and reliability, the output voltage for a given input voltage

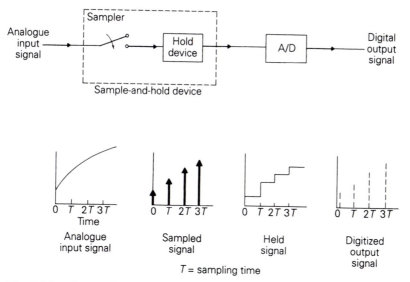

Fig. 2.13 Converting an analogue signal to a digital signal.

will change, or drift, with temperature and age. This is one disadvantage of analogue controls – they need recalibrating to offset the drift over time.

2.7.2 Analogue-to-digital converters

There are several different ways that an analogue signal may be digitized or sliced into a digital signal, but the two commonest analogue-to-digital converters (ADCs) used in a BEMS are **successive approximation** ADCs and **integrating** ADCs.

2.7.2.1 *Successive approximation ADCs*

A successive approximation ADC has the block diagram in Fig. 2.12(a), where the input signal is fed into a comparator (an op-amp circuit) which compares it to successive half-values of a reference signal (Fig. 2.12(b)). The digital-to-analogue converter (DAC) converts the digital signal in the register memory to an analogue value so it can be compared in the comparator. When the two signals are as close together as possible, the conversion is complete.

Although the ADC is successively approximating to the input signal, any change in the input signal would upset the process. So a **sample-and-hold amplifier** can be used to hold the signal constant until the conversion process is complete. For slowly changing signals this is not necessary. Figure 2.13 shows the input signal as it is processed through the sample-and-hold and the ADC. Typical conversion times for these ADCs are around 20 μs.

Figure 2.13(b) shows that the digital signal does not exactly equal the analogue signal, although it comes fairly close. The resolution of the ADC is limited by the word length of the converter.

Example 2.2
A 0–10 V signal comes in from a water temperature sensor. The lower voltage represents 0°C and the upper voltage represents 120°C. What is the resolution of an ADC with an 8-bit word length?

Solution
Resolution may be defined as the change in the input signal for the digital output to change by the least significant bit (the last bit on the right of a binary number). An 8-bit word can be used to represent in binary the numbers from 0 to 255 (2^8 numbers), so the resolution is

$$10 \text{ V}/2^8 = 39 \text{ mV} \quad \text{or} \quad 120°\text{C}/2^8 = 0.469°\text{C}$$

Normally if the signal is above the halfway point between two levels then it is rounded up, and if it is below the halfway point then it is rounded down. So the readings from this ADC are accurate to $0.469/2 = 0.235°\text{C}$. The resolution of the central station monitor will give graphs a stepped appearance due to the number of lines scanned on the screen. This should not be confused with the ADC resolution.

2.7.2.2   Integrating ADC

The second ADC is the integrating type. Here the input signal, in the form of a voltage, is compared to the voltage across a capacitor being charged from a constant current source. A high-speed digital counter, controlled by the clock circuit, counts the time until the two voltages are equal. From this the input signal can be determined in digital form. This is not a precise converter but is suitable for joystick controllers on PCs used for games.

The integrating converter can be made more accurate by altering it to monitor the discharge of the capacitor against a reference negative voltage with a digital counter digitizing the input signal. Figure 2.14 shows the block diagram for this type of ADC, called a dual-slope or dual-ramp converter.

Figure 2.15 shows its operation with the capacitor of the integrator charging up for a fixed time and then discharging at a fixed rate against a negative reference voltage, hence the fixed slope. Once the capacitor has been discharged, the counter stops; the time the counter has been counting pulses, hence the number of pulses, is proportional to the input signal.

Slow but accurate – some input signals take several milliseconds to be converted – this ADC's great strength is that it can integrate a varying input signal and produce an average value without requiring a sample-and-hold circuit. If

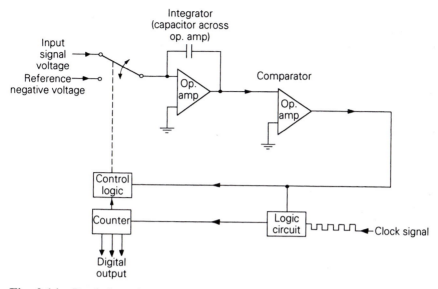

Fig. 2.14 Dual-slope integration ADC.

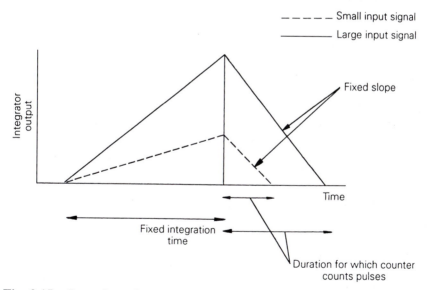

Fig. 2.15 Operation of a dual-slope integration ADC.

the integration is over 20 ms, the period of the 50 Hz mains supply, then any mains interference on the input signal is averaged to zero. As many BEMS sensor signals do not vary very fast (the temperatures of rooms, water, etc., do not vary much in 20 ms), the dual-slope ADC is used in many outstations.

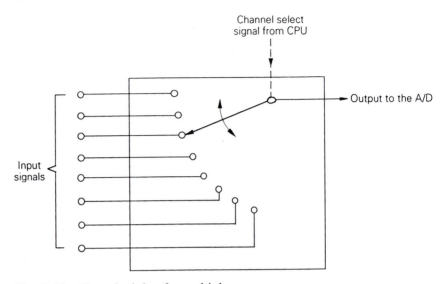

Fig. 2.16   The principle of a multiplexer.

To save having an ADC for each analogue input, a **multiplexer** can switch the sensors in turn to the ADC. The multiplexer (MUX) is a circuit that reads or passes on the information from any selected input channel; Fig. 2.16 illustrates the principle. The early multiplexers, as used on the very first BEMSs, were composed of electromechanical relays. Nowadays CPU outstations have MOSFET multiplexers; MOSFETs are metal–oxide–semiconductor field-effect transistors. The channel is selected by the CPU and the relevant address is sent to the MUX on the address bus [7]. Once the analogue input signal has been converted to a digital signal, it can be stored in a register until the CPU is ready to process it. The CPU will not stay idle while the ADC is converting; it can do many operations in this time. Once the conversion is complete, the ADC could either send an **interrupt signal** to the CPU, or the CPU can regularly poll the multiplexer and other devices, to find any data that is waiting to be processed.

Example 2.3
If an outstation has 8 analogue input channels for sensors and it uses one dual-slope integrating ADC with a multiplexer, how often can each channel be sampled?

Solution
The dual-slope integrating ADC will probably integrate over 20 ms to avoid 50 Hz mains noise, and it is reasonable to add on another 20 ms for interrupt signals or polling to take place before the CPU deals with it. So each channel

could be sampled every $8 \times 40 = 320$ ms, approximately 1/3 s. But the outstation also has to run the control program to produce a control signal as a result of the converted input signal. This program and its sequence table have to be processed before the channel sample can be taken again; a typical sequence cycle time is 5 s. This does not mean that the loop time is also 5 s as typically there are up to 150 sequence steps for a sequence table, and if a loop has five sequence steps then the loop time would be nearer

$$(5/150) \times 5 = 0.17 \text{ s}$$

$$= 170 \text{ ms}$$

2.8 Sampling

Example 2.3 shows that the sampling of the input channels, based on analogue-to-digital conversion and signalling time alone, could be as fast as one-third of a second. Is this too fast or too slow for the average building services plant to be adequately controlled? To answer this, consider **Shannon's sampling theorem**. This states that provided a signal contains no frequency component higher than f_{max} then the signal can be represented by, or reconstituted from, a set of sample values where the sampling frequency is at least $2f_{max}$. In practice the sampling frequency is often 10 times the theoretical limit [8].

Example 2.4
The mixing valve for a hot water heating system takes 30 s to go from fully mixing to fully recirculating. Is a period of 5 s between samples adequate for a temperature sensor just downstream of the valve? Is the analogue-to-digital conversion time satisfactory when using a dual-slope integrating ADC?

Solution
A mixing valve is a three-port valve often used in compensator control of heating systems to mix boiler water with recirculated water returning from the radiators. The time from fully mixing to fully recirculating is half a cycle, if the valve is continuously opening and closing at its greatest frequency. So 30 s from fully open to fully closed is a period of 60 s and a frequency of $1/60 \approx 0.017$ Hz.

Assuming the temperature of the water varies at the same frequency as the valve, the sampling frequency for the sensor should be at least twice 0.017 Hz and in practice 10 times, giving a sampling frequency of $10 \times 0.017 = 0.17$ Hz; this corresponds to a period of 6 s. So a period of 5 s between samples is adequate.

The time taken by the dual-slope ADC can be estimated at 20 ms, as in the previous example; the valve and temperature will not vary significantly during this short time.

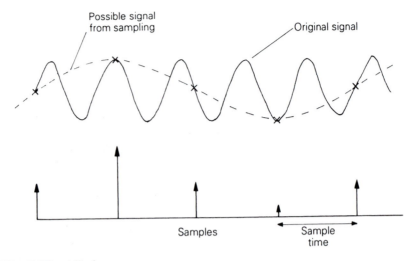

Fig. 2.17   Aliasing.

In practice, to see how the valve or the rest of the plant is operating, selected sampled values from the input sensors can be stored or logged for later evaluation, most likely at the central station. A graph may be constructed by the central station software from this logged data. The outstation will have limited memory, and the sampling rate for this logged data will have to be carefully selected. Too much data and, when the memory becomes full, the later data will simply overwrite the earlier data; the earlier data will be lost. But it will not be lost if it is regularly sent, or downloaded, to the central station for storage, where the memory space is much larger. The frequency of downloading depends on the logging frequency.

There are two regimes for logging data: a short-term regime for tuning control loops at the commissioning stage to see whether the control loop settings provide stability (see later); and a longer-term regime for monitoring. With both regimes it is possible to lose information by having a sampling rate that is too low, then a high-frequency variation may be aliased or interpreted as a low-frequency variation (Fig. 2.17).

Example 2.5
An outstation has a memory set aside for each log that it has; each memory can hold 96 sampled values from inputs. What should the sampling rate be for (a) tuning a loop, (b) long-term monitoring of a temperature?

Solution
For tuning a loop to control a valve which can open in 30 s, Shannon's theorem suggests a sampling time of 5 s should be adequate. But this means the memory can hold data for a period of $(96 \times 5)/60 = 8$ min.

For longer-term monitoring, daily downloading to the central station should present few problems, so 96 samples spread over a day would give a sampling rate of every $(24 \times 60)/96 = 15$ min.

2.9 Output section

Having sensed the temperature of the cylinder via the sensor and the input section, and having put the digitized value through the BEMS control program, the outstation could well have to send out a signal to alter a valve or to switch something. This could be done by simply sending out a pulse. To move a valve by a certain amount, a number of pulses will need to be sent to move the valve for a certain time. This is valve movement on an **incremental** basis, the most popular method.

But besides incremental movement, a valve may require an analogue signal to move it to a definite position, in which case a digital-to-analogue converter (DAC) is required to give a movement on a **whole-value** basis. Figure 2.18 shows a schematic of an 8-bit DAC, where the digital signal is sent through a ladder of resistors with increasing valves; the resultant analogue signal is the output of the summing operational amplifier. This is the reverse of the successive approximation ADC.

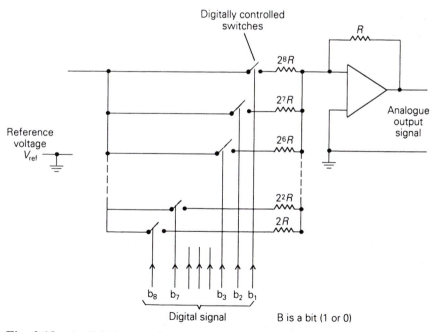

Fig. 2.18 A digital-to-analogue converter.

Example 2.6
If an 8-bit DAC has a digital signal 01000100, what is the analogue output signal?

Solution
Referring to Fig. 2.18, the output signal is

$$V_{ref} = \{b_8(R/2R) + b_7(R/2^2R) + \ldots + b_1(R/2^8R)\}$$

The *R*s cancel to give

$$V_{ref}\{b_8/2 + b_7/2^2 + \ldots + b_1/2^8\}$$

In the given signal only b_7, $b_3 = 1$; the other bits are 0. So the analogue output signal is

$$V_{ref}\{1/2^2 + 1/2^6\} = V_{ref}\{1/4 + 1/64\}$$
$$= V_{ref}\{0.266\}$$

The full-scale analogue output would result from the digital signal 11111111 and it would be

$$V_{ref}\{1/2 + 1/4 + 1/8 + 1/16 + \ldots + 1/256\}$$

which is

$$V_{ref}\{(128 + 64 + 32 + 16 + 8 + 4 + 2 + 1)/256\} = V_{ref}\{255/256\}$$

which is almost the full reference voltage. This voltage can be amplified to match the equipment's requirements, or larger voltages and a.c. voltages can be controlled with relays and contactors. Most valves require a signal in the range 0–24 V d.c.

2.10   Configuration

The outstation control strategy (the relationship between the inputs and the outputs of the outstation, or the program the outstation follows to enable it to control the plant) is set up, or configured, from a library of control elements, or modules, in the EPROM. This means that the library of modules is part of the firmware. There are no standard symbols for these modules; different manufacturers have devised different conventions. Figure 2.19 shows the symbols used in this book; they are used by at least one UK manufacturer.

Later chapters describe the elements in more detail and with relevant control configurations; this section is therefore kept relatively brief. Some manufacturers go directly to a programming language, such as C, C++ or FORTRAN, without the use of configuration diagrams. A control schedule as lines of a program is more compact than configuration diagrams but it can be less easy to understand for the BEMS user.

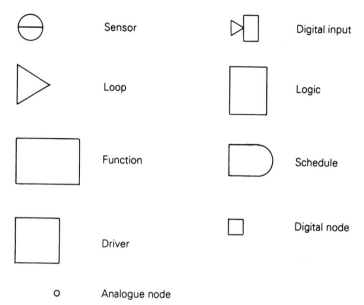

Fig. 2.19 Configuration module symbols.

The **loop module** is the basic control element; it provides an output signal after comparing the input signal, from a sensor, with the required setpoint. Connection to the outstation output is via a **driver module**. Although the loop symbol is a triangle, like the op-amp, it should not be confused with the op-amp. In the following chapters the op-amp is not referred to, so the triangle from here on will only relate to a loop.

The **function module** performs mathematical functions on analogue values: averaging, taking the maximum or minimum from a number of inputs, multiplying, adding, dividing, comparing, etc.

For digital values there are **logic modules** such as the logical AND, NAND, OR, NOR, explained later. They can be used for conditional control, perhaps to switch on the heating at a low level for frost protection if the outside temperature goes below zero. The comparator function module is often used with these logic modules to apply logic to analogue values.

The **schedule module** selects its output depending on the time of day and is effectively a time switch. It is often associated with the optimizer module, which determines the latest time to switch on the heating to get up to temperature by the occupancy time. This is examined in a later chapter.

Sensors and digital inputs are required on **strategy diagrams**, so there are symbols for them shown in Fig. 2.19. Sequence table numbers are placed above each symbol to complete a strategy diagram.

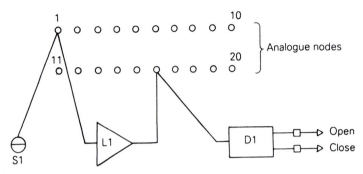

Fig. 2.20   Strategy configuration.

Refer back to the outstation model in Fig. 2.2, which shows the loop and the functions in general. The number of each type of module an outstation can use depends on the size of the outstation. The outstation in Fig. 2.5 has 8 inputs, so it can use up to 16 loop modules, up to 64 function modules plus many other modules.

The strategy required by the user is configured by linking up the required modules to specific storage addresses in the RAM; this is called **soft wiring**. These addressed storage areas, or **analogue nodes**, store the output from a module where it can be read by another module linked to that node. Figure 2.20 shows a temperature sensor connected to a loop, in turn connected to a driver.

The analogue nodes are shown as as a matrix, with certain rows reserved for certain modules. For the sake of illustration, Fig. 2.20 shows a matrix with only 20 nodes; the real matrix has 255 nodes. The top row is for sensors and the second row is for loop ouputs. The driver outputs (in this case to open or close a valve) go to **digital nodes** (the two small squares) which are storage bits in the RAM. Other digital nodes are also stored in a digital matrix similar to the analogue matrix. A digital node may be identified as byte 7, bit 0 or just 7,0. Each byte has 8 bits, numbered from 0 to 7.

An outstation is configured from the central station by going into the configuring part of the software. Here the modules are selected from a **software menu** or library and numbered for identification, then nodes are assigned to the modules. Configuration is a detailed process and some manufacturers do not allow the user to perform it; they will configure the outstations to a client's specifications. Others have standard configurations or **macros** which can be loaded into the outstation, rather than supplying detailed configuration steps. Whatever the manufacturer's procedures, the better the user understands the control strategy, the lower the likelihood of inappropriate or wrong control strategies. Training courses are provided by most manufacturers to impart the necessary understanding.

References

[1] Eyke, M. (1988) *Building Automation Systems – a Practical Guide to Selection and Implementation*, BSP Professional Books, Oxford.

[2] Scheepers, H. P. (1991) *Supporting Technology for Building Management Systems*, Honeywell Europe, Offenbach, Germany.

[3] CIBSE (1986) CIBSE Guide, Volume B, Section B10, Electrical power, Chartered Institution of Building Services Engineers, London.

[4] Diefenderfer, A. J. (1979) *Principles of Electronic Instrumentation*, Holt-Saunders International, Philadelphia PA.

[5] Bentley, J. P. (1988) *Principles of Measurement Systems*, Longman Scientific and Technical, London.

[6] Morris, N. M. (1983) *Microprocessor and Microcomputer Technology*, Macmillan, London.

[7] Stoecker, W. F. and Stoecker, P. A. (1989) *Microcomputer Control of Thermal and Mechanical Systems*, Van Nostrand Reinhold, New York.

[8] Bibbero, R. J. (1977) *Microprocessors in Instruments and Control*, Wiley, New York.

[9] Houpis, C. H. and Lamont, G. B. (1985) *Digital Control Systems: Theory, Hardware, Software*, McGraw-Hill, New York.

[10] Scheepers, H. P. (1991) *Inside Building Management Systems*, Honeywell Europe, Offenbach, Germany.

[11] Money, S. A. (1987) *Practical Microprocessor Interfacing*, Collins, London.

[12] Morris, N. M. (1983) *Control Engineering*, 3rd edn, McGraw-Hill, London.

[13] Chesmond, C. J. (1986) *Control System Technology*, Edward Arnold, London.

[14] Sinclair, I. R. (1988) *Practical Electronics Handbook*, Heinemann/ Newnes, Oxford.

[15] White, R. (1998) *How Computers Work*, 4th edn, Que, Indianapolis.

3
The BEMS central station

The central station, sometimes known as the supervisor or slightly confusingly as the central processing unit (CPU), is the heart of the BEMS and the main communication channel for the operator. It contains the user software and the main storage of data relating to the plant and buildings controlled. In older BEMSs, originating around the late 1970s and early 1980s, the central station was a minicomputer **floor model**, with the electronics chassis robustly encased, standing upright on the floor and integral with the operator's desk, similar to the current **workstations** or **supermicros** but much less powerful. With the advances in chip technology, the central station in most BEMSs is now a standard Pentium-type PC.

3.1   Central station basics

The PC was made possible by the development of VLSI chips and the first microcomputer (the Apple microcomputer) was made in 1977 by Wozniak and Jobs. Because it is based on a microprocessor, the central station PC is very similar to the outstation, except it also has a keyboard, a screen or monitor, a much larger primary memory and a large backing memory.

The qwerty keyboard (the top line of letters on a standard keyboard starts with QWERTY) is similar to a typewriter keyboard, with additional **function keys** (to perform special operations or routines), cursor movement keys and a pad of numeric keys.

Compared to the outstation, the PC has a much larger primary (RAM and ROM based) memory which can be part of the hundreds of megabytes of RAM installed in single-in-line memory modules (SIMMs). Central station PCs also have large backing or hard disk memories for the Microsoft Windows software, so common on most PCs, and the BEMS operator software and data storage; hard disk storage is typically measured in gigabytes or megabytes.

The primary memory consists of the ROM and RAM chips; the backing or secondary memory consists of non-volatile magnetic disks containing the central station's large software library. As with the outstation, the central station power is characterized by its CPU or microprocessor chip, its speed (MHz) and the size of its primary and secondary memories. More basic information on PCs and their operation is given in the literature [1, 2].

3.2   Secondary memory

Personal computers use two types of disk, floppy and hard. Floppy disks (Fig. 3.1) are now invariably 3.5 in in diameter (the 5.25 in disk is virtually redundant); they are made of plastic coated with a magnetic material and contained within a stiff plastic cover. The cover has a shutter for the head of the disk drive to read or write to the disk.

The disks may be either **double density** (DD) or **high density** (HD); high-density disks store 1.44 Mbyte, and double-density disks store about half this amount. The disks are now sold already formatted (i.e. the sectors, tracks and clusters are already set up). The disks are rotated at about 300 revolutions per minute in the disk drive and the information is stored on the disk in sectors of circular tracks on the disk. The size of the sectors is determined when the disk is initially formatted or programmed. Note that the information on an A4 sheet of paper with single-spaced typing would need about 4 Kbytes.

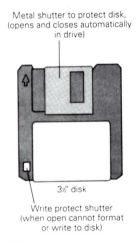

Metal shutter to protect disk,
(opens and closes automatically
in drive)

3¼" disk

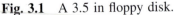

Write protect shutter
(when open cannot format
or write to disk)

Fig. 3.1   A 3.5 in floppy disk.

Iomega have produced a **Zip** disk drive that can take thicker 3.5 in zip disks that can store 100 Mbyte and more. The drive can be installed inside the PC itself or connected externally. This provides an extra drive, a D drive, to add to the storage of the C drive. Such disks can be used for back-up storage, although tapes are often used for this. There are two popular types of tape drive: the **quarter-inch cartridge** (QIC) drive and the **digital audio tape** (DAT) drive. Both have a read head to check the data written onto the tape by the write head. Both are capable of storing a gigabyte or more of data and are therefore useful for back-up.

Hard disks are made from aluminium rather than plastic, and because of their rigidity, they can store more information than a floppy disk and they store it more closely packed. A typical hard disk can hold at least 10 Gbyte and it is increasing all the time. Hard disks can also be rotated faster than a floppy disk and this means they can access information more quickly. The life of the early hard disks was about 10 000 h but modern disks have lives of up to 50 000 h. However, it is always wise to back up a hard disk with zips or tapes just for that awful hour.

A **compact disc read-only memory** (CD-ROM) can store either 74 min of music or 650 Mbyte of data on its 4.75 in diameter surface. A beam of light from a laser diode is focused on the disk. Small pits in the reflective surface scatter the light whereas lands (the area of the reflective surface not cut by a pit) reflect the light back to a light-sensing diode. Hence the binary data can be read. Whereas the tracks on floppy and hard disks are concentric circles, the CD-ROM track spirals out from the centre. Much software is now distributed on CD-ROMs, and they are also available in forms that are writable or rewritable. Write-once read-many (WORM) and write-many CD-ROMs are two options for storing large amounts of data.

All the disks described so far store the data in tracks that are spirals or concentric circles. These tracks are split into sectors so that **files** of data and programs can be located. A **disk directory** of all the files contained on the disk is written on the disk and is updated each time a file is written or saved. The updating is done by the computer **disk operating system**. Examination of the disk directory in Windows Explorer for Windows 98 reveals each file's name, its type, its size in bytes, the date and time it was written and the free space left on the disk in bytes.

Floppy disk drives and hard disk drives are controlled from the microprocessor by controller cards or interfaces on the motherboards themselves. A hard drive that contains most of the control circuitry itself is termed an **integrated drive electronics** (IDE) drive and communication is by the IDE standard. There are also **small computer systems interface** (SCSI) drives. SCSI amounts to an expansion bus (see later) for plugging in CD-ROM drives, scanners and hard drives. Again most of the controller circuitry is on the device, allowing the SCSI to be free to communicate with other devices. Up to seven SCSI devices can be daisy-chained to a single SCSI port.

3.3 Monitors

A **colour monitor** or **display**, sometimes visual display terminal (VDT) or visual display unit (VDU), with at least 256 colours is standard with most modern PCs and the colour is an essential element of BEMS schematics. A typical display is 768 lines by 1024 pixels (picture elements); a pixel is created by several adjoining spots of light. These spots are created when electrons hit the phosphor coating on the inside of the screen. The common display standard, introduced by IBM in 1987 is the **Video Graphics Array** (VGA). The VGA display converts digital data into an analogue signal to vary the brightness of the pixels. Super VGA displays are an enhancement of the standard VGA display, using special chips and requiring a bigger memory for a greater range of colours. Up to 16.7 million colours are available but to use all of them it would require 3.9 Mbyte of video memory.

Notebook computers also have colour displays but here a liquid crystal display (LCD) is used. The liquid crystals, long rod-shaped molecules, can be twisted by an electric charge. Polarized light sent through the crystal changes its plane of polarization. A polarizing filter behind the crystals polarizes the light that is shone through. Another polarizing filter on the other side of the crystals will only pass light that has been twisted by the crystal, and the crystal can only twist the plane of polarization if the crystal itself is twisted by a voltage. This is how the display is controlled. The colour is provided by colour filters placed before the second polarizing filter. The quality is not nearly as good as its electron tube equivalent on the desktop machine, but it is good enough for local use to connect to a BEMS outstation. Some outstations use LCDs as small displays alongside keypads for limited access and information.

3.4 Printers

Although they are now rather dated, there are still many **dot matrix** printers in use with BEMSs. Dot matrix printers and their consumables are cheaper than inkjet or laser printers and their consumables. The dot matrix printhead consists of between 7 and 27 pins arranged vertically. Each pin is controlled by its own electromagnet which moves the pin down when it receives a signal voltage. The appropriate pins are struck against an inked ribbon and the paper to form the characters. This is for **draft-mode printing**. Slower, higher-quality printing requires **letter-quality mode**, where the printhead strikes the pins twice for each character, its position slightly shifted between strikes.

Fig. 3.2 A mouse.

Inkjet printers have up to 50 ink-filled chambers with fine nozzles attached to them; at the bottom of the chamber is a resistance film. When heated by an electric current, the film heats up a thin layer of ink; the ink then boils and forces a small drop out of the fine nozzle onto the paper [3]. The print is of high quality and the printing is fast. One distinct advantage is that a number of coloured inks can be used to print in colour. **Laser** printers are considered the top-quality printers, working like a photocopier which directs toner onto a drum by electrostatic charge; the drum then prints onto a page of paper.

3.5 Mouse

A mouse (Fig. 3.2) is an input device for selecting symbols or **icons** on the screen. Originally developed to minimize the size of the PC keyboard, Apple used it as an essential component of their Macintosh computers. As the mouse proved so useful, especially in making drawings on the screen, it was incorporated by other manufacturers and is now a standard item with PCs.

3.6 Expansion

As some outstations can be expanded by adding extra circuit boards or cards, so most PCs have **expansion slots** on their motherboards for additional cards or for replacing and upgrading existing cards. The expansion slot is a connection to the PC's data, address and control buses, into which boards can easily be inserted. This allows for different video, disk controller and other cards to be inserted at the user's discretion and it allows freedom in designing a PC to suit one's needs. The idea of the PC bus was introduced by IBM in 1981 with an 8-bit bus. This was upgraded to 16 bits in 1984 and became the **Industry Standard Architecture** (ISA) bus. In 1987 IBM introduced the **Micro Channel Architecture** (MCA) bus, a 32-bit bus, but it did not accept the older 8-bit and 16-bit ISA cards and IBM did not release it to other companies. Its competitors, led by Compac, then introduced the 32-bit **Extended Industry Standard Architecture** (EISA) bus, which was downwardly compatible with existing 16-bit ISA cards. EISA buses could transfer data at 33 Mbytes per second and an improved EISA-2 bus ran at 132 Mbytes per second. Being expensive and complex, EISA did not become very popular. However, with faster and faster microprocessors being developed, the relatively slow ISA and MCA buses (operating at 8.22 and 10 MHz compared to the then 33 MHz microprocessor) caused an expansion bus bottleneck, giving slow overall PC performance.

In 1992 the high-speed **local bus** was developed; its two versions give direct access to the microprocessor CPU and they are known as the **Video Electronics Standards Association** (VESA) local bus, or **VL-bus**, and the Peripheral Component Interface (PCI) bus. The local bus is in addition to the ISA, MCA or EISA bus, but it is for the faster devices. The **PCI bus** is a 32-bit expansion bus introduced by the Intel Corporation in 1992 to work with its Pentium microprocessors. It has become the industry standard and is currently being upgraded. The PCI bus also supports **plug-and-play** (PnP), an industry-wide standard for add-in hardware that requires the hardware to identify itself in a standard fashion. This enables the device to communicate with the microprocessor and other devices, without interference from another device and vice versa. Windows 98 supports the plug-and-play standard.

An interesting expansion card is a **network adapter**, a sophisticated circuit board for connecting a PC to other PCs in a **local area network** (LAN). A LAN is the cable connection of two or more PCs so that they can share programs, printers, large high-capacity hard disks, file servers and access to the Internet (see later). This greatly enhances each PC and allows communication between them. The linking of BEMS outstations to a central station is akin to a LAN and it is discussed later on.

3.7   Connections

An **interface** is required to connect the central station PC to an external device, such as a printer or monitor. The interface may be **serial** or **parallel** and it ensures the device is effectively connected direct to the PC bus system.

Under the control of the CPU a parallel interface transmits information, one byte at a time, to and from the relevant device, often a printer. Each bit of the byte is sent simultaneously down its own wire, the wires being in parallel. Actually more than eight wires are needed: eight wires for the bits, some extra wires for controlling the transmission and reception, and one wire for a reference ground voltage. To avoid **crosstalk**, the cable length is limited to about 3 m. Crosstalk is where a voltage in one wire can induce a voltage in a neighbouring wire, just as someone else's telephone conversation can leak into your call.

A parallel interface for connecting scientific instruments together or to a PC is the General Purpose Interface Bus (GPIB), developed by Hewlett-Packard to connect laboratory instruments to a PC; it is a standard interface for 16 parallel wires. In the United States, the IEEE (Institute of Electronic and Electrical Engineers) adopted this standard in 1975 as IEEE-488. It was updated in 1978 and is also known as the IEEE bus. Eight lines are for data and information, five are for general interface management to ensure an orderly flow of information, and three are for data transfer control, or **handshaking**. Handshaking consists of control signals making it possible for two electronic circuits to synchronize their work – it allows them to talk to each other. The receiving device, for instance, has to be ready to receive data from the sending computer.

The original interface standard for sending information to printers was the Centronics standard, based on one of the first printers made by Centronics Data Computer Corporation. A common Centronics connection to the printer has 36 edge-connecting pins and the male connector is held in place by two clips on the edge of the female connector. Improved versions of the parallel port are the **enhanced parallel port** (EPP) and the **extended capabilities** port (ECP); they can transmit data at around 10 Mbps (megabits per second), suitable for connection to PC networks.

A serial interface or serial port transmits data from the PC one bit at a time and one bit after another. This is slower than parallel transmission but it needs fewer connections and less wiring. It uses a 25-pin or 9-pin connector and is commonly used for connecting a mouse or a modem where the speed of data transfer is adequate. The modem is restricted in its speed by the telephone line it uses, and the traditional analogue phone line can only send one message at a time, serially. Very popular for sending digital information serially is the method or **protocol** (set of rules) that complies with the North American Electronic Industries Association (EIA) standard RS-232-C; this also conforms to the internationally recognized V.24 standard except that the on/off pulse shape is different. The V.24 standard was drawn up by the Comité Consultatif International Téléphonique et Télégraphique, an organization superseded by the ITU-TSS (see later). RS-232 was originally introduced in 1960 then revised in 1963 (RS-232-A), in 1965 (RS-232-B) and in 1969 (RS-232-C). The RS-232-C standard specifies the functions and voltage levels of the 25 pins in the standard **D-type** connectors. The D-shape comes from the 13 horizontal pins above the 12 pins. They are also known as **DB** (data bus) connectors; a 25-pin version is called a DB-25 connector. A 0 is represented by a voltage between 5 and 15 V and a 1 is represented by a voltage between −5 and −15 V. Often few of the 25 wires are actually needed. Signals can be satisfactorily transmitted over distances up to 15 m. Transmission over greater distances, up to 1000 m, requires equipment that meets standard RS-423-a.

Most PCs have at least one serial RS-232-C interface or port from the motherboard or expansion slots to a D-type socket on the back of the chassis or case. And most outstations have RS-232-C ports for connecting to a small portable PC, laptop PC or keypad for local programming and adjustment; this is besides any integral display and keypad. A new standard, RS-499, has been introduced to replace RS-232-C [4]. Current PCs also have **Universal Serial Bus** (USB) ports for transferring data at up to 12 Mbps as speed requirements increase. PCs with a USB controller (specialized chips and connections) can handle dozens of devices daisy-chained from a single port using four-wire cables and plugs. Two of the four wires in the cable provide power to the external devices, thereby eliminating independent bulky power supplies. The other two wires, D+ and D−, send data and commands. A high voltage on D+ but not on D− is a 1 bit and a high voltage on D− but not on D+ is a 0 bit. The PC has a **host hub** for connecting the USB sockets, but independent external hubs can also be connected; monitors and other devices may incorporate hubs. This allows the universal serial bus to handle up to 127 devices [2].

Another common serial standard is the 20 mA current loop where voltages can range over ±80 V with the current held constant at 20 mA. But it is mainly due to the RS-232-C standard that many PCs and devices from different manufacturers can use the same serial interface. More importantly, RS-232-C is very similar to a common standard that defines the handshaking signals used to control standard modems by which a BEMS central station can communicate with distant outstations over telephone lines.

3.8   Modems

For communication between the central station PC and a remote outstation in another building, the PC is connected via the RS-232-C port to a modem (Fig. 3.3) which modulates the digital signal into an audio signal suitable for transmission over the public switched telephone network (PSTN). There is a modem at the receiving end which demodulates the audio signal into a digital signal for the outstation microprocessor. The term 'modem' derives from its function of modulating and demodulating signals. The sending modem sends data by switching between 1.07 and 1.27 kHz, and the receiving modem sends data by switching between 2.025 and 2.225 kHz to represent the 0s and 1s.

Fig. 3.3   A stand-alone modem and a modem board to go inside an outstation. (Photo courtesy Trend Control Systems Ltd)

The PC could well have an **internal modem** on its motherboard or as a separate card on an expansion slot. In this case the telephone line simply connects to the internal modem through a connection on the back of the PC. An **external modem** is in its own box; it needs a separate power supply and is connected to the PC via the RS-232-C port or through the USB port. It is connected to the telephone line by an **RJ-11** four-wire telephone connector, also known as a modular jack.

Just like the central station, the outstation needs a modem and a serial interface. Often the outstation has **autodial** facilities to enable it to automatically dial the central station's phone number to raise important alarms or to download sensor logs, preferably overnight to save on phone charges. And the oustation may also be reprogrammed, receive new setpoints, have its data examined, etc., simply over the telephone.

Although ordinary PSTN lines can be used for BEMS communication, it is possible to use an all-digital telephone service called **ISDN** (integrated services digital network). ISDN is a worldwide standard for data and telephone services [5, 6] in three categories: **basic rate interface** (BRI), **primary rate interface** (PRI) and **broadband ISDN** (B-ISDN); B-ISDN is still under development. There are several channels standardized for ISDN, denoted by letters A, B, C, D, E and H. The basic rate ISDN supplies two 64 Kbps (kilobits per second) channels, or B-channels, and a 16 Kbps D-channel for dialling, ringing and caller identification for the other two channels. PRI has 30 B-channels and one D-channel in Europe and 23 B-channels and 1 D-channel in the United States and Japan. This PRI service is for larger organizations whereas BRI can be used by individuals at home or for BEMSs to communicate over a reliable system using the same wires as the standard telephone system. ISDN users have a **terminal adapter**, sometimes wrongly called an ISDN modem. Modems provide a digital-to-analogue conversion before transmission over the standard analogue telephone system, whereas the terminal adapter simply connects digital systems. The terminal adapter can be connected to a PC as well as a fax and telephone. ISDN uses on-the-fly (during transmission) error detection and correction to provide an error-free transmission, ideal for fast communication for PCs or BEMS central stations with data to transmit. Although not yet used extensively by BEMSs, it will become increasingly accessed as communications expand and improve.

The early development of modems and data transmission over the standard telephone lines were hindered by the monopoly power of a small number of telephone companies or utilities dictating what could or could not be connected to the telephone line. An early modem of the 1960s and 1970s, the Bell 103, could operate at 300 bps (bits per second). With the break-up of the telephone monopolies in the late 1970s there was a rapid increase in modem development. Hayes Microcomputer Products took the lead in modem development and pioneered the use of a microprocessor in the modem itself – the **smart modem**. And as they were the pioneers, their control instructions have virtually become an industry standard – the **Hayes command set**.

In 1985 the CCITT introduced a V-dot standard, V.22bis, for 2400 bps modems. CCITT was an international organization that designed standards for analogue and digital communications involving modems, computer networks and fax machines; it has been superseded by the International Telecommunications Union Telecommunications Standards Section (ITU-TSS). Other V-dot standards were introduced later, with V.90 being a standard for 56 000 bps. This probably represents the limit for analogue/voice telephone transmission. ISDN can achieve much higher transmission rates (as shown above).

If it is required that a BEMS central station should communicate with its outstations over the PSTN, not only will it require interface cards and modems, it may also need software modifications if it does not already have communications software. One cannot simply connect BEMSs to modems.

With individual bits of data being modulated, sent over the PSTN and then demodulated, there is the possibility of stray voltages and interference on the line causing corruption of the data. Hence some form of **error checking** is employed. This is often a **parity check** on each byte of data sent. Seven of the byte's eight bits are used for the information and another bit is added for the party bit. The parity can either be even or odd. Consider the following byte:

bit 7	bit 6	bit 5	bit 4	bit 3	bit 2	bit 1	bit 0	.
1	1	0	0	0	1	0	1	

Its first seven bits, running from right to left, have odd parity; if an even parity check is used then the parity bit, bit 7, is made a 1 to make the whole byte even. The error-checking convention could be that all bytes will have odd parity, so with bits 0 to 6 the same as before, the parity bit will now be 0:

bit 7	bit 6	bit 5	bit 4	bit 3	bit 2	bit 1	bit 0
0	1	0	0	0	1	0	1

In **asynchronous transmission** the receiver does not know when the transmitter of the information is going to send information. The line is then left at a voltage indicating a 1 (for RS-232-C this is −5 to −15 V, which is interrupted by a start bit of 0 (for RS-232-C this is 5 to 15 V). A stop bit is used to indicate the end of the byte. So the above byte now looks like Fig. 3.4. **Synchronous transmission** is faster than asynchronous transmission as the transmitter and receiver have synchronized clock signals and do not require the start and stop bits.

The route between the central station and the outstation, via PSTN or otherwise, is called a channel. When a message is sent from the central station to the outstation, the outstation can make a duplicate of the received

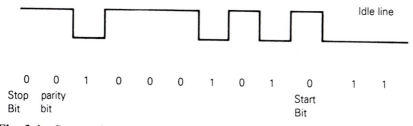

Fig. 3.4 Start and stop bits for transmission.

signal and send it back to the central station; this appears under the original message on the central station screen and it is used to check for successful transmission. A channel like this that can send messages simultaneously in both directions is a **full-duplex** channel. A **half-duplex** channel does not produce this echo and can only transmit in one direction at a time.

For a computer to transmit text as well as numbers, it has to convert them to binary signals. Morse code was the earliest code for electric transmission of information. Although dots and dashes can be deciphered by a human operator, Baudot code is better for teletypewriters, which do the decoding themselves. The Baudot code uses five pulses of equal length to represent letters; it is limited in the number of characters it can represent, upper and lower case. So in 1966 a number of manufacturers developed ASCII (American Standard Code for Information Interchange); this 7-bit code has now become an industry standard.

The central station PC keyboard characters are also converted to ASCII for the CPU to understand them. For instance, the letter E in ASCII is 01000101, which is the decimal number 69. This example has eight bits; the extra bit is a parity bit. If even parity is used, add a 0 when the number of bits is even. If odd parity is used, add a 0 when the number of bits is odd (as in the example). Parity checking is a simple test for validity.

3.9   Networks buses and structured cabling

As BEMSs have been developed using microprocessor and PC technology, it is worth discussing PC networks and buses to give a background to BEMS communications. Many buildings now have extensive IT networks and these can also be used for BEMS [7, 8], with BEMS manufacturers increasingly providing software and connections to these networks in their outstations.

3.9.1   PC networks

Traditionally when one thinks of communications, one thinks of the telephone or television. PCs can be linked to both, to the telephone by a modem and to cable TV by a cable modem and an Ethernet cable (see later). However, with the advent of digital communications, e.g. ISDN, the potential for communication between ordinary PCs and PC networks is great. **Computer telephony integration** (CTI) is now well established, with many organizations giving their employees **voice mailboxes** for their telephone messages to be recorded.

However, **local area networks** (LANs) are more relevant to BEMSs. The standards and protocols for computer **interoperation** (different manufacturers' computer systems working together) emerged in the early 1980s primarily from IBM, the US Department of Defense (DoD) and the Xerox

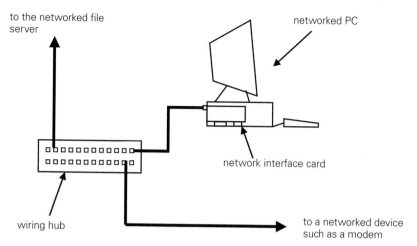

Fig. 3.5 A LAN and some of its components.

Corporation's Palo Alto Research Center (PARC). The DoD was involved as it used computers for its air defence system. In the 1970s, the DoD had many computers that could not be easily linked (they could not interoperate), so it devised the TCP/IP protocol (Transmission Control Protocol/Internet Protocol) for communications between computers over quite long distances. This is now the basic communication protocol used on the Internet. (A protocol is a standard that specifies the format of data as well as the rules to be followed in networking and communications.)

Also in the 1970s computer companies were beginning to link up computers in LANs. The **Ethernet** hardware, communication and cabling standard was conceived and developed at PARC. This standard is for a high-speed single-channel bus LAN with carrier sense multiple access with collision detection (CSMA/CD) protocol for access to the bus (see later). At Datapoint Corporation the **Attached Resource Computer Network** (ARCnet) standard was developed but kept as a proprietary protocol. In 1986 IBM brought out its **token-ring** networking topology. A ring topology is shown in Fig. 3.7 and discussed in Section 3.10. Actually the token ring is in a central hub from which all connections are made, so it appears as a star network, or hybrid star/ring configuration. Ethernet and ARCnet are two of the four specified BEMS networks in ASHRAE's BACnet specification (see later), which can be used for the main backbone LANs. A LAN consists of a **wiring hub** with a variety of cables going out to a number of PCs and other devices, e.g. printers and modems (Fig. 3.5).

All these devices have **network interface cards** or **LAN adapters** added to the PC's expansion bus that control the data flow between the PC's internal data bus and the serial stream of data from the network cable. A **network operating system** (NOS), which consists of a number of programs or networking software, e.g. Novell's NetWare, is required to run the LAN. Windows 98, 95 and NT have built-in networking functions. The PC controlling the LAN is the server, or file server – a dedicated PC often with a large memory and a fast microprocessor.

3.9.2   Cabling and structured cabling

In an Ethernet (or IEEE standard 802.3) LAN, a commonly used system, thick or thin coaxial cables are often used as the bus. Sometimes the cable runs from one PC to another via a T-connector at each PC. If there is a break in the cable, it affects all the devices on that cable. Increasingly used in Ethernet LANs are **unshielded twisted-pair** (UTP) cables, with eight pairs of twisted plastic-coated wires, similar to telephone wires, contained in a plastic outer jacket. These emanate from the wiring hub and connect to a PC by RJ-45 connectors, similar to telephone plugs but with eight connections. Twisted-pair cabling systems are often known as **10baseT** or **100baseT**, referring to 10 Mbps or 100 Mbps using baseband signalling (base) and twisted-pair cabling (T).

The twisting of the wires in UTP cables cancels out **crosstalk**, noise from adjacent pairs, as well as noise from other devices and wires. Like telephone cable, UTP lacks ideal twisting and other electrical characteristics, but it is cheap, okay for reasonably fast data transmission and easy to pull through conduit and trunking. It has five categories, **Cat 1** to **Cat 5**; Cat 1 is telephone cable or RS-232 cable and Cat 5 can handle data at 100 Mbps.

Shielded twisted-pair (STP) wiring is used in the IBM token-ring LAN. Here the cable typically has two twisted pairs of wires, each pair enclosed in foil and then both pairs enclosed in another foil shield with braided copper shielding around that. The whole is enclosed in a plastic outer jacket. This gives extremely good protection from outside electrical interference but makes it costlier than UTP cables and less easy to install.

Fibre-optic cables can also be for very fast data transmission over long distances. Often a cable consists of two glass fibres with their own plastic shields, surrounded by reinforcing Kevlar fibres contained within plastic outer jackets. But connection is difficult and special connectors are employed to produce good optical transmission.

Structured cabling provides a standard way to wire a building for all types of network in an IT system. A main distribution frame, housed in a cabinet, acts as an interface to the outside world, perhaps to the telephone system. A vertical cable, often fibre-optic, then runs from the main distribution frame to distribution cabinets on each floor; the cabinets contain **patch**

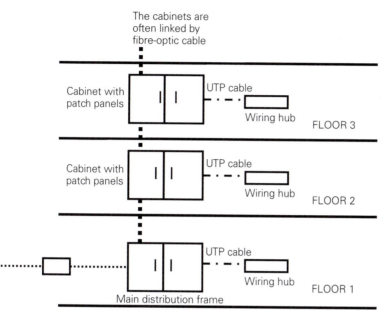

Fig. 3.6 Structured cabling.

panels for connecting horizontal UTP cable to wiring or LAN hubs, from which UTP cables can be run to wall sockets for PCs and devices (Fig. 3.6).

3.9.3 LAN connection

There are limits to the sizes of LANs and LAN cable lengths and also the number of devices that can connect to each run, or segment, of cable. **Repeaters** and **bridges** enable this extension. A repeater receives a signal and retransmits it to other cable segments. A bridge is selective in its transmission and can reduce unnecessary data transmission. A **router** connects a number of networks and decides, using routing tables, the best way to get data from one point to another; it is like a sorting office for letters. Routers may add routing information to the data packets they receive, helping to direct them through the networks. If the LANs do not have the same protocol, maybe they come from different BEMS manufacturers, then a **gateway** with its own microprocessor is required to translate one protocol into another. Figure 3.7 shows some of these devices. Figure 3.8 shows a typical LAN and Fig. 3.9 shows a large building using interconnected LANs (sub-LANs). Here the bridges are in the outstations on each floor.

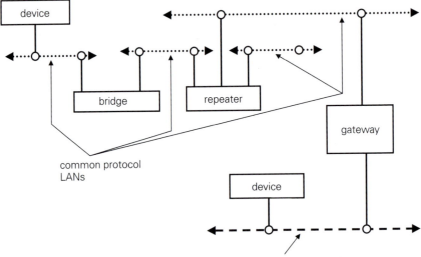

common protocol
LANs

proprietary LAN with different protocol

Fig. 3.7   The connection of LANs.

3.9.4   The Internet

By 1969 the US Department of Defense (DoD) had developed **ARPANET**, a reliable, universal and self-healing communications system for its command and control systems using different computers around the country [3]. This was the test bed for developing TCP/IP (Transmission Control Protocol/ Internet Protocol) used on the Internet. In 1983 ARPANET was split into a high-security military network (Milnet) and an R&D network. In the 1990s the independent **Internet Architecture Board** (IAB) took over developing and regulating the Internet. An internet is a connection of LANs with a common communications protocol using routers. Many internets exist that are not connected to the **Internet**, the enormous, worldwide system of linked LANs used for **electronic mail** (e-mail), **file transfer** and the **World Wide Web** (WWW). There are also **intranets**, which are internets for certain organizations only.

Each computer on the Internet is given an **IP address**, four numbers between 1 and 256 separated by periods, e.g. 192.168.177.155. Connection to the Internet is via an **internet service provider** (ISP) which has high-speed access to the Internet **backbone** network. There are extra protocols for e-mail, where individual PCs are not connected to the Internet but connected to an e-mail server, a larger PC dedicated to controlling the internal LAN and its communication with the Internet. A simple protocol for communicating with the server is by using POP3 (Post Office Protocol 3). This way a user can log in, log out, fetch, store and send messages.

An e-mail address, such as author.thisbook@umist.ac.uk is a series of ASCII strings, not binary numbers. As all PCs only respond to binary numbers, the **domain name system** (DNS) protocol maps an e-mail address or name onto a binary IP address. The domains for the names are generic, e.g. ac (academic organization), uk (the country domain) in the address above; com would be a commercial organization, org a non-profit organization.

Many organizations with structured cabling around their buildings can easily connect PCs to an e-mail server. Other devices, such as BEMS outstations or a BEMS bus, can also be linked to the organization's intranet or to the Internet server, thereby saving separate BEMS communications cabling. A number of BEMS manufacturers are now providing bridges to intranets and the Internet.

3.9.5   The Web

The World Wide Web, or simply the Web, is a framework for accessing linked documents that has an easy-to-use colour graphical interface. It began at CERN, the European centre for particle research, where a number of researchers in different countries needed to discuss and design experiments and analyse the results. Berners-Lee, a physicist at CERN, proposed a web of linked documents in 1989. The first graphical interface, or **browser**, was released in 1993 under the name **Mosaic**. Its author, Marc Andreessen, then formed his own company, Netscape Communications Corp., which produced the Netscape Navigator browser. In 1994, CERN and MIT set up the World Wide Web Consortium to develop the Web, standardizing protocols and encouraging interoperability between sites. MIT runs the US part of the consortium and the French research centre INRIA runs the European part. The consortium's web page is http://www.w3.org.

The Web uses Hypertext Transfer Protocol (HTTP) for exchanging information; http://www.w3.org is a **universal resource locator** (URL) where http is the protocol and www.w3.org is the PC where the page is located. The web is a worldwide collection of documents, usually called pages, which may contain links to other related pages anywhere in the world. This linking is done using **hypertext**. Hyperlinks are strings of text that are underlined, highlighted or coloured, and clicking on them moves the user to the linked page. The software that enables this to happen is the browser combined with the HTTP protocol. The browser sends an HTTP request to the Web server, the Web server sends back a stream of **Hypertext Markup Language** (HTML) embedded with graphics and other scripts, and the PC then recreates the page from the HTML.

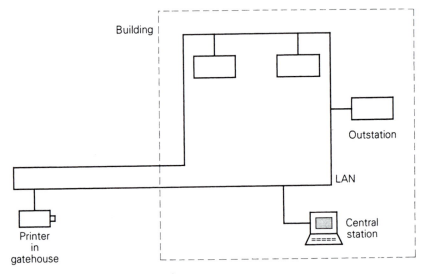

Fig. 3.8   A local area network.

A number of organizations actually display their BEMS data and graphics on a Web page for the world to see, not just the maintenance people and operators from a distant site. Links to maintenance databases and plant and building drawings can make Web pages linked to the BEMS very useful. Occupants in buildings can leave comments, complaints, etc., using a Web page and receive replies by e-mail. General information can also be posted on the Web. Nevertheless, it may be some time before telephone calls to the site operative are totally replaced.

3.9.6   BEMS LANs

BEMS central stations, outstations, sensors and actuators can be linked together on a LAN or a hierarchy of LANs in a big building or buildings. Figure 3.8 shows a building with the BEMS LAN extending to a printer in the security gatehouse for alarms out of hours when the building is unoccupied. The printer would have to have its own network adapter card.

Figure 3.9 shows a a large multi-tenanted office block of three floors. Small outstations, each with say four inputs and four outputs, will control the individual items of plant (often air-conditioning fan coil units or variable air volume terminal units) on each floor. Each floor would have its own separate LAN, a **sub-LAN**, for its small outstations. A larger outstation would be master of the sub-LAN and would be joined into the main LAN of the building, which carries the central station and the other sub-LAN masters (Fig. 3.9).

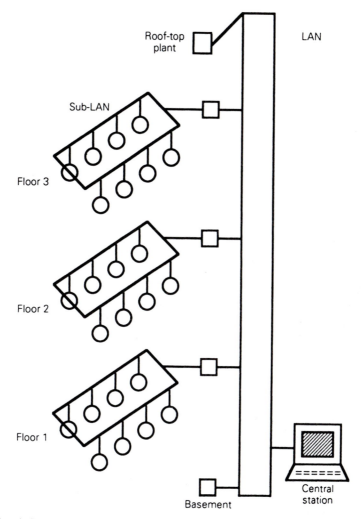

Fig. 3.9 A large building with sub-LANs.

Each floor's sub-LAN would control that floor's heating and air-conditioning system and monitor its energy consumption for that particular tenant. To help with identification, each outstation could be given useful **attributes**, such as the floor it is on and whether it is situated in a zone facing south, north, east or west. This would allow for better control, such as reducing the heating in synchronization with the sun's movement around the building when there is significant solar gain.

Small outstations are now being produced which control individual items of plant, so that the plant itself becomes intelligent. The original equipment manufacturers (OEMs) can then put the outstations on their plant at the factory and test and commission them before they are sent to site. At the site, the outstation just has to be connected to the LAN. Even temperature sensors, switches and relays have become intelligent (smart sensors, switches, etc.), with their own microprocessor and communications link onto a bus around the building.

There is the possibility of fire detection systems and security systems also linking into BEMSs, although failures on one system must not affect another system so that the building becomes unsafe in an emergency.

With this potential for communication between systems, and with outstations becoming smaller and cheaper so that individual items of plant can become intelligent and communicate on the BEMS LAN, intelligent buildings may soon be a reality. However, there is still the problem of communication between different manufacturers' equipment and different BEMSs. At present, BEMS manufacturers can have widely different LAN topologies, as well as other incompatibilities. This is discussed later on.

3.10   Network topology

The way that the BEMS outstations and PCs are positioned on a LAN is called the topology. There are three basic LAN topologies: star, bus and ring (Fig. 3.10). In the star topology all outstations are connected to the central, message-switching station (CS). The advantage of this topology is that it requires most of the communication intelligence to be at the central station and less at the outstations. However, it does depend crucially on the station at the centre of the star.

With the bus topology all stations have the potential to communicate independently with each other. Information is sent along the bus in **packets** or **frames**, often 8 bits long, with a receiver identification 'tag' on it. All stations are 'listening' for messages but the tag will identify where the message is going. One advantage of this topology is that new stations can be inserted by simply tapping into the bus cable. The disadvantage is that information between various stations can travel in opposite directions and collide. So a **carrier sense multiple access** (CSMA) protocol is needed that requires each station to 'listen before speaking' and to transmit when there are no other messages being sent. The protocol is much simpler if the central station controls the bus and it also saves hardware and software at the outstations.

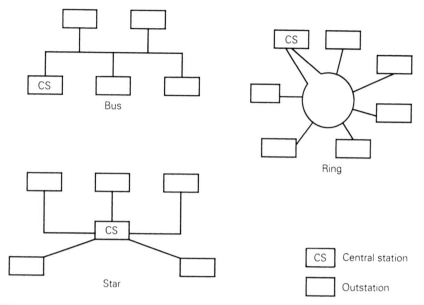

Fig. 3.10 Network topology.

In a ring topology the network cable is connected to each station in turn and the information travels round the ring in one direction only. Ring topologies often use the **token-passing** protocol, where a token (a pattern of bits) is transmitted around the ring from one station to the next. Only the station with the token is allowed to transmit if it wishes, and it does so by attaching a target address and its information to the token. The token and information is then passed to the next station on the ring, which copies it and retransmits it. Only the addressee station will make use of the information. Figure 3.11 shows the message format for a BEMS network using the ring topology for a message from station A to station B.

The message can be up to 100 ASCII characters long. The outstations, modems, central stations, printers and other devices are physically connected to the ring through their node controllers or network adapter cards. The ring itself is a two-wire 20 mA current loop. If the sending node sees its message go round the ring 10 times without being accepted, it removes the message and tries again later. The LAN in Fig. 3.8 is an example of a ring type.

Just as sub-LANs are connected to a main LAN via a larger master out-station (Fig. 3.9), so LANs themselves can be connected together to form a **supernet** or **internet**. If two LANs have different protocols, they can be connected together by a **gateway** which converts them. With a single manu-facturer's BEMS, a gateway would only be necessary to connect to another manufacturer's system.

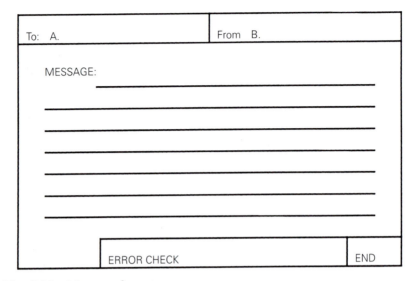

Fig. 3.11   Message format.

3.11   Software

Like all computer systems, a BEMS needs a set of instructions to tell it what to do. The actual chips, circuit boards and the physical parts of the BEMS are called the hardware, and the instructions are called the software. Programs stored in ROM are known as firmware; although they are sets of instructions, they are contained within hardware and their description reflects this compromise.

The microprocessor, the chip at the heart of all BEMSs outstations and the central station, only operates with digital signals. These signals form a **low-level programming language**, or **machine language**, which consists of binary numbers representing instructions for the microprocessor as well as data and memory addresses. Machine language instructions consist of two parts: the **operation code**, or opcode, is the operation to be performed, and the **operand** is the number or address on which the operation is to be performed. Most CPU instructions can be summarized as fetch and execute operations. There are surprisingly few instructions for a CPU. The Intel 8088, used on many of the early IBM PCs, has a set of about 100 basic instructions. The instruction to add is simply 00000101.

Assembly code, or **assembly language**, is slightly easier than machine code as easy-to-remember mnemonics are used for the instructions instead of binary digits. For instance, the add instruction would now be ADD and the subtract instruction would be SUB. The mnemonic instructions still have to be converted into machine code, and a program written in assembly code has to be run through another special program, called an **assembler**, to translate it into machine code. Before translation the program is known as the **source code** and after translation it is known as the **object code**. When an outstation has a keypad on it, a code similar to assembly code is used to communicate with the outstation CPU.

Both machine code and assembly code deal with the CPU at its closest or lowest level, hence they are known as low-level languages. Low-level languages are difficult to use and are often specific to particular PCs. Therefore, easier **high-level languages**, such as BASIC, Visual Basic, Pascal, C++ and FORTRAN were developed for writing programs, with a short **statement**, instruction, or object in the language representing a lot of machine-coded instructions.

The high-level languages have to be translated to machine code, just like the low-level languages. But this does not mean that a BEMS user has to learn a programming language such as C++, because most BEMSs use application software or an application package, often based on Microsoft Windows, where the programming is hidden and a non-specialist can use it. An example of application software is a word processing package. Application software is what has made PCs so popular. Many BEMSs have software which can display graphics and plans of plant rooms; the plant status is displayed and the temperatures are dynamically updated (Fig. 3.12).

Figure 3.12 is an example of Windows software using a **windowing environment**, where the screen can be divided into a number of different boxes or windows and a separate program or diagram can be shown on top of the main window. With this type of program the basic method of presenting information is to use schematic diagrams with colour graphics. Associated with these programs are **icons**, small pictorial symbols that often appear at the bottom of the screen. For instance, the icon for an alarm program is a bell, and the icon for the password program is a padlock; the padlock opens when the password is entered correctly. A mouse is the primary input device, moving a pointer around the screen to select the required command or a small box on a diagram, called a **click box**. Clicking the mouse on a click box, often with a double click, produces another window of graphic information.

Microsoft Windows, regularly upgraded to keep pace with hardware developments, is now used on many BEMSs. Microsoft Corporation, founded in 1975 by Bill Gates and Paul Allen, has written the **operating system** software used by many PCs; perhaps the best known are MS-DOS and Windows. The operating system is a set of programs that control the PC's hardware and manage its use of software, making the PC easier to use. It acts as an interface between the user and the PC and sets up the PC for use when it is switched on.

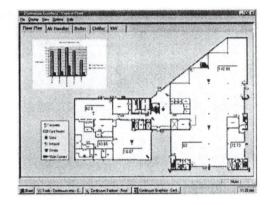

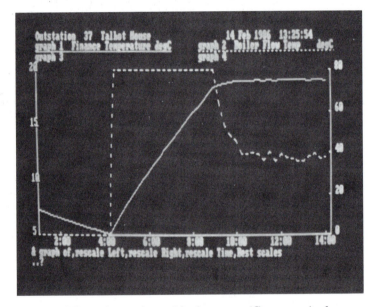

Fig. 3.12   Building plan and graphical output. (Courtesy Andover Controls Ltd and Trend Control Systems Ltd)

BEMS software may include a **computer-aided drawing** (CAD) package for drawing plant schematics such as Fig. 3.13, a word processing package for writing reports, and a **spreadsheet package** for manipulating data, doing numerous calculations and producing various graphs. Spreadsheets are rows and columns of data that can be manipulated by formulae and easily turned into graphs by the spreadsheet software. These packages can be very useful for monitoring and targeting energy consumption; they are discussed in more detail in Chapter 10. Examples of spreadsheet packages are Lotus 1–2–3,

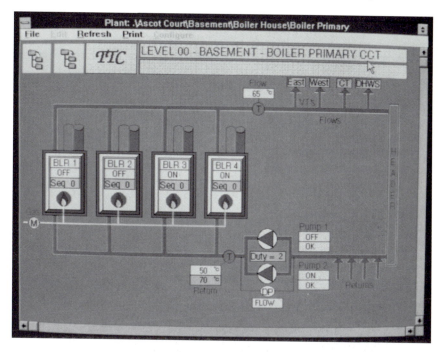

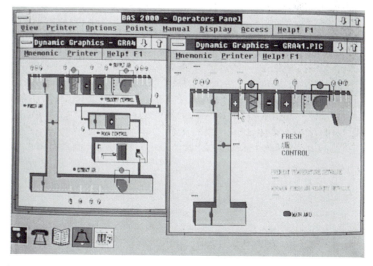

Fig. 3.13 Plant room schematics. (Courtesy Satchwell Control Systems Ltd and Trend Control Systems Ltd)

| Name; | Atkins House | | | | | | |
| Address; | Sabiri Way, Milestown | | | | | | |

| Account No; | 53476 | | Oil | 35 second | | Supplier; | Fiddlers |

Date	Stored	Delivered	Price p / litre	Cost of delivery £	Litres used	Oil Yearly Running Total, (YRT)	Delivery cost YRT £
03-May-99	30594	40922	15.08	6171.04	55651	645300	97324.51
31-May-99	55301	54532	14.63	7978.03	29825	629120	97358.27
05-Jul-99	44868	13640	15.63	2131.93	24073	625304	97104.6
02-Aug-99	26957	0	15.63	0.00	17911	623281	93043.12
06-Sep-99	32549	27280	15.63	4263.86	21688	621450	93237.71

Fig. 3.14   Spreadsheet output.

Quattro Pro and the well-known Microsoft Excel. Figure 3.14 shows a spreadsheet of oil consumption from a large site.

3.12   Compatibility, communications and standards

One of the great drawbacks of a BEMS is that once one has bought a system, other manufacturers' equipment (outstations, sensors, software, etc.) cannot be connected to it without a great deal of programming effort and without the relevant information from the manufacturers. This may seem like a wicked ploy by the manufacturers, but it simply arises because there are so many possible ways of writing a piece of software (including the different languages), of storing data in an outstation (including the different memory sizes and the relevant area of memory for say temperature sensor logs), and there are many different types of equipment. In other words, one manufacturer's BEMS will not communicate with another's.

Part of this communication problem stems from the computer industry, where there are many different protocols (sets of rules for governing communication) for PCs; this means that PCs from different manufacturers cannot easily communicate. Standards are being developed to try to resolve this. An all-encompassing communications standard for the many different types of computers and equipment will be very complex, but to try to describe proposed standards and what the network hardware and software do, the **International Organization for Standardization** (ISO) developed the reference model of **open systems interconnection** (the OSI model) [9, 4].

To understand the model, it is worth considering our own methods of communicating. When we speak to someone near to us, we have to establish initial contact, probably by looking at them and saying their name. Then we send the message, which consists of words in the language we are using, ordered to obey the rules of grammar and syntax. If the person is a foreigner who does not speak our language, they will indicate in some way that they do not understand and that the communication has not been understood. Likewise PCs and BEMS have different languages and grammars, although they share binary signals, as we often share the same alphabet as French and German people. For a group of people at a meeting, there has to be a protocol for speaking; we cannot all speak at once. This is similar to outstations on a LAN.

If we phone someone then we have to dial their number. For a local number in inner London it could be 928 8298. If we are in inner London then this suffices. But if we are in outer London then we have to add 0207 so that the phone system can route our call to inner London. If we are in another country then we have to add further numbers to route the call to the United Kingdom. Our phone conversation has to have various layers of numbers, or protocols, sent immediately before contacting the other person. We do not have to understand how the call is routed or how the message is sent (analogue or digital). The system is transparent to us, and as far as we are concerned, the person we are talking to could be almost standing next to us. We are virtual users.

The OSI model defines the various layers of protocol which have to be added to messages for sending between computers and computer equipment that are communicating. The model is split into seven layers called protocol levels. Figure 3.15 shows two computer systems, X and Y, which could well be two manufacturers' central stations connected together, perhaps on a LAN. Each system has hardware and software layers conforming to the OSI model, so that they can communicate.

The layers divide the model into smaller, more manageable segments. Each layer isolates the lower layers from the higher layers and adds values to the services provided by the lower layers. Figure 3.16 shows this addition of protocol **headers** and **tailers** to the basic message, similar to the addition of digits in our telephone example. Layers 1 to 4 are called the transfer service as they are responsible for moving information from one point to another. Layers 5, 6 and 7 are called user layers as they give the user access to information on the network.

Layer 1 is the physical layer and defines the physical connection between the computer and the network, defining the actual connectors, cables and signalling voltages. This is like the RS-232-C standard. LANs for PCs with large-scale data transmission would use coaxial cables, proprietary cables such as Ethernet cables or fibre optics.

Layer 7 is the application layer; in a BEMS it is the central station application software, perhaps producing plant diagrams and temperature graphs in a windows environment. Many BEMS manufacturers are now using Microsoft Windows as the basic central station software, but there is variation in its application and the underlying software.

For more details on the OSI model and computer communications the reader is referred to the literature [2–4]. Examples of BEMS incompatibility

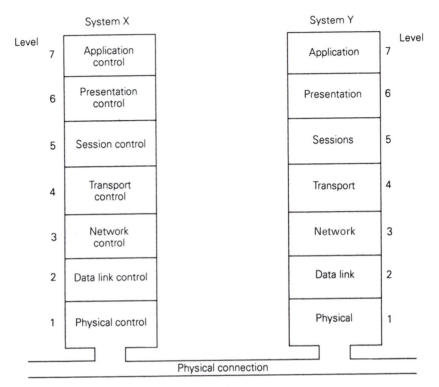

Fig. 3.15 The OSI seven-layer model.

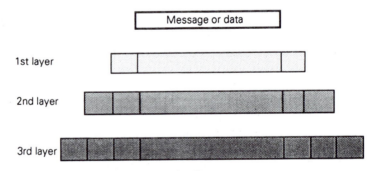

Fig. 3.16 Addition of headers and tailers.

are (1) how the outstation software stores sensed data from outstation sensors; and (2) how information on the actual sensors is stored, plus their role in the control configuration. The fact that manufacturers will use different sensors is another complication. Consider the transmission of data from an outstation to a central station.

The central station may need the last day's outside temperature readings from an outstation's sensor. Once it has established contact with the outstation, the central station needs to know the address for this information in the outstation's memory and the command or code to retrieve it. The temperature data may be stored as voltage readings to be converted when required, or it may already have been converted into temperatures. The outstation may store the data from all temperature sensors in one memory block and have a protocol that sends the whole block of data to the central station. The error-checking routines (e.g. a parity check) may also differ from manufacturer to manufacturer. So there are many possibilities for the protocols and each manufacturer's protocol is effectively unique. Of course this protocol, as well as the other BEMS software, is of great commercial importance to the manufacturer, as it may well have cost many staff-years to develop. In fact, there will be few people in the company who actually know and have access to the protocol.

However, protocols have been released to a software house which produces monitoring software for a number of different BEMSs. Having obtained the protocols, much time and effort had to be expended in writing software to interface the protocols. In a very few isolated cases, clients have been given access to BEMS protocols and software, and the clients have unwisely made alterations with rather disastrous consequences.

West Sussex County Council has successfully used BEMSs from three different manufacturers. It needed a mainframe computer and considerable programming effort to get the BEMS data and commands into one common format. This enabled all three BEMSs to be accessed from one computer terminal.

At the London Stock Exchange an older-generation central station was replaced by a new **head-end computer** which communicated with the old outstations and additional outstations supplied by a different manufacturer. Communication was facilitated by using the public host protocol developed by US BEMS manufacturer American Autometrix.

Instead of adapting the head-end computer, another approach is to go to the sensor end of the system [10] and use a common bus with an open protocol to connect together different manufacturers' smart or intelligent equipment. The smartness comes from employing common interface chips. A number of head-end and bus systems are described below.

3.12.1 Development of a BEMS communications standard

The BEMS Centre at BSRIA produced a report on the development of a communications standard for BEMS [11]. Although the proposal did not evolve very far, it is interesting to see the industry-independent models it used. The standard related to the applications software of BEMS and was just

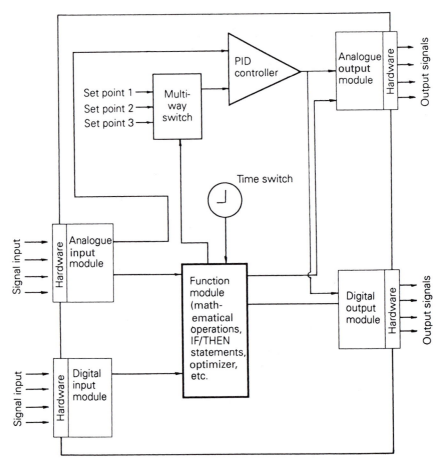

Fig. 3.17   BEMS centre model of an outstation.

concerned with layer 7 of the OSI model. The proposed standard considered a model of a typical BEMS outstation (Fig. 3.17) and the data required to define its control configuration, inputs and outputs, and the data it stored.

It was proposed that the data should be stored on a spreadsheet basis with sheets for outstation details, sensors, sensed data, control loops, control configurations, etc. (Fig. 3.18). The main point was that the spreadsheet would be a common format for all manufacturers' BEMSs and this would greatly ease the accessing of data between different systems.

3.12.2   The FND standard

Much work on a BEMS communications standard has been done in Germany; FND (firm neutral data transmission) is promoted by German public

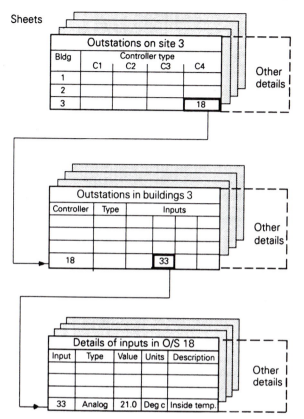

Fig. 3.18 Spreadsheet basis for storing BEMS data.

authorities, who own 50% of all large buildings in Germany, and Profibus (process fieldbus) is sponsored by the German government.

FND is effectively a gateway between BEMS LANs. Its specification was published in 1988 and conformance tests were published in 1990 [12]. In relation to the OSI model, the FND standard has layers 4 to 6 empty and layer 1 uses the CCITT X.21 standard. Layers 2 and 3 of FND use the CCITT X.25 standard [9]. Layer 7 contains the FND protocol. Communications between different BEMSs in the FND concept is like communications between islands, each containing a particular manufacturer's central station, and an overall central station centre. Each island (IZ) has an island central standard interface adapter (IZ-SSA), which communicates with the central station through the central interface adapter (LZ-SSA). Figure 3.19 shows this communication between two different manufacturers' BEMSs.

The manufacturer's communication system is converted to FND through all seven layers, starting at the physical layer and moving up to the application

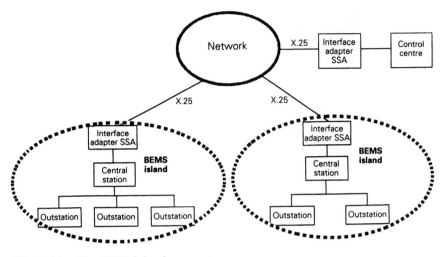

Fig. 3.19   The FND island concept.

layer, where the manufacturer's data and messages are also converted to FND for transmission. The FND protocol is centred on data points: digital inputs are message points, analogue inputs are measuring points, analogue outputs are setpoints, digital outputs are switch points. There are also high and low limits and totalizer points.

In December 1989 a Landis and Gyr Visonik 400 was connected to a Honeywell Delta 2000 and a Honeywell Excel, all controlling exhibition halls in Berlin. This involved a lot of engineering effort in setting up the FND data points for the IZ and LZ SSAs, and the island central and main central parts of the system [12]. Figure 3.19 shows the FND concept of linking BEMS islands as realized in Berlin. FND has not been adopted much outside Germany and Switzerland.

3.12.3   The Profibus fieldbus

Profibus is an example of a **fieldbus** and is relatively low-tech, intended for small and medium-sized buildings. Fieldbus is the name given by the International Electrotechnical Commission (IEC) to a low-level industrial data bus [9]. It has wider application to the manufacturing industry besides BEMSs. Fieldbus networks are intended to connect actuators, sensors, controllers and similar devices at a low level of communication rather than allow for distributed computation and large data communications as in a computer LAN. Hence fieldbuses do not need the full seven-layer architecture of the OSI model [13].

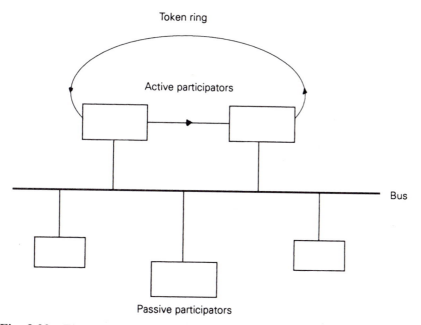

Fig. 3.20 The Profibus structure.

Whereas FND is for complete BEMSs to communicate, Profibus is for outstation-to-outstation communication or controller-to-controller communication. Layer 1, the physical layer, is based on the RS-485 standard using shielded twisted-pair cables. The maximum length between outstations is 1200 m, and up to 4800 m with repeaters. Layers 3 to 6 are empty. Layer 7, the application layer, is divided into two sublayers. The upper layer is a fieldbus messaging service (FMS) and is tailored to suit other fieldbus applications.

The Profibus project was sponsored by the German government with 13 companies and 5 research institutions participating. It was started in November 1987. A Profibus User Group was founded in 1989 and took over the running of the project. Figure 3.20 shows the structure of the fieldbus, with **active** and **passive participators**.

The active participators are on a decentralized token ring and the passive participators are on a centralized polling bus. When one of the active participators has the token, it may poll the passive (slave) participators for information as well as transmitting to other active (master) participators. Once an active participator has the token, it is master of the whole bus. The passive participators need less hardware and software and can be used to keep the cost of the system down.

Table 3.1   Some objects for a typical BEMS

Control loop
Analogue input
Analogue output
Analogue value
Binary input
Binary output
Binary value
Schedule
Group

Table 3.2   Properties of the binary input object

Object identifier	Polarity
Object type	Inactive text
Present value	Active text
Description	Change-of-state time
Status flags	Elapsed active time
Reliability	Change-of-state count
Override	Time of reset
Out-of-service	

3.12.4   The ASHRAE BACnet

In 1987 the American Society of Heating, Refrigerating and Air-Conditioning Engineers (ASHRAE) formed a standards committee to develop BACnet (Building Automation and Control Network), a standard communications protocol for building energy management and control systems. In 1995 BACnet was adopted as an American National Standard, ANSI/ASHRAE 135–1995.

The BACnet protocol may be understood by considering it as two separate but closely related parts: a model of the information in the BEMS, in terms of **objects**, and the functions or 'services' used to exchange it [14]. Like the FND standard, BACnet will interface between BEMSs from different manufacturers. It will model the design with 20 standard **object types**; some examples are given in Table 3.1. These are similar to the 'points' often used to describe the size of an outstation. For instance, an analogue input could be a temperature sensor input, an analogue output could be a control signal output, and an analogue value could be a control setpoint. These objects can then be described by up to 123 properties. Properties for the binary input object are given in Table 3.2 [14].

The functions or **services** provide commands for accessing and manipulating information. Table 3.3 lists the five groups of services [15]. The object access services provide the means to read from and write to the properties of

Table 3.3 The five groups of services

Alarm and event
File access
Object access
Remote device management
Virtual terminal

objects and to create or delete them. The virtual terminal services allow a monitor and keyboard to interact with the system as if it were directly connected to the system, the communication and protocols all being dealt with 'unseen'.

BACnet specifies four types of LANs, and a serial EIA-232 interface, that can be used [16]: the high-cost but high data rate Ethernet (standard ISO/ IEC 8802–3), the medium-cost lower-rate ARCNET (standard ATA/ANSI 878.1), and a low-cost **MS/TP** (Master–slave/token-passing) LAN (standard ANSI/ASHRAE 135–1995) using EIA-485 (formerly RS-485) signalling over twisted pairs. LonTalk is also specified. In terms of the OSI seven-layer model, BACnet has four of the layers: application, network, data link and physical. By mixing these different LAN types, a large building can have a low-cost BEMS communication system. In a very large building in San Francisco, MS/TP LANs are used to communicate with unit controllers, each LAN controlling one or two floors with routers to an Ethernet backbone to a file server and BEMS central station. There is also a gateway to a proprietary LAN [6]. This is similar to Fig. 3.7.

There are now thousands of BACnet systems installed in at least 14 countries, although an industry certification program is still being developed [17].

3.12.5 Echelon's LonWorks

While standards committees were discussing open standards for BEMSs, Echelon produced the cheap, programmable Neuron chip which can handle network and input/output functions. This chip would enable even small devices, such as sensors and switches, to be connected to a **fieldbus** and so communicate with other BEMS devices and central stations. The term 'fieldbus' is used as its data transmission rate is low, 4.8 to 1250 kbps, but sufficient for the small 'field' devices. However, Echelon released the protocol it developed, **LonTalk**, so that it became an open protocol for other manufacturers to use, along with incorporating the Neuron chip. It has subsequently become the basis of one of the four BACnet-specified LANs. A number of BEMS and controls companies, as well as manufacturers of access controls, lifts, fire and security equipment, HVAC systems, lighting

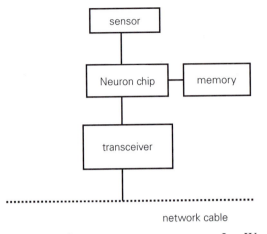

Fig. 3.21   Components of a temperature sensor on a LonWorks network.

and metering, now use Neuron chips and the LonWorks network. The LonMark Interoperability Association establishes standard specifications for the functionality of devices which are to be connected to the LonWorks network; it also carries out conformance testing. This enables interoperability to become more of a reality. Products conforming to the standards can be issued with a LonMark logo as a warranty for interoperability [7].

Devices are connected to the bus at **nodes**. A transceiver must be used in each device to connect its Neuron chip to the bus. Different transceivers are available for different bus systems, such as twisted-pair cables, mains or power cables, and wireless or infrared. Each device is treated as an object with a defined set of properties, similar to BACnet. Figure 3.21 shows the components of a temperature sensor on a LonWorks network.

Often the LonWorks fieldbus network is connected to a faster backbone network to which BEMS central stations and file servers are connected. Integration of the fieldbus with the TCP/IP protocol of the Internet/intranet will enable existing IT systems to be used for the BEMS.

3.12.6   ElBus

Another small bus system is the European Installation Bus (EIB or EIBus), originally developed by Siemens of Germany but now pan-European with a European standard administered by the **European Installation Bus Association** (EIBA) in Brussels. EIBA is independent of any one company. Any manufacturer whose devices conform to EIBA standards and tests may use the EIB mark. The bus was developed so that it could be installed by ordinary electricians and contractors adjacent to main power cables in trunking.

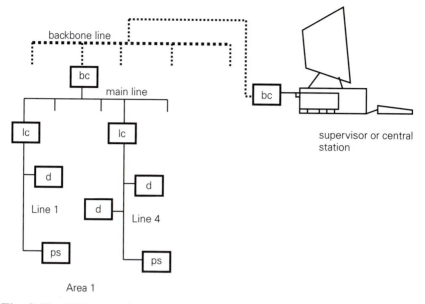

Fig. 3.22 EIBus topology.

The bus is a simple unshielded twisted-pair cable carrying both power (a nominal 28 V d.c.) and signals, at 9.6 kbps, to the connected devices. A device, or **application unit** (AU), is connected to the bus by a **bus coupling unit** (BCU), programmed during commissioning by a PC with EIB Tool software. Each bus, or **bus line**, has a maximum length of 1000 m onto which may be connected up to 64 devices (Fig. 3.22). Up to 12 such bus lengths may be connected using **line couplers** to form an **area**. A device is then characterized by its address of area, line and device number.

Line-crossing 'telegrams' can be sent between lines but all other messages are kept within the line. This reduces the amount of data transmission. The BEMS central station, or **head-end supervisor**, is a PC connected to the bus by a gateway.

3.12.7 Batibus

Batibus is another small system bus similar to EIBus, and it too was developed by a company making electrical switchgear and control equipment for electrical panels, Merlin Gerin. There is a Batibus Club International, a standard for the protocol and product certification for devices from different manufacturers. Although it originated from work on a bus for communication between electrical components, it was then extended to a bus for

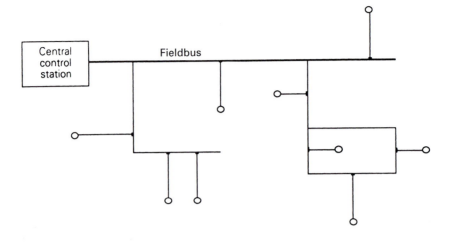

O Intelligent sensors and actuators

Fig. 3.23   The Batibus system.

building control. The French designers took the word 'batiment' and called their system Batibus.

The concept is that Batibus should be multifunctional; easy to install by general electrical contractors; able to control heating, lighting, intruder and fire detectors; and easily configured by building users [18]. It has been designed to be economic for buildings up to 10 000 m². The Batibus cable is a single unshielded twisted pair operating at 4.8 kbps. It can be laid in a bus, star or ring topology [19]. If the bus is installed next to mains power cables then a shielded insulated twisted-pair cable is recommended. Figure 3.23 shows a system with a combination of these topologies.

A communications module is required for each device to communicate via the bus. Communication between devices on the bus is operated under CSMA/CA control (carrier sense multiple access with collision avoidance). Up to 1000 devices can be connected to the bus with up to 75 powered directly from the bus, at 15 V d.c. Adjustment of two code wheels or DIP switches on each device gives 240 different settings of the microprocessor circuits to give devices unique addresses. If there are more than 240 devices on the bus then similar devices will presumably have to share an address. Although up to 1000 devices may be connected to the bus, with a twisted-pair cable using wires of diameter 0.75 mm² the maximum length of cable is 1500 m and the maximum distance between the central control station and the furthest device is 200 m.

Table 3.4 Modules that can be connected to the R-Bus

Room temperature controller
Local set point
Control output
Analogue output
Unit control
Local command
Variable air volume

South Link Business Park in Oldham has 10 office blocks with the Batibus system controlling the electric heating. The blocks and their heating systems are designed to the Electricity Association's Energy Efficient Design (EED) guidelines. One office block is split into two Batibus networks, each with its own central control unit and two output modules to switch the heater contactors.

In December 1999 a specification was issued on a standard European building bus that combines elements of Batibus, EIBus and the European Home System (a bus system for domestic applications developed with EU funding). It is hoped this will be adopted as a European standard to allow more interoperability and allow the European system to compete better with LonWorks and CEbus in the United States [20].

3.12.8 Intelligent room control

Honeywell has developed its own small bus system, the Intelligent Room Bus or R-Bus. It uses a three-wire 1.5 mm^2 unshielded cable providing 24 V d.c. and signals to and from actuators and sensors. The maximum cable run is 200 m. Typical modules that can be connected to the bus are given in Table 3.4.

The system uses a current loop bus limited to 3.5 mA current with a data transfer rate of 1200 bps. Up to 16 modules can be connected to an R-Bus, controlled by a **multicontroller module**, which contains the application software. The multicontroller has interfaces to control two buses. A PC can be connected to a multicontroller as a central station.

3.12.9 Summary

Much work has been done on generating open standards for communications, and gradually interoperability will become more of a reality. This does not mean there will be one communication medium in a building, especially a large one. It is more likely that a number of buses will be used for different levels in the BEMS [7], as shown in Table 3.5.

Table 3.5   Levels of communication within a building (CEN TC247, WG4)

Level	Example	Proposed draft standards
Management	Central station to central station	BACnet, (Ethernet, ARCNET) FND
Automation	Outstations, DDC controllers	BACnet, (MS/TP, ARCNET) FIP, Profibus
Field	Sensors, actuators	LonTalk, EIBus, Batibus, EHS

References

[1]   Pilgrim, A. (1995) *Build your own Pentium processor PC and save a bundle*, 2nd edn, McGraw-Hill, New York.

[2]   White, R. (1998) *How Computers Work*, 4th edn, Que Corp., Indianapolis IN.

[3]   White, R. (1995) *How Computers Work*, 2nd edn, Ziff-Davis, Emeryville CA.

[4]   Tanenbaum, A. S. (1996) *Computer Networks*, 3rd edn, Prentice Hall, Englewood Cliffs NJ.

[5]   Derfler, F. J. and Freed, L. (1998) *How Networks Work*, 4th edn, Que Corp., Indianapolis IN.

[6]   Hodson, P. (1997) *Local area networks, including internetworking and interconnection with WANs*, 3rd edn, Letts, London.

[7]   CIBSE (1999) CIBSE Applications Manual, Automatic controls, Chartered Institution of Building Services Engineers, London.

[8]   Anon (1998) The hows and why of buying open systems. *M&E Design*, Nov.

[9]   Houldsworth, J. (1990) OSI Handbook, International Computers Ltd, London.

[10]   Wilkins, J. and Willis, S. (1990) What are smart sensors? *Building Services Journal*, **12**.

[11]   BSRIA (1989) The development of an application level communications standard for building management systems, BSRIA, Bracknell.

[12]   Fischer, P. (1990) FND and PROFIBUS: standard communications in building automation systems communication standards for BEMS. Paper presented at the BEMS Centre Colloquium, Birmingham.

[13]   Pimentel, J. R. (1989) Communication architectures for fieldbus networks, *Control Engineering*, Oct.

[14]   Bushby, S. T. and Newman, H. M. (1991) The BACnet communication protocol for building automation systems. *ASHRAE Journal*, April.

[15]   Swan, W. (1996) The language of BACnet. *Engineered Systems Magazine*.

[16] Swan, W. (1997) Internetworking with 'BACnet'. *Engineered Systems Magazine*.

[17] Applebaum, M. A. and Bushby, S. T. (1998) 450 Golden Gate Project, BACnet's first large-scale test. *ASHRAE Journal*, July.

[18] Joseph, D. (1990) BATIBUS communications standards for BEMS. Paper presented at the BEMS Centre Colloquium, Birmingham.

[19] Joseph, D. (1990) Catching the right bus. *Electrical Design*, Oct.

[20] Anon (1999) Single European building bus is revving up. *M&E Design*, Jan.

4
Sensors and their responses

A BEMS outstation is basically a microprocessor processing digital electrical signals. But most plant equipment controlled by a BEMS is controlled on temperature, pressure or flow, rarely on electrical signals. So these non-electrical parameters must be measured and converted to electrical signals, invariably a voltage. This measurement and conversion is the function of the **sensor**, **transmitter** and **transducer**. The sensor responds to the change in the measured parameter (e.g. the temperature), the transducer changes the sensor signal to an electrical signal (e.g. a pressure into a voltage) and the transmitter is the electronic circuitry to enable a suitable strength voltage proportional to the sensed parameter to be sent to the outstation. Often the sensor transmitter and transducer are integral in a measuring device and it is difficult to separate them, so that sensors are commonly known as transducers.

Many of the details of measuring devices relate to their electronic circuitry and the use of operational amplifiers. Such aspects are not dealt with here, but they are covered in the literature [1–4].

4.1 Binary and pulse inputs

Although detailed devices are required for an outstation to measure temperature, say, some inputs to the outstation are just open/closed, high/low or on/off signals (i.e. binary signals) which do not need analogue-to-digital converters (ADCs) but which may need to be conditioned to the appropriate voltage level for the outstation and, in the case of pulses, to be of sufficient duration for the outstation to detect them during the sequence cycle time.

One form of binary input is from a **status sensor**, usually an electrical relay or switch that is open or closed. Status sensors can monitor, for instance, whether boiler burners are operating or electric motors are running. Although energized by the burner or motor to its relevant state, the sensor must have volt-free contacts, so that no large voltages and currents are connected to the outstation. This is done either by using volt-free auxiliary contacts on the appliance or by using a relay energized by the appliance, as explained in Chapter 2.

A second form of binary input is from a **pulse sensor** or **event counter** which sends a pulse to the outstation each time an event occurs. A common event that is sensed is the rotation of a utility's energy meter, so that the BEMS can count the rotations and hence monitor energy use. The utilities now provide meters which are serially interfaceable to BEMSs, which produce pulsed output as the meter moves. Event counting is also useful for **condition-based maintenance** by monitoring the number of times that plant and equipment is switched on and used. (This is discussed further in Chapter 14.)

4.2   Temperature sensors

By far the most used sensor for a BEMS is the temperature sensor. There are several different types and they have different cost and accuracy implications. Fielden and Ede [5] point out that sensors are likely to provide the greatest source of failure and maintenance problems in a BEMS.

4.2.1   Platinum resistance thermometer

Platinum resistance thermometers are the more expensive and accurate temperature sensors used in BEMSs. They rely on the fact that platinum, like other metals, increases its resistance as it gets hotter; the relationship is of the form

$$R_t = R_0(1 + \alpha t + \beta t^2 + \Gamma t^3) \tag{4.1}$$

where R_t = metal resistance at temperature $t°C$ (Ω)
R_0 = metal resistance at $0°C$ (Ω)
t = temperature ($°C$)
α = temperature coefficient (K^{-1})

β and Γ are constants which are very small and usually neglected, so that equation (4.1) becomes

$$R_t = R_0(1 + \alpha t) \tag{4.2}$$

Table 4.1 shows the values of α for some common metals. Due to its highly linear, repeatable relationship of resistance against temperature over a wide temperature range and its chemical inertness, platinum is the most widely used metal for resistance thermometers. The metal in the form of a wire is traditionally wound on a mandrel and the resistance determined by passing a

Table 4.1   Values of α for some common metals

Metal	$\alpha/10^{-3}\ (K^{-1})$
Nickel	6.7
Iron	4
Copper	4.3
Platinum	3.91
Silver	4.1

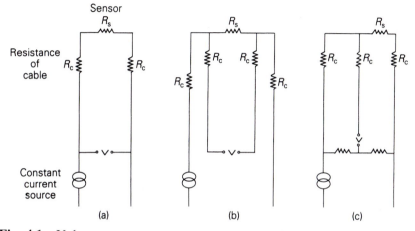

Fig. 4.1   Voltage measurement across a sensor using (a) two-wire, (b) four-wire and (c) three-wire cable.

constant current through it and measuring the voltage across it. Due to the purity of platinum that can be achieved, there is a low statistical variation in the resistance between similar samples, and there is a British Standard for platinum resistance thermometers (PRTs), BS 1904. A typical platinum element has the following properties [1]:

$$R_0 = 100.0\ \Omega$$

$$R_{100} = 138.5\ \Omega \quad \text{(resistance at 100°C)}$$

$$R_{200} = 175.83\ \Omega \quad \text{(resistance at 200°C)}$$

$$\alpha = 3.91 \times 10^{-3}\ K^{-1}$$

$$\beta = -5.85 \times 10^{-7}\ K^{-2}$$

A Wheatstone bridge circuit is often used to determine the sensor's resistance to great accuracy [7, 4].

To avoid voltage drops through the two wires connecting the sensor to the outstation, four wires are used (Fig. 4.1). Since four wires are rather expensive, a compromise of three wires is sometimes employed.

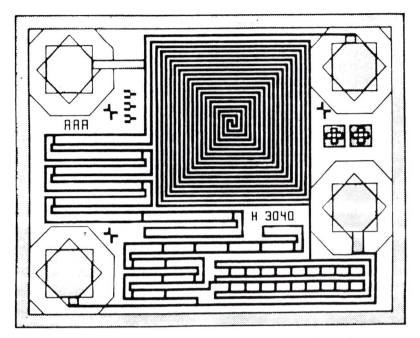

Fig. 4.2 Layout of 1.0 mm × 1.3 mm laser-trimmed chip with photochemically etched resistor network. The sensing element is the 10 μm thick spiral square.

Tolerances of ±0.075 Ω at 0°C are specified in BS 1904 for class 1 elements and ±0.1 Ω for class 2. Bentley [1] states that in a typical element 10 mW of electrical power dissipated in the element as I^2R heating causes a temperature rise of 0.3°C.

Modern BEMS platinum resistance thermometers (PRTs) use platinum films on a substrate about 1 mm^2. The thin film of platinum is accurately trimmed by laser. Often-used PRTs are standard PT100 PRTs, which conform to the BS 1904 class B standard. A chip sensor [6], which actually uses a nickel–iron alloy film, 10 μm thick, to avoid the expense of platinum is shown in Fig. 4.2, and it still provides a linear and stable sensor. The sensing element is the laser-trimmed square spiral on the silicon substrate, with a resistor network around it.

The sensor chip, as with other types of sensor elements, is installed in a suitable plastic case for internal and external air temperature measurement and in a probe for ducted air and piped water measurement (Fig. 4.3).

The duct sensor element consists of a platinum element sensor positioned at the end of the probe, with four wires connecting it to the transmitter in the

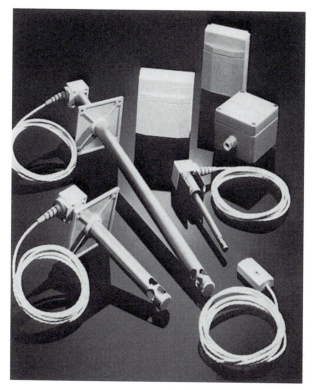

Fig. 4.3 A variety of sensors.

box at the probe end. The transmitter is an electronic circuit supplied with 24 V d.c. from the outstation which amplifies and conditions the sensor signal for sending to the outstation; the outstation could be quite distant from the sensor element. The signal from the transmitter is sent along a two-wire cable as a signal current of 4–20 mA. The current is proportional to the temperature sensed, with an offset of 4 mA.

4.2.2 Thermistors

A cheaper sensing element is the **thermistor**, a ceramic material made by sintering mixtures of metal oxides into whatever shape is required: bead, disc or rod. The size can be quite small, which gives it a distinct advantage. Copper leads are connected to the thermistor element, encapsulated in a vitreous material.

Thermistors are a form of semiconductor and their resistance depends on their composition. A minority have positive temperature coefficients (PTCs),

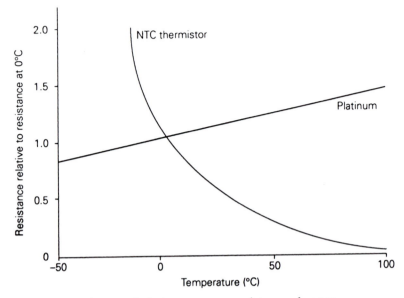

Fig. 4.4   Thermistor and platinum sensor resistance changes.

where resistance increases with increasing temperature. The majority have negative temperature coefficients (NTCs), their resistance falling with increasing temperature. Although the change in resistance with temperature is larger than for platinum, the resistance changes non-linearly with temperature, as shown in Fig. 4.4.

The relationship between resistance and temperature for an NTC thermistor is given by

$$R_t = R_\infty \exp (\beta/T) \qquad (4.3)$$

where R_t = thermistor resistance at temperature $t°C$ (Ω)
R_∞ = resistance that thermistor tends to at high temperatures (Ω)
β = thermistor material constant (K)
T = absolute temperature (K)

A more practical and commonly used alternative to equation (4.3) is

$$R_t = R_{25} \exp\left(\beta\left(\frac{1}{T} - \frac{1}{273 + 25}\right)\right) \qquad (4.4)$$

where the resistance at 25°C is taken as the reference point. A typical thermistor has an R_{25} of 12 kΩ with β = 3750 K. The tolerance on such a thermistor is ±7% of R_{25} and I^2R heating of 7 mW raises the read temperature by 1°C, so thermistors cannot be expected to be as accurate as PRTs. In fact,

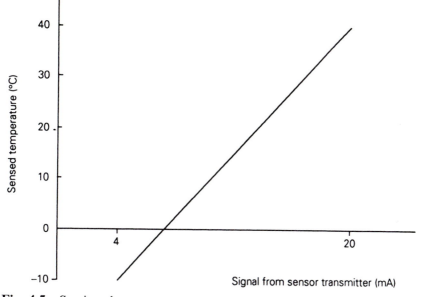

Fig. 4.5 Semiconductor sensor output.

the **field accuracy** of one BEMS manufacturer's thermistor is ±1°C, whereas the thermistor element itself has an accuracy of ±0.2°C.

4.2.3 Semiconductor circuit sensors

Silicon semiconductor diodes and transistors are sensitive to temperature changes and the junction voltage of these devices changes by about 2.2 mV per kelvin over a temperature range −55°C to 150°C. These can be made up into integrated circuit devices, effectively current sources which produce a current proportional to temperature and very linear. The temperature sensitivity of 1 μA K^{-1} is amplified by an operational amplifier circuit in the transmitter to a level suitable for the 4–20 mA standard signal. A resistor in the transmitter circuit can be adjusted to vary the temperature range. Figure 4.5 shows the output from the transmitter of a BEMS semiconductor sensor.

The current signal to the BEMS from the sensor can be converted to a voltage signal for processing in the BEMS by simply putting an accurate resistor in series with the semiconductor sensor (Fig. 4.6).

It is interesting to note that semiconductor circuits in parallel yield the average sensed temperature, and when in series the minimum sensed temperature. Most BEMS manufacturers who use semiconductor sensors, however, would not complicate their sensor range with series and parallel sensors,

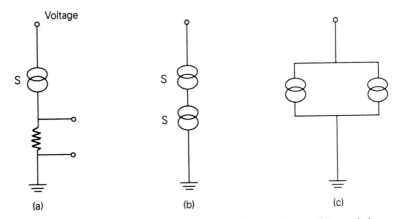

Fig. 4.6 Semiconductor sensor, S, to give (a) a voltage, (b) a minimum sensed temperature and (c) the average sensed temperature.

but would use separate sensors and calculate the average, minimum or maximum by using outstation logic function modules.

4.3 Setting up sensors

When a sensor is measuring a temperature, its transmitter is sending a signal, proportional to the sensed temperature, to the BEMS outstation. The outstation needs to know what range of temperature the sensor is measuring. Is it a room temperature range of 0–40°C or a hot water temperature range of 0–100°C? Often different sensors are used with different ranges, e.g. immersion sensors for water pipes, duct sensors, inside air temperature sensors, outside air temperature sensors and surface contact clamp-on sensors (Fig. 4.3).

In the EPROM of the outstation (Chapter 2) there is a **sensor module**, and a sensor-type module, that reads the signals from the sensor. First, the signal (for this outstation 4–20 mA) is converted to a voltage of 1–5 V by a **signal conditioning circuit** and it is then converted to a digital signal by an ADC. All this is done in the input unit by the **analogue input interface**.

Initially, the outstation has to be **configured** as to the type of sensor connected to a particular input and its corresponding analogue node, say node 1 for sensor 1 at input 1, node 2 for sensor 2 at input 2, etc., and the relevant alarm levels set. This is done through the sensor module.

The **sensor-type module** converts the standard digital signal, from the sensor module, to the relevant engineering units, degrees Celsius for a temperature sensor. A number of different sensor types can be set up in the outstation RAM to accommodate different sensors: room, duct, water, etc.

Bottom of range =

Top of range =

sensor Upper limit to working range =

sensor Lower limit to working range =

Fig. 4.7   Information from the sensor-type module.

But once a room temperature sensor type has been set up, then it defines the temperatures from the corresponding conditioned voltages for a standard room temperature sensor.

To set up the outstation modules in RAM, one has to get into the **configuration mode** of the outstation. This is not a routine operation for a site caretaker or boilerperson, so entry to the configuration mode will be password-protected at a high security level. Entry, for most outstations, will be via the central station or a local portable PC. Figure 4.7 shows the kind of information that has to be entered in to the central station to set up the sensor-type module in the outstation's RAM.

The capital letters in Fig. 4.7 indicate the key to type on the central station keyboard to enter the relevant value, e.g. T is for the top of the sensor temperature range corresponding to 20 mA (or 5 V), and B is for the bottom of the sensor temperature range, corresponding to −5 and −20 mA. The upper and lower limits to the working range set limits to the signal to be expected, normally the temperatures corresponding to 4 mA and 20 mA. Outside these limits, an **out-of-limits** alarm is raised, indicating that the sensor may be malfunctioning.

Example 4.1
For an outside air temperature sensor $B = -151°C$ (−5 V), and $T = +51°C$ (+5 V). What typically are the signals sent back to the outstation from the sensor, and what is the corresponding range of temperature (U and L)?

Solution
Knowing that the signal from the sensor will most likely be 4–20 mA, it is better to consider currents than voltages. So −5 V corresponds to −20 mA and +5 V to +20 mA. Assuming that the sensor is linear, then B and T are two points on a straight-line graph:

$$y = mx + c \qquad\qquad (4.5)$$

where y = the temperature axis (°C)
   x = the current signal axis (mA)
   m = the slope (°C mA^{-1})
   c = the intercept (°C)

Substituting in the two values gives

$$-151 = -20\,m + c \qquad\qquad (4.6)$$

$$51 = 20\,m + c \qquad\qquad (4.7)$$

From equations (4.6) and (4.7) we have

$$y = 5.05x - 50 \qquad\qquad (4.8)$$

The sensor's transmitter sends a signal between 4 and 20 mA, so using equation (4.8):

$$L = -29.8$$

$$\approx -30°C$$

$$U = 51°C$$

To configure the outstation input for this sensor, one would type the following responses:

$$? \, B \, -151$$

$$? \, T \, 51$$

$$? \, U \, 51$$

$$? \, L \, -29.8$$

The example data defines the outside air temperature type of sensor and stores it in the outstation RAM. This type of sensor we can define as Y1, using the capital notation, and the sensor-type module has been set up for a type Y1 sensor.

Still in the configuration mode, the sensor module is then selected to set the relevant outstation input channel, say the third input channel, as the input for the outside air temperature sensor, consequently defined as S3. Typing in the relevant commands to the central station allows the outside air temperature sensor S3 to be set up.

S3 will be configured as a type Y1 sensor, as defined in the sensor-type module set up above, but also the desired high and low sensor alarms can be set, the sensor labelled, e.g. 'outside air temp.', and a **sensor alarm delay** of a few minutes set to ensure that the sensor alarms will not operate unless it is a genuine alarm, not just a fluke, transient reading from a sensor, perhaps due to interference. Sensor alarms should not be too tight otherwise alarms may inundate the central station. On one user's BEMS, which had no alarm delay software, alarms were reported via a modem to the maintenance depot printer and staff sent out to rectify the problems. Often just as someone had been dispatched, the printer would report that the alarm was no longer present; it had been of short duration!

Once the sensor type and sensor modules have been set up, the outstation has all the information it needs to define the sensor and calibrate the milliamp signals to temperatures for control and logging.

4.3.1 Non-linear sensors

The sensor type considered above was a linear sensor, where the voltage and temperature were related by a simple, straight-line equation. Unfortunately, a commonly used sensor is the thermistor, a non-linear device. When the thermistor's output is converted to a digital voltage for the outstation CPU to process, due to the thermistor's non-linear characteristic, the digital voltage will also be non-linear.

There are two methods of tackling this problem, either linearize the thermistor over a small temperature range by placing an accurate resistor in series or parallel with the thermistor or use a **look-up table** of temperatures corresponding to voltage signals. The linearization method uses a standard sensor-type module; the look-up method replaces the sensor-type module with a look-up table.

Example 4.2
It is proposed to use a thermistor for an air temperature sensor, for both internal and external use. Compare the two methods for using this for a BEMS outstation, employing another resistor and using a look-up table. The data on the thermistor resistance is given below.

Temperature (°C)	Resistance (Ω)
−10	55 340
0	32 660
10	19 990
20	12 490
25	10 000
30	8 058
40	5 326

Solution
The first method is to put a resistor in series or parallel with the thermistor. Looking at the temperature range, at 25°C the resistance of the thermistor is 10 000 Ω. It would be quite easy to obtain an accurate resistor of this value and it would make the calculations easy if the thermistor were linearized around this temperature. With a 10 000 Ω resistor in series with the thermistor, then the voltage across it is

Table 4.2   Resistance ratios

Temperature (°C)	$\dfrac{R}{R + R_{th}}$	$0.5\left(\dfrac{R_{25}}{R_{th}}\right)$
−10	0.15	0.09
0	0.23	0.15
10	0.33	0.25
20	0.44	0.4
25	0.50	0.50
30	0.55	0.62
40	0.65	0.94

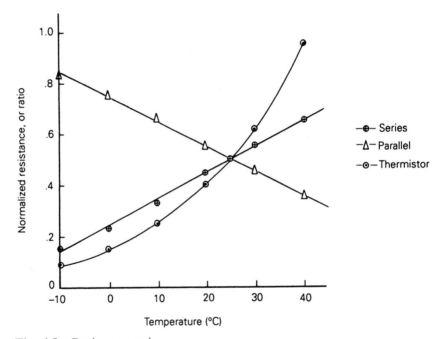

Fig. 4.8   Resistance ratios.

$$V_{meas} = V_{supp}\frac{R}{R + R_{th}}$$

where R = constant resistance (Ω)
V_{meas} = measured voltage signal across R (V)
R_{th} = thermistor resistance (Ω)
V_{supp} = voltage supply to the sensor (V)

The term $R/(R + R_{th})$ is much more linear than R_{th} itself, as shown in Table 4.2 and Fig. 4.8. This has normalized R_{th} around the R_{25} value (10 000 Ω), at a value of 0.5 to compare with the combined term.

If a 10 000 Ω resistor is placed in parallel with the thermistor, then when a voltage is applied the current can be measured (or the voltage across another resistor in series). The current signal from the parallel resistors is

$$I_{meas} = V_{supp} \frac{R + R_{th}}{R\,R_{th}}$$

Again, the resistance ratio produces a reasonably linear relationship, as the normalized values in Fig. 4.8 show.

The second method, the look-up table, could be done by applying a constant current source to the thermistor and measuring the voltage across the thermistor. There would then be a direct non-linear relationship between temperature and voltage. A look-up table could then be calculated and put in the EPROM. But rather than have a table of every possible voltage and temperature, a more compact way is to approximate the non-linear relationship to a number of straight lines. Five straight lines, each spanning 10 K, from $-10°C$ to $40°C$ would be satisfactory. Then intermediate voltages can be converted into temperatures by a sensor-type module, as used for linear sensors.

Reference [8] examines interfacing a thermistor to a microcomputer, and determines a look-up table, as well as the ADC, the program for conversion and the display of a temperature.

4.4   Block diagrams of the sensor

This configuration process can be seen in terms of the block diagram, much used in control theory. The sensor and transmitter can be represented by the block diagram shown in Fig. 4.9.

The semiconductor sensor changes its current flow in relation to the temperature change it senses. The temperature–current relationship is described by the **static gain**, K_s, which is the slope of the temperature current line shown in Fig. 4.10 and the intercept c_s:

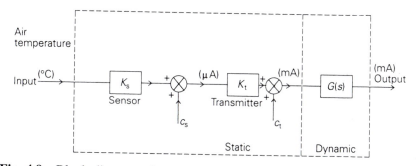

Fig. 4.9   Block diagram of sensor and transmitter.

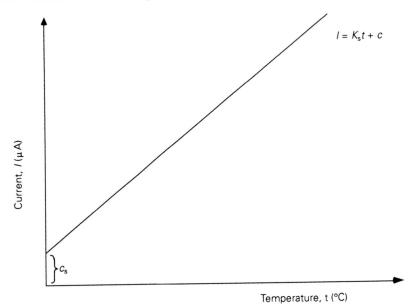

Fig. 4.10   The static gain of a sensor.

$$I = K_s t + c_s \qquad\qquad (4.9)$$

where I = sensor current (μA)
$\quad K_s$ = slope or dI/dt (μA K^{-1})
$\quad t$ = temperature (°C)
$\quad c_s$ = intercept (μA)

The circular symbol with the cross inside is used in block diagrams to represent addition (+) or subtraction (−). The static gain is that gain when the current and temperature are not varying with time (time does not appear in the temperature – current equation) and the sensor is in equilibrium; the term **steady-state gain** is sometimes used instead. The **dynamic** or time-dependent characteristic of the sensor and transmitter is represented by the $G(s)$ block, where s is a complex variable:

$$s = \sigma + j\omega$$

where $j = \sqrt{(-1)}$
$\quad \sigma$ = real part of s
$\quad \omega$ = imaginary part of s

This dynamic block and the operator s are discussed in more detail in Chapter 5. Note that sensors can be non-linear (like the thermistor) and they can also suffer from hysteresis and age [1], which makes the sensor relationship and block diagram of Fig. 4.9 more complicated.

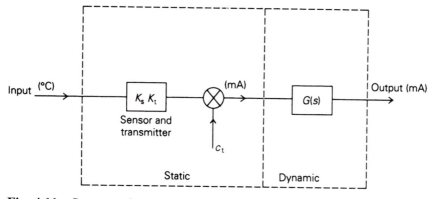

Fig. 4.11 Sensor and transmitter combined.

If the sensor has a negligible value of c_s, then the sensor and transmitter blocks can be combined (Fig. 4.11). The configuration is simply the process of defining the static gains, $K_s K_t$, and the intercept, c_t, of the sensor–transmitter combination. This is done by inputting two prescribed points of the linear relationship into the outstation. It also shows the elementary beginnings of manipulating block diagrams to simplify control analysis.

4.5 Interference

Interference is defined as the picking up of a **deterministic signal**, such as a sinusoidal signal at 50 Hz from a power cable, affecting the required transmitted signal. There are also **random signals** due to random motions in electronic devices, and these are called **noise**.

The greatest problem for BEMSs is external interference and noise, often from other plant in the plant room. Nearby a.c. power circuits operating at 240 V or 415 V (50 Hz) can produce sinusoidal interference signals known as mains **pickup** or **hum**. D.c. Circuits are less likely to produce interference, but when d.c. or a.c. circuits are switched, then transient noise can be produced. Lighting from fluorescent and other discharge lamps is another common cause of interference, with the arc of the lamp occurring twice per cycle. However, lamp interference occurs only with malfunctioning of the luminaire or poor design.

Interference is picked up by the BEMS equipment by **inductive coupling, capacitive coupling** or **multiple earths**, or combinations of all three [1]. The first two refer to the interfering circuit and the BEMS circuit having a mutual inductance or a mutual capacitance. Multiple earthing arises from connected pieces of equipment being separately earthed and the local earth not being at zero potential. This is often due to large electrical equipment upsetting the potential locally.

As mutual inductance and capacitive coupling are inversely proportional to the distance between the circuits, it is sensible to keep the BEMS and its sensors and cabling as far away from interfering equipment as possible.

But this is not always practical, especially for sensor cables which are often quite long. Here the inductive coupling can be reduced by twisting the sensor cables in pairs to produce the required number of wires in the cable as **twisted pairs**.

To avoid capacitive coupling, the BEMS equipment can be contained in an earthed metal screen or shield. Ideally the equipment should be insulated from the screen. This is a small problem for the outstation in its electrical cabinet, but it is not so easy for the sensor and its cable. The cable can be screened by having a light metal lattice around the twisted pairs, just under the plastic sheath, similar to a coaxial cable.

In general, BEMSs manufacturers take additional precautions against interference as such small signals are used by the microprocessor, which can easily be contaminated by interference. Such measures can be the use of op-amps as diffrential amplifiers for signals to reject a lot of the interference. The **common mode rejection ratio** (CMMR) is a measure of an amplifier's performance; it should be as high as possible, a typical value being 10^5 or 100 dB.

Another subtle ploy is to use an integrating ADC which integrates over 20 ms, the period of the 50 Hz mains cycle. Being a sine wave, the positive part of the interference signal cancels out the negative part during integration. Filtering and averaging of sensor readings are other methods to obtain better signals.

4.6   Sensor dynamic response

We have considered the static gain of the sensor, but the sensor also has a dynamic response, represented in Fig. 4.9 by the $G(s)$ block. This dynamic response comes from the sensor having its own thermal mass which takes time to heat up and cool down as the air or water whose temperature is being measured changes. This is described by the following equation:

$$\text{Rate of change of heat into sensor} - \text{Rate of change of heat out of sensor} = \text{Rate of change of heat stored in sensor}$$

(4.10)

A useful way of understanding equation (4.10) is to consider a couple of analogies: a cylinder with a hole in it (or more exactly an open pipe and valve in the bottom), and an electrical resistor capacitor circuit with a battery and switch (Fig. 4.12). Table 4.3 shows the analagous quantities. From the table it can be seen that the previous equation can be rewritten as:

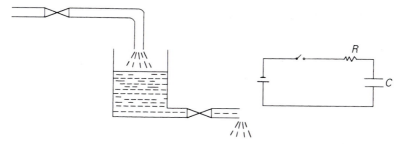

Fig. 4.12 Water and electrical analogies.

Table 4.3 Electrical, thermal and liquid analogies

Quantity	electrical charge (C)	thermal heat (J)	liquid volume (m^3)
Potential	voltage (V)	temperature (K)	height (m)
Flow	current (A)	Watt (W)	(m^3 s^{-1})
	(charge per second)	(heat per second)	
Resistance[a]	(Ω)	(K W^{-1})	(s m^{-2})
Capacity[b]	(F)	(J K^{-1})	(m^2)

[a] Resistance = potential/flow
[b] Capacity = quantity/potential

$$\frac{\text{Heat flow}}{\text{into sensor}} - \frac{\text{Heat flow}}{\text{out of sensor}} = \frac{\text{Rate of change of}}{\text{heat stored in sensor}} \qquad (4.10)$$

which is analogous to the flow of water in the cylinder and the flow of charge in the capacitor. It is interesting to note that temperature is analogous to voltage for electricity and to height for water. Temperature is the driving potential for heat transfer.

The best way of examining the response of a sensor is suddenly to change the temperature of the fluid it has been measuring for some time and with which it is in equilibrium. One could effect the same response by plunging a BEMS water temperature sensor into a tank of hot water from a tank of cold water. This instantaneous step change in the water temperature causes the sensor temperature to change, but as the sensor takes time to heat up or cool down it does not instantaneously achieve the new water temperature. Figure 4.13 shows this for a sudden increase in the fluid temperature.

When the sensor is suddenly plunged into a hotter fluid, it will gain heat and there will be no heat loss term. It is assumed that the sensor takes a negligibly small amount of heat from the bulk of the fluid, so the fluid's temperature is unaltered. Equation (4.10) becomes

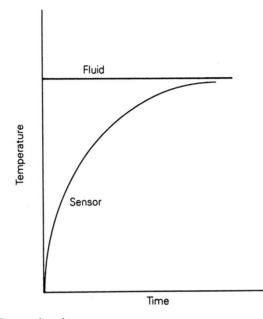

Fig. 4.13   Sensor heating up.

$$h_s A(t_{\text{fluid}} - t_s) = MC_p \frac{dt_s}{dT} \tag{4.11}$$

where M = mass of the sensor (kg)

C_p = specific heat capacity of sensor (J kg^{-1} K^{-1})

t_s = sensor temperature at time T (°C)

t_{so} = sensor temperature at time $T = 0$ s (°C)

t_{fluid} = fluid temperature (°C)

T = time (s)

h_s = heat transfer coefficient to sensor (W m^{-2} K^{-1})

A = area of sensor (m^2).

Rearranging equation (4.1) yields

$$\frac{dT}{\tau} = \frac{dt_s}{(t_{\text{fluid}} - t_s)} \tag{4.12}$$

where τ is the sensor time constant (s), given by

$$\tau = \frac{MC_p}{h_s A} \tag{4.13}$$

The τ equation can be verified by dimensional analysis:

$$\left[\frac{MC_p}{h_s A}\right] = \left[kg\frac{J}{kg\ K}\frac{m^2 K}{W}\frac{1}{m^2}\right] = [s]$$

The larger the mass of the sensor, the larger the time constant and the slower the response. The sensor casing also adds to the mass, and so does a sensor pocket or well, as shown in Table 4.3 for a thermistor.

Integrating equation (4.12):

$$\int_0^T \frac{T}{\tau} = \int_{t_s\ initial}^{t_s} \frac{dt_s}{(t_{fluid} - t_s)} \tag{4.14}$$

where it is assumed that the step change in fluid temperature has occurred at time $T = 0$ s, after which the fluid temperature immediately becomes t_{fluid} and the sensor temperature rises from its original value of $t_{s\ initial}$ at time $T = 0$ s to t_s at time T s; equation (4.12) becomes

$$T = \tau\{-\ln(t_{fluid} - t_s) - [-\ln(t_{fluid} - t_{s\ initial})]\}$$

Note that the minus sign is due to the $-t_s$ in the denominator of the integral. Rearranging the equation gives

$$T = -\tau\ \ln\left[\frac{(t_{fluid} - t_s)}{(t_{fluid} - t_{s\ initial})}\right] \tag{4.15}$$

Another common form of this is

$$\frac{(t_{fluid} - t_s)}{(t_{fluid} - t_{s\ initial})} = \exp(-T/\tau) \tag{4.16}$$

Example 4.3
A hot water temperature sensor in a pipe has an initial temperature of 10°C. A valve then opens up and allows water through the pipe from the boiler at 80°C. What is the sensor temperature 1 min later? The sensor time constant is $\tau = 1$ min.

Solution
The sensor temperature, t_s, is obtained explicitly from equation (4.16) as

$$t_s = t_{fluid} - \exp(-T/\tau)\{t_{fluid} - t_{s\ initial}\}$$

subtracting $t_{s\ initial}$ from both sides, it simplifies to

$$t_s - t_{s\ initial} = \{t_{fluid} - t_{s\ initial}\}[1 - \exp(-T/\tau)] \tag{4.17}$$

Here $t_{s\ initial} = 10°C$
$t_{fluid} = 80°C$
$T = \tau = 60$ s

Table 4.4   Typical sensor time constants

Sensor element	Time constant (min)
Thermistor in water	0.2
Thermistor in pocket in water	0.6
Thermistor in still air	2.4
Resistance thermometer bulb in water	1.0

So that

$$t_s - 10 = \{80 - 10\}[1 - \exp(-1)]$$

$$t_s = 10 + 70[1 - 0.368]$$

$$t_s = 54.2°C$$

4.6.1   Time constants

Example 4.3 shows that it takes some time for the sensor temperature to rise; the time constant of the sensor is the determining factor. This was defined by the sensor parameters in equation (4.13) but it can also be measured as the time for the sensor temperature to rise to 63.2% of its full temperature rise. Equation (4.17) can be used to demonstrate it. When $T = \tau$, equation (4.17) is

$$t_s - t_{s\ initial} = \{t_{fluid} - t_{s\ initial}\}[1 - \exp(-1)]$$

$$= \{t_{fluid} - t_{s\ initial}\}[0.632]$$

The rise in t_s from its initial value, $t_{s\ initial}$, is 63.2% of the temperature difference between the fluid temperature, t_{fluid} (the temperature the sensor is approaching) and the initial sensor temperature, $t_{s\ initial}$.

Typical time constants for various sensors are shown in Table 4.4, which has been taken from reference [9]. The velocity of the fluid in which the sensor is placed also affects the sensor time constant as the heat transfer coefficient reduces at low velocity. For instance, when the water velocity around a sensor is below 0.2 m s^{-1}, the time constant increases significantly [9].

From a log of the water temperature sensor in a pipe in Example 4.3, a BEMS would produce a graph like Fig. 4.13 when the sensor was plunged into the hot water; this is a graph of a **first-order system**.

If a tangent is drawn on the first-order curve at time T_1 and it crosses the final temperature which the sensor is trying to achieve, t_{fluid} in the example, at T_2, then

$$T_2 - T_1 = \tau$$

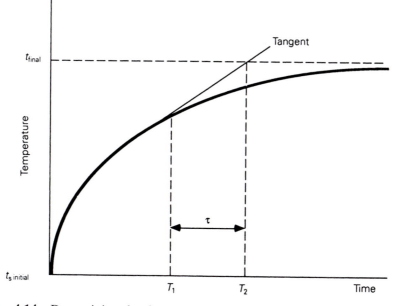

Fig. 4.14 Determining the time constant of a first-order system.

as is shown in Fig. 4.14. This can be demonstrated from equation (4.17):

$$t_s - t_{s\ \text{initial}} = \{t_{\text{final}} - t_{s\ \text{initial}}\}[1 - \exp(-T/\tau)] \qquad (4.17)$$

where t_s = sensor temperature (°C)
  $t_{s\ \text{initial}} = t_s$ at time $T = 0$
   t_{final} = final temperature which the sensor is trying to achieve
    τ = cylinder time constant (s)
Differentiating equation (4.17) to obtain the slope of the tangent gives

$$\frac{dt_s}{dT} = \frac{1}{\tau} \exp(-T/\tau)\{t_{\text{final}} - t_{s\ \text{initial}}\} \qquad (4.18)$$

The equation of the tangent line at the point T_1 is

$$t_s = \left[\frac{1}{\tau} \exp(-T_1/\tau)\{t_{\text{final}} - t_{s\ \text{initial}}\}\right]T_1 + c \qquad (4.19)$$

where c is the intercept of the line (°C).
  The line has the two points on it at t_s and $T = T_1$ and t_{final} and $T = T_2$. So from equation (4.19):

$$T_2 - T_1 = \left[\frac{1}{\tau} \exp(-T_1/\tau)\{t_{\text{final}} - t_{s\ \text{initial}}\}\right]^{-1}\{t_{\text{final}} - t_s\} \qquad (4.20)$$

From equation (4.16) it can be seen that

$$\frac{t_{final} - t_s}{t_{final} - t_{s\ initial}} = \exp(-T_1/\tau)$$

which when substituted into equation (4.20) gives

$$T_2 - T_1 = \tau \qquad (4.21)$$

Note that t_{final} is the hot water temperature, and that the sensor temperature will approach it very closely, although specifically it will not exactly equal it. However, in practical terms, when $T = 2\tau$ then

$$t_s - t_{s\ initial} = 0.86\{t_{final} - t_{s\ initial}\}$$

and when $T = 3\tau$ then

$$t_s - t_{s\ initial} = 0.95\{t_{final} - t_{s\ initial}\}$$

and when $T = 4\tau$ then

$$t_s - t_{s\ initial} = 0.98\{t_{final} - t_{s\ initial}\}$$

In other words, after 2τ the sensor temperature has risen to within 86% of t_{final}, and to within 95% and 98% after 3τ and 4τ, respectively.

4.7   Measurement errors

The dynamic response of a sensor affects how quickly it approaches the measured fluid temperature. Until the sensor is in equilibrium with the fluid, there will be some error in the sensed temperature.

   An even greater factor in the error of a sensor's temperature measurement, especially for water temperature measurements, is how well the sensor is in contact with the fluid it is measuring. To measure the temperature of hot water in a pipe, for instance, the sensor could be inserted as part of a probe into the pipe, with a mechanical seal to allow removal.

   Another more popular method is to place the sensor into a **pocket**, or well, which has already been screwed or welded into the pipe. The sensor can then readily be taken out for maintenance or replacement.

   An easier way, especially for BEMS installations on existing plant, but more prone to error, would be simply to strap the sensor to the pipe. This method is frequently employed on domestic hot water cylinders and calorifiers, the sensor casing being held in place by a band around the calorifier. Such sensors are referred to as **strap-on sensors** or **surface contact sensors**, or simply **contact sensors**.

   Wherever sensors are used to measure water temperature, it is unlikely they will be in direct contact with the water. Pockets or wells will be installed

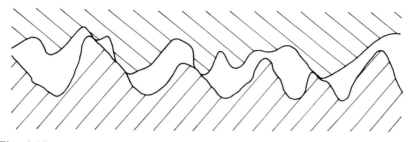

Fig. 4.15 Two surfaces in contact.

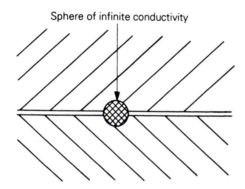

Fig. 4.16 Holm's model of a constriction resistance.

into which the sensors can be placed. The sensor will then be measuring the inside surface temperature of the pocket, whereas the outside surface of the pocket will be in contact with the water. Provided that the pocket is made of a material of good conductivity (brass and stainless steel are suitable and commonly used materials), the thermal resistance will be small.

Unfortunately, there will be more resistance between the surface of the sensor and the inside surface of the pocket. This is due to the two surfaces being rough on a microscopic scale and making contact at relatively few points of small area. Microscopically it is like putting the profile of Switzerland on top of Scotland. Figure 4.15 shows this contact between two surfaces.

Where contact is made, the area is often small and it forms a **constriction resistance** to heat flow, like three lanes of motorway traffic being funnelled into one lane, with the resultant hold-up or resistance to flow. Holm describes the original work on constriction resistances in relation to electrical contacts in his seminal book [10], but the theory is very similar for thermal contacts. Holm's model is of a contact point being replaced by a small sphere of radius r, where πr^2 is equal to the contact point area (Fig. 4.16).

The sphere has infinite conductivity and the resistance to heat flow through one of the contacting surfaces is

$$R_{\text{th constr}} = \frac{1}{\pi r k}$$

where $R_{\text{th constr}}$ = thermal constriction resistance (m² K W⁻¹)

where $R_{\text{th constr}}$ = thermal constriction resistance (m^2 K W^{-1})
k = conductivity of the contact member (W m^{-1} K^{-1})
r = sphere radius (m)

A more elaborate model gives the more correct

$$R_{\text{th constr}} = \frac{1}{2ak} \tag{4.22}$$

where a is the radius of the circular contact area (m).

For two surfaces in contact, each will have a constriction resistance of $1/4ak$, and the combined resistance will be the two constriction resistances in series, $1/2ak$. There will be convection and radiation heat transfer across the gaps between the surfaces [16], but it is assumed that the resistances of these two modes is much greater than the resistance of the constriction.

So for a sensor in a pocket, there is the small resistance due to the metal of the pocket itself and a larger constriction resistance due to the contact area of the sensor and pocket. To reduce the constriction resistance, the area of contact needs to be increased. This can be done by putting some silicon grease or oil into the pocket to fill in the gaps and so increase the area of contact. The conductivity of the fluid is much larger than the air it replaces and it therefore shorts out the constriction resistances. To retain the oil or grease, the pocket must be positioned vertically.

The need to put oil and grease in pockets is an additional maintenance requirement which inevitably may be forgotten. So, to ensure that there is good thermal contact between a sensor and its pocket, springs can be put at the sensor tips to ensure there is adequate pressure, and consequently good contact, with the pocket. Even with this, there is still the influence of the pipe wall and its heat loss that can affect the sensor's reading [2]. With all these factors, errors of 20% or more below the fluid temperature can be made.

Compared to the sensor in a pocket, the strap-on sensor is even more subject to error. This is because it is measuring the outside surface temperature of the pipe, or calorifier, not the water temperature itself. The pipe surface temperature can be considerably lower than the water within it due to radiation and convection losses, depending on the level of insulation. However, the insulation has to be stripped off around the sensor location for access to the pipe surface. Conduction can then occur through the sensor, its case and wires. Constriction resistances are invariably higher as well, due to the curved surface of the pipe. Hence serious errors can occur with strap-on sensors.

Example 4.4

A surface-mounted sensor has only 10^{-5} of its surface area making contact with a 25 mm copper hot water pipe.

(a) What effect has this on the temperature read by the sensor?
(b) What happens when a heat transfer fluid is inserted between the sensor and pipe?

Assume there are no convection or radiation losses.

Solution

(a) The actual effect is difficult to quantify as little work has been done to measure thermal constriction resistances [16]. The fraction of the actual area of contact used in this example has been taken from measurements between flat surfaces in contact [17], under a pressure of ~ 9 kPa.

Assuming the constriction resistance to be due to a single circular contact, then if the surface area is A, the radius of actual contact, r, is

$$r = \left[\frac{10^{-5} \times A}{\pi}\right]^{0.5}$$

So the constriction resistance is

$$R_{th\ constr} = \frac{1}{2k}\left[\frac{\pi}{10^{-5} \times A}\right]^{0.5} \tag{4.23}$$

and with $k = 390$ W m^{-1} K^{-1} [19],

$$R_{th\ constr} = 0.72(A)^{-0.5}$$

This has units of K W^{-1}, not the usual units.

Taking the thickness of the copper pipe to be 1.2 mm [18], and the sensor head to which the thermistor or PRT is attached to be of equal thickness and conductivity, the resistance of the pipe (and the sensor), R_p, is

$$R_p = \frac{1}{kA}$$

$$= \frac{1.2 \times 10^{-3}}{390A}$$

$$= 3.1 \times 10^{-6}\ A^{-1}$$

which is substantially below the constriction resistance. For instance, if the nominal contact area is 4 cm^2, the ratio of the constriction resistance to the pipe and sensor resistance will be

$$\frac{R_{th\ constr}}{2 \times R_p} = 2.3 \times 10^3$$

So if the water in the pipe is at 80°C and the sensor reads 70°C, the temperature drop across the constriction resistance will be

$$\frac{R_{\text{th constr}}}{2R_p + R_{\text{th constr}}} \times 10 = 9.99 \text{ K}$$

(b) Taking the conductivity of the heat transfer grease to be $0.1 \text{ W m}^{-1} \text{ K}^{-1}$, and the separation of the sensor and pipe surface to be 2 μm (dictated by the surface roughness), the resistance of the grease, R_g, is

$$R_g = \frac{2 \times 10^{-6}}{0.1 \times 4 \times 10^{-4}}$$

$$= 0.05 \text{ W K}^{-1}$$

This compares with a constriction resistance of 35.85 W K^{-1}. As the constriction resistance and the grease resistance are in parallel, the combined resistance is predominantly due to the grease, i.e. 0.05 W K^{-1}. So if the heat flow through the resistances to the sensor remains the same with the grease, then without the grease the temperature drop reduces from 10 K to

$$\frac{10}{R_{\text{th constr}}}(R_g + 2R_p) = \frac{10}{35.85}(0.05 + 0.015)$$

$$= 0.018 \text{ K}$$

Effectively the temperature difference, and error, has disappeared with the introduction of the grease. But this is a simplified, idealized example, to illustrate the importance of good contact between the sensor and pipe.

4.8   Comfort

Constriction resistances do not affect air temperature sensors as they are surrounded by air, but there are two other problems. Both relate to inside temperature sensors. The first is how representative the sensor reading is of the bulk air temperature in the room. Warm air will come from an emitter which may well be the other side of the room to the sensor and there will be a **distance–velocity** lag before the sensor senses the warm air. Also the sensor may be in a stagnant area with little air movement, unrepresentative of the bulk of the air. This is not a sensor error, but a positional error.

The more important problem is that an air temperature sensor is not measuring comfort, even though a principal aim of a BEMS is to control the building services plant, so that the building is reasonably comfortable for the occupants. A measure of the thermal comfort of a room is the **dry resultant temperature** in that room. The dry resultant temperature takes account not only of the air temperature, but also of the radiant temperature. One only has to sit near a large single-glazed window in winter or on a sunny day to appreciate the radiant component of comfort. The dry resultant temperature for a typical room with air speeds of the order of 0.1 m s^{-1} is approximated as follows [20, 11]:

$$t_{res} = \tfrac{1}{2}t_r + \tfrac{1}{2}t_{ai} \qquad (4.22)$$

where t_r is the temperature measured by a thermometer at the centre of a blackened globe, 100 mm in diameter. The CIBSE Guide refers to t_{res} at the centre of a room, giving it the symbol t_c [20, 12, 13]. Both sections state that at the centre of the room t_r approximates to the mean surface temperature of the room, t_m:

$$t_m = \frac{\Sigma(At_{surf})}{\Sigma A}$$

where A = area of a surface of the room (m^2)
   t_{surf} = temperature of a surface of the room (°C)
It would be impractical for a BEMS sensor to be housed in a 100 mm blackened globe in the centre of a room. Also most BEMS room temperature sensors predominantly measure the air temperature. There is not much published data to support this statement, but there is a little more data on room thermostats [14, 17], which have similar housings to BEMS room temperature sensors and serve a similar purpose. Letherman [7] quotes the steady-state response of a thermostat to air temperature and radiant temperature:

$$t_t = \Gamma t_{ai} + (1 - \Gamma)t_{surroundings} \qquad (4.23)$$

where t_t = thermostat sensor temperature (°C)
         t_{ai} = air temperature (°C)
   $t_{surroundings}$ = mean surface temperature of surroundings (°C)
         Γ = dimensionless factor relating thermostat sensitivity to air and radiant temperatures
For a ventilated casing to the thermostat, a mean value for Γ is quoted as 0.9 and for an unventilated casing it is 0.67. Presumably the second term on the right of equation (4.23) includes the effect of the influence of the internal wall on which the thermostat is positioned. An internal wall is often chosen for positioning thermostats and BEMS sensors as opposed to the inside surfaces of external walls as the latter can be quite cold and much lower than the room air temperature.

To relate the air temperature to the dry resultant temperature, the type of heating has to be considered. Different types of heating give out different proportions of radiant and convective heat, leading to different values of t_{ai} and t_m. The CIBSE Guide relates t_{ai} and t_c by the ratio F_2 and provides tables of values of F_2 (modified to F_{2cu} in reference [20]) for various heating systems (radiators, convectors, radiant panels, etc.) [13]. The ratio is defined as

$$F_2 = \frac{t_{ai} - t_{ao}}{t_c - t_{ao}}$$

where t_{ao} is the outside air temperature (°C).

Example 4.5
A factory is heated by forced warm-air heaters. The system is designed to provide a dry resultant temperature of 19°C when it is −1°C outside; F_2 is 1.17 and t_m is 15.8°C. What is the BEMS room sensor reading if it is in a ventilated case?

Solution
Equation (4.23) for the thermostat can be used with t_m put in as an approximate value for $t_{surroundings}$. In fact, $t_{surroundings}$ will probably be slightly higher than t_m as it is closer to a warm inside wall. The value of t_{ai} is required for equation (4.25), and this can be obtained from the F_2 ratio:

$$1.17 = \frac{t_{ai} - (-1)}{19 - (-1)}$$

giving

$$t_{ai} = 22.4°C.$$

Putting the values into equation (4.23):

$$t_t = (0.9 \times 22.4) + (0.1 \times 15.8)$$

$$= 20.16 + 1.58 = 21.74°C$$

So to maintain t_c at 19°C and t_{ai} at 22.4°C, the BEMS sensor must control to 21.7°C.

To alleviate the problem of obtaining a representative inside temperature of a building, where some rooms will inevitably be colder than others, due to the heating system being unbalanced or having wrong emitter sizes, a number of sensors can be employed. Their outputs are then addressed to an **averaging function module**, which produces a signal of the average value for the relevant control loop. Typically one average module can have up to four sensors, but a number of modules can be combined in parallel to increase the number of sensors.

Minimum and **maximum function modules** can also be used in parallel with the average module; they identify the extreme rooms to indicate how unbalanced the heating system is.

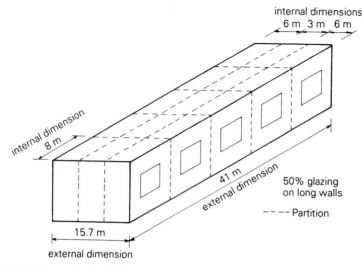

Fig. 4.17 Example building.

4.8.1 Monday morning blues

Most BEMS internal temperature sensors are predominantly air temperature sensors. So when the building is warming up on a Monday morning, after the heating has been switched off over the weekend, the BEMS will switch off or modulate the heating when the required air temperature is met. This could mean that the fabric, which takes much longer to heat than the air, has not got very warm. The radiant temperature, and hence the dry resultant temperature, would therefore be low and the building still uncomfortably cold.

Chapter 9 on optimizer control deals with preheating, but there is an interesting example there that demonstrates the difference in time to heat the air and the fabric. Figure 4.17 shows the single-storey office building used in that example.

Table 4.5 shows the heat stored in the various elements of the building, and Table 4.6 shows the individual time constants of the elements (the time constants are derived in Chapter 9 on optimizer control). Notice that the air can heat up much quicker than the fabric elements. Although there is heat exchange between the air and the fabric, which will limit large temperature differences, there is still a distinct possibility that the air temperature will rise well ahead of the fabric temperature. The BEMS sensor will respond to the air temperature and control the heating accordingly. This gives a low value for the dry resultant temperature, especially on a Monday morning when the fabric has lost heat over the weekend, hence the term 'Monday morning blues'.

Table 4.5   Heat stored in the elements of the building

Element	Heat stored	
	MJ	*Fraction of total (%)*
Window	13	1
Air	59	4
Brick	100	8
Block	229	17
Roof	340	25
Partitions	614	45
Total	1355	100

Table 4.6   Time constants of the elements of the building

Element	Time constant, τ (h)
Brick	5.8
Block	5.1
Window	0.7
Roof	4.5
Partitions	6.7
Air (one air change per hour)	1.0

Some BEMS manufacturers recommend that a sensor is placed within an internal wall to control the preheating on the internal fabric temperature. However, this will effectively give the sensor a long time constant, necessitating the use of another sensor to measure the air temperature inside the room for controlling the heating when the building is up to temperature.

4.9   Smart sensors

A laser-trimmed thin film sensing element on a silicon substrate is shown in Fig. 4.2. It would be possible to add on to the same chip a CPU to make it intelligent. With an ADC, a communication section and some memory to store look-up tables and the sensor-type module then one has the basics of a **smart sensor**. The ADC would allow a digitized signal to be transmitted to the outstation and the CPU would enable more sophisticated signal processing to be carried to avoid interference. A smart sensor would also allow direct control of a valve actuator or relay. Figure 4.18 shows the concept of a smart sensor, where the boxes do not represent areas on the chip, but merely discrete functions of the chip in its DIL package.

It is expensive to manufacture a chip that contains these specialized functions for just one sensor type. However, by including a limited range of sensor elements, the chip becomes multi-purpose, widening its market and

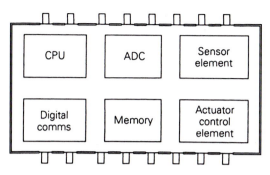

Fig. 4.18   Concept of a smart sensor.

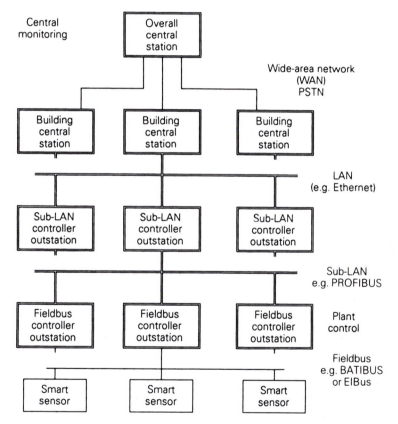

Fig. 4.19   Possible architecture for a BEMS with smart sensors.

reducing its cost [15]. Users then select the sensing element they require. Using smart sensors would tie in with small outstations for controlling individual items of plant, and also the use of fieldbuses. Figure 4.19 shows a probable architecture for a BEMS using smart sensors.

References

[1]   Bentley, J. P. (1988) *Principles of Measurement Systems*, 2nd edn, Longman, London.

[2]   Stoecker, W. F. and Stoecker, P. A. (1989) *Microcomputer Control of Thermal and Mechanical Systems*, Van Nostrand Reinhold, New York.

[3]   Chesmond, C. J. (1982) *Control System Technology*, Edward Arnold, London.

[4]   Diefenderfer, A. J. (1979) *Principles of Electronic Instrumentation*, 2nd edn, Holt-Saunders, Eastbourne.

[5]   Fielden, C. J. and Ede, T. J. (1982) *Computer-based Energy Management in Buildings*, Pitman, London.

[6]   Hencke, H. (1989) The design and application of Honeywell's laser-trimmed temperature sensors. *Measurement and Control*, **22**, October.

[7]   Letherman, K. M. (1981) *Automatic Controls for Heating and Air Conditioning Principles and Applications*, Pergamon, Oxford.

[8]   Martin, G. (1982) *Microprocessor Appreciation Level III*. Technician Education Council, Hutchinson, London.

[9]   CIBSE (1986) Guide, CIBSE Volume B, Section B11, Automatic control, Chartered Institution of Building Services Engineers, London.

[10] Holm, R. (1958) *Elektrische Kontakte*, Springer–Verlag, Berlin.

[11] CIBSE (1986) CIBSE Guide, Volume A, Section A1, Environmental criteria for design, Chartered Institution of Building Services Engineers, London.

[12] CIBSE (1986) CIBSE Guide, Volume A, Section A5, Thermal response of buildings, Chartered Institution of Building Services Engineers, London.

[13] CIBSE (1986) CIBSE Guide, Volume A, Section A9, Estimation of plant capacity, Chartered Institution of Building Services Engineers, London.

[14] Fitzgerald, D. (1969) Room thermostats, choice and performance. *Journal of the Institution of Heating and Ventilating Engineers*, **37**.

[15] Wilkins, J. and Willis, S. (1990) What are smart sensors? *Building Services* (CIBSE) **12**(12).

[16] Holman, J. P. (1989) *Heat Transfer*, McGraw-Hill, New York.

[17] Bowden, F. P. and Tabor, D. (1971) *Friction and Lubrication of Solids: Part 1*, Oxford University Press, Oxford.

[18] CIBSE (1986) CIBSE Guide, Volume B, Section B16, Miscellaneous Equipment, Chartered Institution of Building Services Engineers, London.

[19] CIBSE (1986) CIBSE Guide, Volume C, Section C7, Units and Miscellaneous Data, Chartered Institution of Building Services Engineers, London.

[20] CIBSE (1999) CIBSE Guide A, Environmental Design, Chartered Institution of Building Services Engineers, London.

5
Basic and fuzzy logic control

Having considered the basic elements of a BEMS in previous chapters, in this chapter we examine the basics of control. This involves the system that the BEMS is controlling and its interaction with the BEMS.

5.1 The need for control

It is initially worth considering the need for control to get an appreciation of its importance. Consider a boiler and heating system for a building. The system is sized on the required values of the inside dry resultant temperature, t_c (also t_{res} but the subscript c refers to the centre of the room) and the design outside air temperature t_{ao}. The CIBSE Guide gives design values for both t_c and t_{ao} [1, 2, 22].

Most heating systems are not sized to the exact building heat loss but are oversized to get the building quickly up to temperature in the morning. Typically systems are 20% oversized, although the CIBSE Guide suggests a larger margin [3, 22].

The guide indicates that the design value of t_{ao} should be −3°C in London for a lightweight building; the term 'lightweight' refers to the thermal capacity of the structure (Chapter 7). Between 1976 and 1995, in London, on only one occasion a year on average did the 24 h mean t_{ao} fall below this temperature [22]. For heavyweight buildings with 20% oversizing, the recommended t_{ao} is −1.9°C and there was only one occasion a year on average in which the 48 h mean t_{ao} fell below this temperature. The longer time average for the heavyweight building reflects the fact that it will not be so greatly influenced by short-term temperature fluctuations.

Example 5.1
What is the average heat output from a heating system during the heating season for a lightweight office in London. The system is 20% oversized. The heating system produces little difference between t_c and t_{ai}, i.e. $F_2 = 1.0$.

Solution
For an office the CIBSE Guide recommends t_c should be 21°C to 23°C, say 21°C here. From the CIBSE Guide [2] the average t_{ao} for the 30-year average period 1941–70 (October to April) is 7°C. The design inside-to-outside temperature drop is

$$t_{c\ design} - t_{ao\ design} = 21 - (-3)$$
$$= 24°C$$

However, the heating system is 20% oversized, so the temperature drop from inside to outside that it can maintain is

$$24 \times 1.2 = 28.8°C$$

Therefore the heating is capable of maintaining 21°C inside under favourable conditions when it is −7.8°C outside. The heating system produces t_c close to t_{ai}, so the heat output from the heating system can be taken as proportional to the temperature difference. The ratio of the average heat output to the design heat output is

$$\frac{21 - 7}{21 - (-7.6)} = 49\%$$

This calculation, like the heating design calculation, does not include the internal heat gains (heat from people, equipment, lighting and the sun [4, 22]) that occur and will further reduce the heat requirement during the day. So the control of the heating system must match the heat requirement, otherwise there will be potential waste of heat and energy. The energy consumption through the heating season should be approximately proportional to

$$t_c - t_{ao} = 21 - 7$$
$$= 14\ K$$

A 1 K rise in t_c through the heating season would increase the consumption by

$$\frac{1}{14} = 7.1\%$$

It should be noted that the heating system should not raise t_c above 19°C to save energy. Internal gains can raise it to 21°C [22].

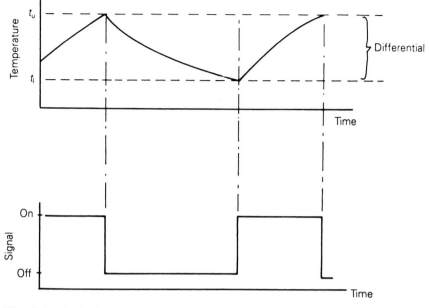

Fig. 5.1   On/off control.

5.2   On/off control

The simplest and most common control is **on/off** control. As shown in Fig. 5.1, an on/off controller sends an 'off' signal when the temperature reaches the upper level, t_u, and an 'on' signal when the temperature drops to the lower level, t_1.

Using the water analogy (Chapter 4) of the cylinder being filled with water while water flows out from a pipe in the bottom, the supply valve is shut off at the upper level, h_u, and opened at the lower level, h_1, as shown in Fig. 5.2.

The water level is measured by a sensor in the base of the cylinder, which is pressure sensitive and sensitive to the level of water above it. This sends a signal to the BEMS outstation which then controls the valve. The valve is either fully open or fully closed; there is no modulation.

The gap between t_u and t_1 (also h_u and h_1) is the **differential**, sometimes called the **overlap** or **hysteresis**. Figure 5.3 shows that halving the differential (from $2D$ to D) will halve the period (from $2\tau_p$ to τ_p) and double the frequency (from f to $2f$) as

$$f = \frac{1}{\tau_p} \tag{5.1}$$

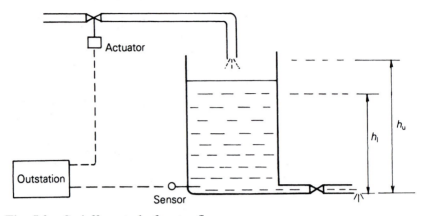

Fig. 5.2   On/off control of water flow.

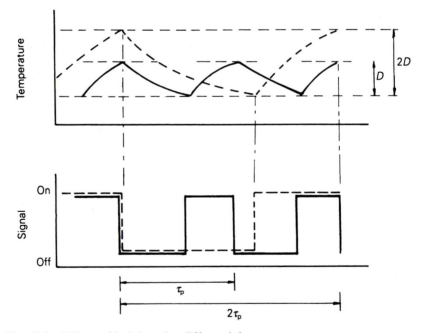

Fig. 5.3   Effect of halving the differential.

Accurate control requires a small differential. Too small a differential and the frequency of cycling will be high; the controlling device (the valve in the water cylinder example) will get increased use and wear. So a compromise is required between close control with low device life and looser control with longer device life. Often, if on/off control is used, the accuracy of control needed is not high, otherwise a more sophisticated and expensive control would be used. Also the application determines the accuracy of control.

For instance, a boiler supplying hot water to a radiator heating system will have controls, such as room thermostats and even perhaps thermostatic radiator valves, controlling the heating system. So the boiler thermostat does not have to have too small a differential as there is control further along in the system. A differential of 5 K is generally considered adequate for the boiler, although at low loads a wider differential is more efficient. However, the room thermostat needs a closer differential than 5 K, otherwise the room temperature could vary from hot to cold, e.g. 22°C to 17°C. A differential of 1 K to 2 K would be more appropriate. Although the differential of thermostats cannot easily be altered, on/off control with a BEMS, employing a temperature sensor and an output channel for sending on/off signals, does have an easily adjusted differential.

5.3   Binary drivers

An example of on/off control with a BEMS is shown in Fig. 5.4. Here hot water is pumped around a radiator heating system. A temperature sensor in one of the heated rooms is connected to an outstation which monitors the temperature. When the temperature reaches the upper temperature, say 21°C,

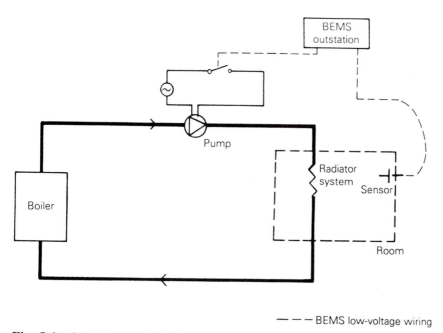

— — — BEMS low-voltage wiring

Fig. 5.4   On/off control of a heating pump.

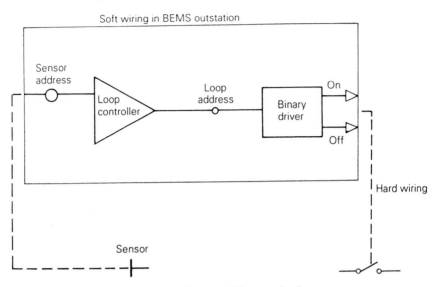

Fig. 5.5   BEMS configuration for on/off control of a pump.

then the outstation switches off the pump. This is a simple example used for illustrative purposes but is a typical control system for a house, except that a thermostat would be used instead of a BEMS.

The BEMS configuration diagram for this control strategy is shown in Fig. 5.5. The hardware, such as the ADC, has been missed out for simplicity. The **loop controller**, the triangle in the diagram, converts the sensor reading, already conditioned by the sensor and sensor-type module, to a value between 0% and 100% depending on how near or far the temperature is to the required setpoint. If the setpoint is 21°C then the signal will be 0%. (Further details on the loop controller and its operation are given in Chapter 6.)

The loop controller, or loop module, sends this output to an address where the **binary driver** is programmed to take the value at this address as an input. A driver, like the loop module, is a software module, or program, that is used to drive plant in response to an input signal. The binary driver is a driver where the on (t_1) and off (t_u) signals can be individually set by the operator to correspond to any value of input signal between 0% and 100%, as shown in Fig. 5.6. In other words, the on/off differential can be fully adjusted. If the differential were set to zero, then noise on the input would cause the driver to operate randomly. The binary driver sends a signal to the relay in the pump circuit to close and start the pump if the value is at or below t_1. If the value is at or above t_u, a signal is sent to open the relay to stop the pump. In between these two values no signal is sent.

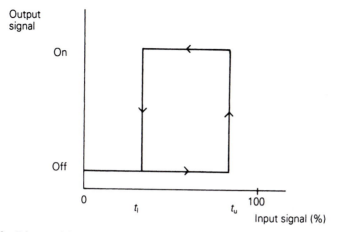

Fig. 5.6 Binary driver signals.

5.4 System dynamics

To be able to set up on/off differentials, as well as more sophisticated controls, it is useful to understand the response of the system being controlled. Figure 5.7 shows a simplified block diagram of the on/off control considered above.

All the variables in the block diagram are functions of time, shown by (T); $r(T)$ is the setpoint, t_u, at which the pump is switched off; and $y(T)$ is the output from the system, the temperature in the room. The sensor measures this temperature and sends back a signal to the outstation which gives a temperature in the outstation of $\hat{y}(T)$, an approximation or estimate of $y(T)$. The estimate is often very close to the actual value but measurement and approximation errors may occur (Chapter 4). At the summing

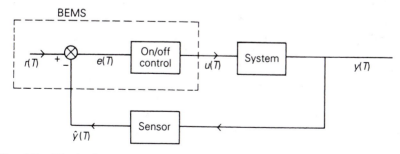

Fig. 5.7 Block diagram of BEMS on/off control.

point (the circle), the **error**, $e(T)$, is determined by subtraction from the setpoint:

$$e(T) = r(T) - \hat{y}(T) \tag{5.2}$$

Although it is commonly used for $e(T)$, the term 'error' is a misnomer, in that it is not the inaccuracy of a poor sensor, but simply the difference between the sensor reading and the required setpoint.

When the error is zero, the on/off control in the BEMS (the loop controller and the binary driver) sends a signal, $u(T)$, which switches off the pump. The pump will not come on again until the error has increased to equal the differential:

$$e(T) = t_u - t_1$$

The sensor block in Fig. 5.7 represents the static sensor and transmitter blocks of Fig. 4.9, as well as the dynamic block $G(s)$. Included in the sensor block of Fig. 5.7 is the outstation hardware, such as the ADC, so that $\hat{y}(T)$ is the measured room temperature. If the sensor and outstation are accurate and of good quality, the static gain terms of the sensor block in Fig. 5.7 have a combined value close to unity. But the dynamic response block of Fig. 4.9 will still be present in the sensor block of Fig. 5.7, whatever the quality and accuracy of the sensor to describe the sensor response.

The loop containing the sensor is called the **feedback loop**, feeding the result of the control action of the on/off control in the BEMS back to the BEMS, so that the error of the action can be determined. If the sensor and BEMS hardware are fast-acting with good accuracy, the loop feeds back virtually the true value and is a **unity feedback loop**.

The system block in Fig. 5.7 represents the pump and the rest of the heating system that goes to produce the resultant temperature output in the room. Note that the arrowed lines in the block diagram represent signals or information flow, not heat flow, although heat is present in the system. If the influence of the outside temperature, t_{ao}, were included in Fig. 5.7, then it would be as a **disturbance**, shown as another block introducing a disturbing signal after the system (Fig. 5.8).

The system block represents the pump, pipework, hot water, radiators and boiler. These devices have their own dynamic responses. If all the device responses can be described mathematically by expressions like equations (4.10) and (4.11), the system is a **first-order system**. There is a first-order differential term but no higher differential terms. The resistor–capacitor circuit of Chapter 4 is an example of a first-order system with a time constant of RC. The time constant for equation (4.11) is (MC_p/h_sA).

The radiators in our heating system can be approximated to first-order devices. Adams and Holmes [5] give the time constants for radiators, and a few values are shown in Table 5.1.

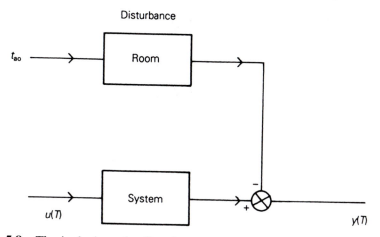

Fig. 5.8 The inclusion of a disturbance.

Table 5.1 Time constants for radiators (min)

Water flow rate (kg s^{-1})	Single-pipe circuit	Two-pipe circuit
0.014	5.5	8.0
0.032	2.9	3.6
0.068	1.5	1.9

5.5 Laplace and z-transforms

When there are a number of components comprising a system, as in our pumped heating system, there will be numerous differential equations to describe the dynamic response of the system. So the Laplace transformation is used to simplify and transform the differential equations into algebraic equations. The Laplace transformation is defined by

$$\mathscr{L}\left[f(T)\right] = \int_0^\infty f(T)\exp\left(-sT\right)dT \equiv F(s)$$

where $\mathscr{L}$ is the symbol for the operator to give the Laplace transformation of a function. The Laplace transformation is used for transforming functions such as $f(T)$, which are functions of time T, into functions $F(s)$, which are functions of the complex variable s. There is a limitation that $f(T)$ is zero for $T < 0$, so the transformation is useful for examining sudden changes to the system such as a step change at $T = 0$. The complex variable s is

$$s = \sigma + j\omega \tag{5.3}$$

where $j = \sqrt{(-1)}$

σ = real part

ω = imaginary part

Using the Laplace transformation, differentiation becomes

$$\mathcal{L}\left[\frac{df(T)}{dT}\right] = sF(s) - f(T)|_{T=0} \tag{5.4}$$

where $f(T)|_{T=0}$, or $f(0)$, is the value of $f(T)$ at $T = 0$. Unless there is any discontinuity in the function at $T = 0$ and the initial conditions are zero, then $f(0) = 0$, so equation (5.4) is simplified to

$$\mathcal{L}\left[\frac{df(T)}{dT}\right] = sF(s) \tag{5.5}$$

With the time constant τ in equation (4.11), it becomes

$$t_{\text{fluid}} - t_s = \tau\frac{dt_s}{dT} \tag{5.6}$$

Performing a Laplace transformation on equation (5.6), with the initial conditions at zero, gives

$$t_{\text{fluid}} - t_s = \tau s t_s$$

or

$$t_s(s)(\tau s + 1) = t_{\text{fluid}}(s) \tag{5.7}$$

Here $t_s(s)$ has not been changed to the normal notation of $T_s(s)$ for the Laplace transform of $t_s(T)$, to avoid confusion with T representing time. The dynamic response of the sensor in Fig. 4.9 was given as $G(s)$. The s here is the Laplace complex variable and $G(s)$ is the **transfer function** of the sensor, the ratio of the Laplace transform of the output from the sensor to the Laplace transform of the input, with the initial conditions at zero, i.e.

$$G(s) = \frac{F_{\text{output}}(s)}{F_{\text{input}}(s)} \tag{5.8}$$

where $F_{\text{output}}(s)$ is the Laplace transform of the output, and $F_{\text{input}}(s)$ is the Laplace transform of the input. For our sensor, described by equation (5.7), the transfer function is

$$G(s) = \frac{t_s(s)}{t_{\text{fluid}}(s)}$$

$$= \frac{1}{1 + \tau s} \tag{5.9}$$

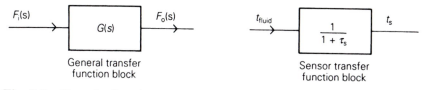

Fig. 5.9 Transfer functions.

Table 5.2 Laplace transforms of some common functions

Function of time	Laplace transform
$f(T)$	$F(s)$
$\dfrac{df(T)}{dT}$	$sF(s) - f(T)\mid_{T=0}$
Unit impulse, $\delta(T)$	1
Unit step, $u(T)$	$1/s$
Unit ramp, T	$1/s^2$
Sine wave, $\sin wT$	$1/(s^2 + w^2)$
Exponential decay, $\exp(-aT)$	$1/(s + a)$
Exponential growth, $1 - \exp(-aT)$	$a/s(s + a)$
Time delay of time T	$\exp(-sT)$

This lends itself well to the block diagram in Fig. 5.9, where $F_i(s)$ and $F_o(s)$ are the input and output, respectively. Table 5.2 shows the Laplace transforms of some common functions.

If a unit step change in fluid temperature, t_{fluid} in our example, with a Laplace transform of $1/s$,

$$\mathscr{L}(t_{\text{fluid}}) = \frac{1}{s} = F_{\text{input}}(s)$$

is applied to the sensor with its transfer function

$$G(s) = \frac{1}{1 + \tau s}$$

then

$$\mathscr{L}(t_s) = F_{\text{output}}(s) = G(s)F_{\text{input}}(s)$$

$$= \frac{1}{s}\frac{1}{1 + \tau s} \tag{5.10}$$

This is converted back to the **time domain** by taking the inverse Laplace transform:

$$\mathcal{L}^{-1}\left(F_{\text{output}}(s)\right) = t_s \qquad (5.11)$$

From Table 5.2 we can see that

$$\mathcal{L}(1 - \exp[-aT]) = \frac{a}{s(s + a)} \qquad (5.12)$$

Equations (5.12) and (5.10) are equivalent if $a = 1/\tau$. So the output from the sensor, $\mathcal{L}^{-1}\{F_{\text{output}}(s)\}$ or t_s, is

$$1 - \exp[-T/\tau] \qquad (5.13)$$

Equation (5.13) is not exactly the same as equation (4.17):

$$t_s - t_{s\,\text{initial}} = \{t_{\text{fluid}} - t_{s\,\text{initial}}\}[1 - \exp(-T/\tau)]$$

This is because the Laplace transformation required the initial conditions to be zero. So the **boundary conditions** (i.e. at the boundary $T = 0$) have to be fitted to the inverted equation. One of the boundary conditions in Example 4.3 was that the actual step change was not unity or 1 K, but ran from 10°C to 80°C, a step change of 70 K. The Laplace transform of this step change is $70/s$. For the sensor $\tau = 60s$, so the transfer function of the example sensor is

$$G(s) = \frac{1}{1 + 60s}$$

As can be seen in Table 5.2, the Laplace transform of a unit impulse is 1, so the response of a system to a unit impulse is simply the transfer function:

$$G(s) = \frac{F_{\text{output}}(s)}{F_{\text{input}}(s)} \quad \text{and} \quad F_{\text{input}}(s) = \mathcal{L}\{\delta(T)\} = 1$$

so

$$F_{\text{output}}(s) = G(s)$$

The overall transfer function of the sensor must also include the steady-state gains. As the transfer function is concerned only with the dynamic response, when things change at $T = 0$, then the change in time of the steady-state gain, or the slope of the steady-state equation is also required, although it is of little significance to the dynamics. As the sensor and transmitter constants, c_s and c_t, do not change with time, they become zero when the steady-state equation is differentiated with respect to time to determine the slope. So the overall transfer function is $K_s K_t G(s)$ mA K^{-1}.

Being derived from the ratio of the output to the input, $G(s)$ has the units of inverse time for the first-order system, so the steady-state gains effectively act as scaling factors. Often these steady-state gain units are not shown in block diagrams as, working through a feedback control loop, one starts at the required temperature, works through the blocks and comes back with the feedback temperature. The block units have cancelled each other out.

It is also often assumed that the sensor introduces no error or approximation into the measurement, i.e. that $\hat{y}(T) = y(T)$. Further details of the Laplace transform and transfer functions, and their use in control theory, can be found in Marshall [6] and DiStefano *et al.* [12]. The Laplace transformation is useful for simplifying and solving differential equations but the algebraic transformation equation itself can be used for determining the stability of a controller without transforming back to a function of time, T, as will be shown later.

But before any transfer function can be determined, one must first have a differential equation, or equations, describing the system. Where these are unknown, one can try to identify the system empirically by switching it on and off, or by changing the setpoint, $r(T)$, to simulate step functions and observing the response. In Chapter 6 the response of a system to various control changes is used to set up the controller for optimum performance. Another method of identification is to examine the system's response to a sine wave. For a linear system, the output will also be a sine wave, but compared to the input wave, it will be of different **amplitude** (i.e. height of the wave), and **phase** (i.e. the peak of the output wave appearing later or earlier than the input wave). The changes in amplitude and phase are plotted as **Nyquist**, **Bode** or **Nichols plots** [6]. However, it is difficult to subject heating systems and buildings to sinusoidal changes, although small electrical sinusoids of suitably long periods can be injected into actuators once the BEMS's control has been disconnected (the loop opened), but with the BEMS's sensor logging the response. Needless to say, there is little such data from building services plant, although the mathematics of Laplace transforms and these various plots make good exam questions!

So far, we have considered the use of Laplace transforms to solve differential equations, where the functions are analogue and continuous in time. But BEMSs and microprocessors deal with discrete, sampled signals. For these there are **difference equations** rather than differential equations, and the z-transform rather than the Laplace. (It is not intended to deal with them in great depth and the interested reader is referred to the literature [13, 14].)

The difference equation arises as the digital data is sampled, perhaps every T seconds. So the analogue signal $y(T)$ becomes

$$y(0), y(1), y(2), \ldots, y(kT), \ldots y(\infty T)$$

where k represents an integer. Usually the function is not of interest up to infinity, but to some general time kT. Hence the **discrete-time signal** is referred to as $y(kT)$, or simply $y(k)$. To appreciate a difference equation, consider an expression like equation (5.6) for a first-order system:

$$x - y = \tau \frac{dy}{dT} \tag{5.14}$$

where for simplicity x and y have replaced t_{fluid} and t_s respectively. For a discrete-time function, the differential can be replaced by the **backward difference** term:

$$\frac{dy}{dT} = \frac{y(kT) - y((k-1)T)}{T} \tag{5.15}$$

Substituting equation (5.15) into equation (5.14), along with $x = x(kT)$ and $y = y(kT)$, gives

$$y(kT) = y((k-1)T) - \frac{1}{\tau} T\{y(kT) - x(kT)\} \tag{5.16}$$

This is a difference equation, derived from the backward difference method. The forward difference method, rarely used in practice [14], employs $y((k+1)T)$ and $y(kT)$ instead of $y(kT)$ and $y((k-1)T)$ in equation (5.15).

The z-transform can be obtained from equation (5.16):

$$Y(z) = z^{-1}Y(z) - \frac{1}{\tau} T\{Y(z) - X(z)\} \tag{5.17}$$

where Y and X indicate that they are the z-transforms of y and x, as is the similar notation for the Laplace transforms. By comparing equations (5.16) and (5.17), notice that z^{-1} acts like an operator on $Y(z)$, shifting it back to the previous sample. The discrete-time transfer function for the first-order system is

$$G(z) = \frac{Y(z)}{X(z)} = \frac{T}{\tau(1 - z^{-1}) + T} \tag{5.18}$$

Comparing this with the Laplace transfer function in equation (5.9), z is related to s, for backward difference equations, by

$$s = \frac{1 - z^{-1}}{T} \tag{5.19}$$

One distinct advantage of the z-transform is that the coefficients for a more complex transfer function than a first-order system (most real items of plant are more complex than first-order systems) can be identified directly by numerical techniques on a computer. One such technique is **recursive least squares** (Chapter 14); the literature describes how to use it for identification [15, 16].

5.6   Dead time and distance–velocity lag

So far we have considered simple first-order systems like the sensor in Chapter 4. However, the heating system in Fig. 5.4 is more complex. The block diagram method implies that the block and transfer function for each system element, such as the pump, pipework and radiators, can be combined to form the overall system transfer function (Fig. 5.9). Indeed the simplification of complex systems into cascaded first-order systems can be a useful step in analysing complex systems. Stoecker and Stoecker illustrate this by examining a temperature sensor bulb containing a liquid [11].

Combining and cascading transfer functions and blocks is generally satisfactory, provided there is no **interaction** or **loading** between the blocks. In thermal systems the interaction is relatively small but one must be aware of its possibility. Nagrath and Gopal [7] illustrate interaction by considering two *RC* circuits (Fig. 5.10). One would expect the combined transfer function to be

$$\frac{1}{1 + \tau s} \frac{1}{1 + \tau s}$$

whereas more correctly it is

$$\frac{1}{\tau^2 s^2 + 3\tau s + 1}$$

Gille *et al.* [8] advocate using transfer matrices where interaction is considered likely.

To examine a more complex system, but not as complex as a heating system, consider a domestic hot water (DHW) storage calorifier, heated by primary hot water from a boiler flowing through a coil in the calorifier to heat up the secondary stored water (Fig. 5.11).

If the stored DHW is initially cold and the valve is opened to let in primary hot water from the boiler at time $T = 0$, then the primary hot water takes a short time to get to the coil and the coil temperature, t_c, does not

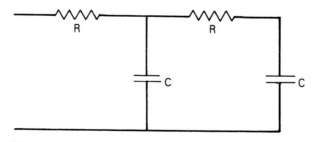

Fig. 5.10   Two electrical circuits.

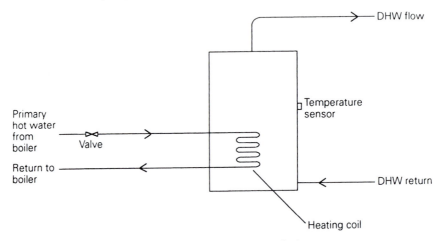

Fig. 5.11   Temperature measurement of domestic hot water.

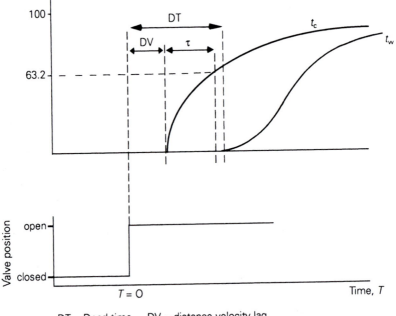

DT = Dead time   DV = distance-velocity lag

Fig. 5.12   Responses and time lags.

immediately respond (Fig. 5.12). This is an example of a **distance–velocity lag**. The coil then has to heat up, and as it can be regarded as a first-order system, it gives rise to the first-order response or **exponential lag**. The water in the calorifier then has to heat up and eventually the sensor begins to warm

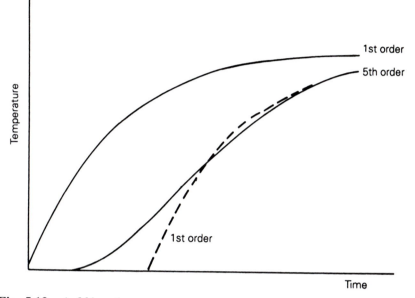

Fig. 5.13 A fifth-order system.

up and register a temperature change. As shown in Fig. 5.12, there is quite a time lag between the hot water valve opening and the BEMS temperature sensor responding. There is a **dead time** before the sensor shows any perceptible change in temperature.

On a practical note, the temperature that the coil is trying to heat the water to is around 80°C, the primary water temperature from the boiler. This temperature is too high for domestic hot water and would waste energy. A temperature of 55°C is more appropriate [9]; CIBSE Technical Memorandum 13 gives information on minimizing the risk of legionnaires' disease [10]. Consequently, the BEMS would cut off the supply of hot water from the boiler by closing the valve when t_w reached 55°C.

The overall response of the system (calorifier, valve, primary and secondary water and sensor), as measured by the sensor, is the response of n first-order systems interacting to form an nth order system. An nth order system has n time constants and is described by a differential equation with differential terms up to $d^n t$.

As Fig. 5.12 shows, there is a very slow initial rise in temperature which contributes to the dead time. However, the fifth-order system in Fig. 5.13 can be approximated to a first-order system with a dead time.

All these dead times and distance–velocity lags in systems contribute to poor control (see later). Here it suffices to say that, for on/off control, the BEMS will not switch on or off at the required limits as the sensor is lagging

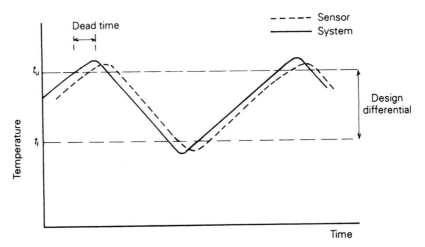

Fig. 5.14   The effect of time lags.

behind the controlled system due to system and sensor lags and responses. This will produce a wider differential than intended (Fig. 5.14).

5.7   Reaction rate and cycling

At low loads the dead times and lags can produce considerable overshoot of the upper differential limit, as the rise in temperature is very steep. This rise in temperature is the **process reaction rate**, or simply the reaction rate.

This reaction rate can be observed with a BEMS by logging the appropriate temperature sensor and displaying the logged data as a graph. The reaction rate is simply the slope of the on/off temperature curves displayed. The steepness of these slopes gives an idea of the heat load and the heat supply. This may be quantified by considering a first-order system, and using the familiar first-order equation:

$$\text{Heat flow into system} - \text{heat flow out of system}$$
$$= \text{rate of change of heat stored in system}$$

$$= MC_p \frac{\mathrm{d}t}{\mathrm{d}T} \qquad (5.20)$$

where M = mass of the system (kg)
$\quad\quad C_p$ = specific heat capacity of the system (kJ kg^{-1} K^{-1})
$\quad\quad t$ = temperature of the system (°C)
$\quad\quad T$ = time (s)
If we consider the DHW storage calorifier shown in Fig. 5.15 and discussed earlier, then equation (5.20) becomes

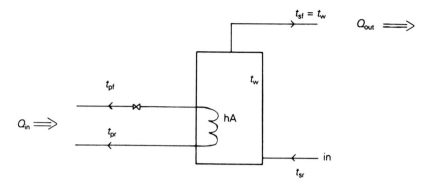

Fig. 5.15   DHW cylinder.

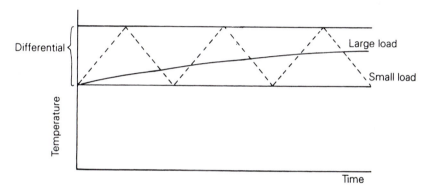

Fig. 5.16   Process reaction rates.

$$MC_p \frac{dt_w}{dT} = Q_{in} - Q_{out} \tag{5.21}$$

where t_w = DHW temperature (°C)
$\quad\quad Q_{in}$ = heat supplied from the boiler (W)
$\quad\quad Q_{out}$ = heat requirement of the DHW load (W)

During heating the slope is given by dt_w/dT. This is often known as the process reaction rate, a term which originated from the process control industry.

If Q_{in} and Q_{out} are almost equal (i.e. there is a large DHW load) then dt_w/dT will be small and there will be a shallow temperature rise. If t_w is within the differential, there will be little cycling on and off (Fig. 5.16). When Q_{in} is much greater than Q_{out}, dt_w/dT and the process reaction rate will be high; this could lead to significant cycling (Fig. 5.16).

If there is no DHW load (i.e. no outflow of DHW), and ignoring the heat losses, then $Q_{out} = 0$ and all the heat gets stored in the cylinder. Putting in more detail about the heat exchange from the cylinder coil gives

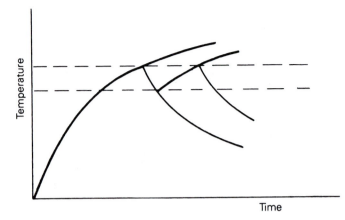

Fig. 5.17   First-order system cycling.

$$MC_p \frac{\mathrm{d}t_w}{\mathrm{d}T} = hA\left[\frac{t_{pf} + t_{pr}}{2} - t_w\right] \tag{5.22}$$

where h = heat transfer coefficient for coil (W m^{-2} K^{-1})
             A = area of the coil (m^2)
             t_{pf} = temperature of primary hot water flow from the boiler (°C)
             t_{pr} = temperature of primary hot water return to the boiler (°C)

$\dfrac{t_{pf} + t_{pr}}{2}$ = the mean temperature of the coil (°C)

The temperature rise will be a first-order curve (Fig. 5.17) with a time constant of $\tau = MC_p/hA$. The slope of the temperature rise will be

$$\frac{\mathrm{d}t_w}{\mathrm{d}T} = \frac{t_m - t_w}{\tau} \tag{5.23}$$

where $t_m = \dfrac{t_{pf} + t_{pr}}{2}$ (°C)

Once the temperature of the stored water reaches $t_w = 55$°C, the valve shuts off the primary hot water. Although heat losses from the calorifier were ignored earlier, there will be losses and they will make Q_{out} non-zero. The slope of the subsequent cooling-down process will be

$$\frac{\mathrm{d}t_w}{\mathrm{d}T} = \frac{Q_{out}}{MC_p} \tag{5.24}$$

The form of Q_{out} will be similar to the heat exchange term from the coil, in that it will be proportional to the temperature difference between the hot water, t_w, and the ambient inside air temperature, t_{ai}:

$$Q_{out} = h_{loss} A_{cal}(t_w - t_{ai})$$

where h_{loss} = heat transfer coefficient for the heat losses from the calorifier (W m^{-2} K^{-1})

A_{cal} = area of the calorifier (m^2)

So the calorifier water will cool down as a first-order curve with a time constant of $\tau = MC_p/h_{loss}A_{cal}$ (Fig. 5.17).

The valve will open for the primary hot water to flow through the coil when t_w has dropped through the on/off differential, t_{diff}. And t_{diff}, typically 5 K, is small in comparison to the overall temperature rise of the calorifier water from cold, say 10 K, to 55°C. Hence the heating-up and cooling-down curves, or process reaction rates, can be approximated to straight lines over the small range of t_{diff}. The lines have constant slopes and dt_w/dT has two constant values, one for the on state and the other for the off state.

The time that the primary hot water is flowing, or the heating is on, is

$$T_{on} = \frac{dT}{dt_w} t_{diff}$$

$$= \frac{t_{diff} MC_p}{Q_{in} - Q_{out}} \tag{5.25}$$

Here the outflow and heat losses, Q_{out}, have been included for completeness. Q_{out} will vary according to the load; the water drawn off from the secondary system and the supply of cold water. Q_{in}, however, will be relatively constant as the mean primary hot water temperature in the coil, t_m, will be fairly constant and t_w will be limited by the differential. Relating Q_{in} and Q_{out} by a load factor, Φ, with Q_{out} regarded as the load, we have

$$Q_{out} = \Phi Q_{in} \tag{5.26}$$

where $0 \leqslant \Phi \leqslant 1$. Then equation (5.25) can be recast as

$$T_{on} = \frac{t_{diff} MC_p}{Q_{in}(1 - \Phi)} \tag{5.27}$$

The time that the hot water flow is off, T_{off}, is

$$T_{off} = \frac{t_{diff} MC_p}{\Phi Q_{in}} \tag{5.28}$$

The period of cycling is given by

$$T_{cycle} = T_{on} + T_{off}$$

$$= \frac{t_{diff} MC_p}{Q_{in}} \left[\frac{1}{1 - \Phi} + \frac{1}{\Phi} \right]$$

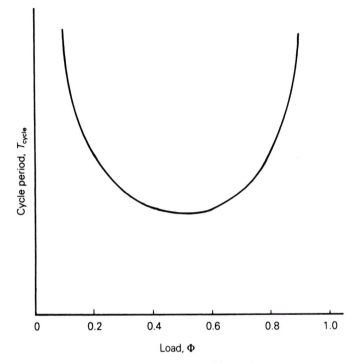

Fig. 5.18   The variation of cycle period with load.

$$= \frac{t_{\mathrm{diff}}\, MC_p}{Q_{\mathrm{in}}} \left[\frac{1}{(1 - \Phi)\Phi} \right] \qquad (5.29)$$

A graph of this equation is given in Fig. 5.18. Note that the graph is flat-bottomed, with a minimum cycling period at $\Phi = 0.5$ of

$$T_{\mathrm{cycle}} = 4 \frac{t_{\mathrm{diff}}\, MC_p}{Q_{\mathrm{in}}} \qquad (5.30)$$

It is interesting that the cycling is not proportional to the load, with increasing cycling at low loads, as is commonly thought.

With a BEMS one has the capability to observe the cycling of plant and to set the optimum differential. The effects of dead times and lags will show up and may lead to remedial work to reduce excesses.

Example 5.1
The DHW calorifier considered above would probably have the secondary DHW being circulated and used. This was ignored in the calculations above. How would it affect the response of the calorifier?

Solution

Assume the heat loss from the calorifier to be negligible compared with the heat in the used DHW. Then

$$Q_{out} = \dot{m}C_p(t_{sf} - t_{sr}) \tag{5.31}$$

where $\dot{m}$ = mass flow rate of DHW (kg s^{-1})
  C_p = specific heat capacity of the system (DHW) (kJ kg^{-1} K^{-1})
  t_{sf} = temperature of secondary DHW flow water = t_w (°C)
  t_{sr} = temperature of secondary DHW return water (°C)

So the heating-up process is given by equation (5.21), now including the loss term in equation (5.31):

$$\frac{MC_p\,dt_w}{dT} = hA(t_m - t_w) - \dot{m}C_p(t_w - t_{sr})$$

or

$$\frac{dt_w}{dT} = \frac{1}{\tau}(t_m - t_w) - \frac{\dot{m}}{M}(t_w - t_{sr}) \tag{5.32}$$

If t_{sr} is the temperature of the feed water to the calorifier and if t_{sr} is constant, taking the Laplace transform of equation (5.32) gives

$$st_w(s) = \frac{1}{\tau}(t_m(s) - t_w(s)) - \frac{\dot{m}}{M}t_w(s)$$

$$t_w(s)\{s\tau + 1 + \beta\} = t_m(s) \tag{5.33}$$

where $\beta = \dot{m}\tau/M$.

This can be cast into the same form as a first-order equation:

$$t_w(s)\{s\tau^* + 1\} = \frac{1}{(1 + \beta)}t_m(s) \tag{5.34}$$

where $\tau^* = \tau/(1 + \beta)$.

In effect, the time constant τ has been reduced and t_w will not tend towards t_m, but a reduced value, $t_m/(1 + \beta)$. If the DHW cylinder had diameter D and height h and the pipe through which the water flowed had diameter d and the water flowed with a velocity v, then

$$\beta = \frac{\dot{m}\tau}{M} = \frac{d^2 v}{D^2 h}\tau$$

The electrical equivalent is an *RC* circuit with another resistance across the capacitor (Fig. 5.19).

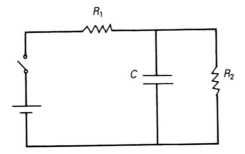

Fig. 5.19   Electrical analogy of DHW calorifier with flow.

5.8   Condition and logic control

Apart from on/off control, BEMS outstations can use the logic potential in their CPUs to perform useful control functions. Consider the heating system in Fig. 5.4 with a boiler and a pump. Control is by switching off the pump when the room temperature sensor reads a higher temperature than the setpoint.

This could be done by a **conditional** statement quite easily in a BEMS which has programmable soft points, as opposed to the function modules mostly discussed up to now. The statement would take this form:

$$\text{IF } t_{ai} > 21 \text{ THEN boiler off}$$

In plain English this means that if the inside air temperature is greater than 21°C, then switch off the boiler. A differential could be introduced with the following statement:

$$\text{IF } t_{ai} < 20 \text{ THEN boiler on}$$

Although this is much easier than having a loop module (a driver or interface to the actuator is still required), there is a problem of the initial heat-up of the room. When t_{ai} first rises through 20°C, before it has yet reached 21°C, the boiler is already on and the absence of a condition may create problems. Another conditional statement could solve this by indicating whether or not the boiler was already on.

But whichever method is used, a loop module or a programmable soft point, a condition statement can independently add to the heating control by switching off the boiler when it is warm outside, on a warm spring or autumn day. The existing control just switches off the pump when the room is up to temperature, leaving the boiler to idle, keeping its own thermostat warm. With an external air temperature sensor, the boiler can be switched off when it is mild:

$$\text{IF } t_{ao} > 15 \text{ THEN boiler off}$$

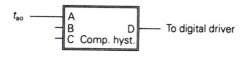

if $A < (B - C/2)$ then $D = 0$

if $A > (B + C/2)$ then $D = 1$

Fig. 5.20   Comparator hysteresis module.

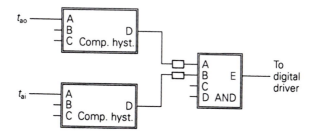

Fig. 5.21   Internal and external control conditions.

For a BEMS using function modules, the conditional control is performed by a **hysteresis comparator** module and a digital driver, similar to the binary driver but receiving a digital input instead of an analogue input. A hysteresis comparator module is shown in Fig. 5.20: A is the input from the sensor analogue node, and C is the hysteresis band either side of B, the setpoint.

The hysteresis band stops any oscillation in output when the temperature is exactly at the setpoint. For our heating example, B could be set to 14°C and C to 2 K. The boiler would then be switched off when $t_{ao} = 15$°C and on again when $t_{ao} = 13$°C.

It is assumed in this control that when t_{ao} is 15°C, then with internal gains in the room (lighting, occupants, equipment and solar), t_{ai} will be around 21°C. To make sure this is so, another hysteresis comparator can be configured to the address node of the t_{ai} sensor and the outputs from the two comparator modules given as the inputs to a logical **AND module** (Fig. 5.21).

Only two of the AND module's four inputs are used here. The operation of the AND module is described by a **truth table** (Table 5.3). The two redundant inputs, C and D here, would have to be set at default values of 1 for the truth table to work.

The truth table is so called as it evolved from **Boolean algebra**, or Boolean logic, where 1 is associated with a statement being true and 0 represents a false statement. In Boolean algebra the AND function is represented by •, so Table 5.3 as a Boolean algebraic equation would be

$$E = A \cdot B \qquad\qquad (5.35)$$

Table 5.3 Truth table for AND module

Inputs				Output
A	B	C	D	E
0	0	1	1	0
0	1	1	1	0
1	0	1	1	0
1	1	1	1	1

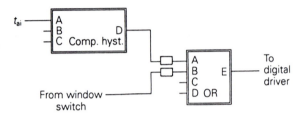

Fig. 5.22 Internal temperature and window switch combined.

Table 5.4 Truth table for OR module

Inputs		Output
A	B	E
0	0	0
0	1	1
1	0	1
1	1	1

Another way of considering the truth table for an AND function is as switches in series; if a switch is open, 0, then no current can pass. A current only passes when all switches are shut.

With our heating control for the boiler, it is possible that the room temperature could be low although t_{ao} is above 15°C because the window is open. A trip switch could be fitted to the window to send a binary signal to the BEMS outstation. This could be combined with the t_{ai} hysteresis comparator digital signal in an **OR** logic module (Fig. 5.22). In Boolean algebra the OR function is represented by +, so the truth table (Table 5.4) as a Boolean algebraic equation would be

$$E = A + B \qquad (5.36)$$

As with the AND function, another way of considering the truth table for an OR function is as switches in parallel; if either switch is closed, i.e. 1, then a current can pass.

Other Boolean logic function modules also exist in most BEMS outstations, such as NOT, NOT AND (NAND) and NOR, but they are seldom as useful for controlling plant.

These logic functions are powerful control elements and for industrial switching there are **programmable logic controllers** (PLCs), a lower form of BEMS outstation. In fact, PLCs have been used to control some building services plant but they have limited capabilities, even compared to smaller BEMS outstations of similar cost.

5.9 Fuzzy logic control

Fuzzy logic relates to how we think and speak. Consider the height of people. In standard logic we could define tall as being over 1.8 m. In set theory this would be two sets, tall and not tall, with a dividing line between them (Fig. 5.23). These are **crisp sets**. However, someone 1.75 m tall would be very close to being tall, closer than someone 1 m tall. The fuzzy set on the right of Fig. 5.23 better describes tall people with a curved or fuzzy dividing line. With the crisp set, the probability of being in one set or the other is 1. If a person is 1.9 m tall then they have probability 1 of being in the tall set. For the fuzzy set, probability is replaced by the **membership function** μ and someone 1.85 m tall would have a membership function of 0.75 for tall and 0.25 for not tall. If the heights were divided into 0.05 m steps then the fuzzy set would be

$$\left\{ \frac{0}{1.7}, \frac{0.25}{1.75}, \frac{0.5}{1.8}, \frac{0.75}{1.85}, \frac{1}{1.9} \right\}$$

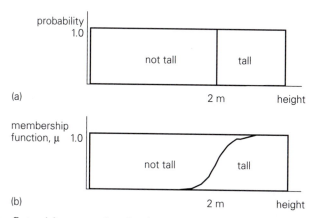

Fig. 5.23 Sets: (a) conventional crisp sets and (b) fuzzy sets.

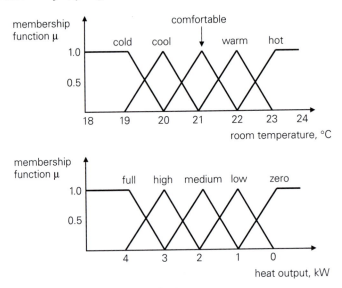

Fig. 5.24   Fuzzy sets for contol of a heater.

In mathematical notation a standard crisp set is

$$\text{Tall} = \{x \mid x > 1.8\}$$

where x is the person's height. The fuzzy set is

$$\text{Tall} = \{x, \mu_{\text{Tall}}(x) \mid x \in X\}$$

where X is all the values of height considered, sometimes called the **universe of discourse**, or U and $\in$ means 'is an element of'. The universal set, sometimes $\mathscr{E}$, is the heights 1.7 m to 1.9 m in this example. Other heights can be considered but those beyond 1.9 m have a membership of 1 and those below 1.7 m have a membership of 0. The universe of discourse could be extended to other heights and other fuzzy sets, such as short and medium. The definition of the set and its shape is set by the expert setting up the system.

   Different shaped sets can be used but often, for mathematical simplicity, straight-edged triangular sets are used. Figure 5.24 shows comfort related to temperature: hot through comfortable to cold. Notice that the sets overlap. If they did not overlap, there would be no definition for the temperatures between the non-overlapping sets. For control based on these sets, there would be no defined control output.

   To control a system we need some rules. Suppose we have a linear, 4 kW electric heater heating a room with an internal air temperature sensor for control, and the heater can have its output varied continuosly from 0 to 4 kW.

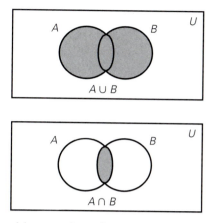

Fig. 5.25 Union and intersection of sets *A* and *B*.

Some rules for controlling the heater could be that when it is comfortable the heater output is 2 kW, when hot the output is zero, when cold the output is 4 kW, when cool the output is 3 kW and when warm the output is 1 kW (Fig. 5.24). This can be represented as follows:

IF room is cold THEN set heater to 4 kW

The phrase 'room is cold' is the **antecedent** and the phrase 'set heater to 4 kW' is the **consequent**. In the MATLAB fuzzy logic toolbox [17] this is written

IF room == cold THEN heater = 4 kW

To see how to interpret these rules mathematically, we have to understand some basic fuzzy set operations. Fuzzy set operations relate to crisp set operations. Figure 5.25 shows the **union** of two crisp sets (symbol ∪) and the **intersection**, (given the symbol, ∩) of two crisp sets.

The union of the fuzzy sets *C* (comfortable) and *W* (warm) (Fig. 5.24), where

$$C = \left\{ \frac{0}{18}, \frac{0}{18.5}, \ldots, \frac{0}{20}, \frac{0.5}{20.5}, \frac{1}{21}, \frac{0.5}{21.5}, \frac{0}{22}, \ldots, \frac{0}{24} \right\}$$

$$W = \left\{ \frac{0}{18}, \frac{0}{18.5}, \ldots, \frac{0}{21}, \frac{0.5}{21.5}, \frac{1}{22}, \frac{0.5}{22.5}, \frac{0}{23}, \ldots, \frac{0}{24} \right\}$$

gives

$$C \cup W = \left\{ \frac{0}{18}, \frac{0}{18.5}, \ldots, \frac{0}{21}, \frac{0.5}{21.5}, \frac{1}{22}, \frac{0.5}{22.5}, \frac{0}{23}, \ldots, \frac{0}{24} \right\}$$

which is shown in Fig. 5.26.

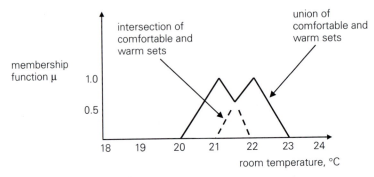

Fig. 5.26   Union (—) and intersection (---) of sets C and W.

This can also be interpreted as a logical **OR** operation. By taking the maximum membership function, this union may be referred to as a **maximizing**, or **max**, function:

$$C \cup W = \left\{ \ldots \max \frac{0,\,0}{20},\, \max \frac{0.5,\,0}{20.5},\, \max \frac{1,\,0}{21},\, \max \frac{0.5,\,0.5}{21.5},\, \max \frac{0,\,1}{22}, \right.$$

$$\left. \max \frac{0,\,0.5}{22.5},\, \max \frac{0,\,0}{23} \ldots \right\}$$

or

$$\mu_{C \cup W}(t) = \max\{\mu_C(t),\, \mu_W(t)\}$$

The intersection of C and W is

$$C \cap W = \left\{ \frac{0}{18},\, \frac{0}{18.5},\, \ldots,\, \frac{0}{21},\, \frac{0.5}{21.5},\, \frac{0}{22},\, \frac{0}{22.5},\, \frac{0}{23},\, \ldots,\, \frac{0}{24} \right\}$$

which is also shown in Fig. 5.26. This can be interpreted as a logical AND operation. The intersection process can be considered as a **minimizing**, or **min**, function:

$$\mu_{C \cup W}(t) = \min\{\mu_C(t),\, \mu_W(t)\}$$

For control by the IF . . . THEN rules, the rules need interpretation for fuzzy sets. Consider the room temperature to be 20.25°C. This temperature falls in both the cool set and the comfortable set, so it **fires** two rules:

IF room is cool THEN set heater to high
IF room is comfortable THEN set heater to medium

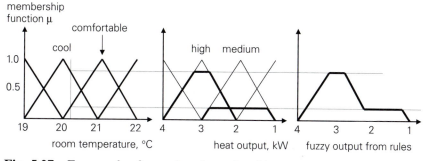

Fig. 5.27   Fuzzy rules for cool and comfortable.

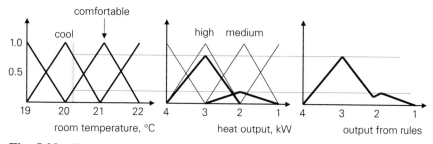

Fig. 5.28   Rule output from squashed sets.

But it is more in the cool rule than the comfortable rule. So the temperature sets truncate the heat output sets in proportion to the temperature membership functions (Fig. 5.27) and produce an output fuzzy set. This method follows the early work of Mamdani [18] based on the work of Zadeh [19, 20]. There is another method of squashing the heat output sets in relation to the membership functions of the temperature fuzzy sets (Fig. 5.28) but this is less common [21].

The rule implementation process in Fig. 5.27 is in effect a **minmax** process. Each rule is applied as a min process and the rules are combined as a max process. Consider IF cool THEN high, with the high fuzzy set given by

$$\text{high} = \left\{ \frac{0}{4}, \frac{0.25}{3.75}, \frac{0.5}{3.5}, \frac{0.75}{3.25}, \frac{1}{3}, \frac{0.75}{2.75}, \frac{0.5}{2.5}, \frac{0.25}{2.25}, \frac{0}{2}, \frac{0}{1.75}, \ldots, \frac{0}{0} \right\}$$

and the cool fuzzy set given by

$$\text{cool} = \left\{ \frac{0}{19}, \frac{0.25}{19.25}, \frac{0.5}{19.5}, \frac{0.75}{19.75}, \frac{1}{20}, \frac{0.75}{20.25}, \frac{0.5}{20.5}, \frac{0.25}{20.75}, \frac{0}{21}, \frac{0}{21.25}, \ldots, \frac{0}{23} \right\}$$

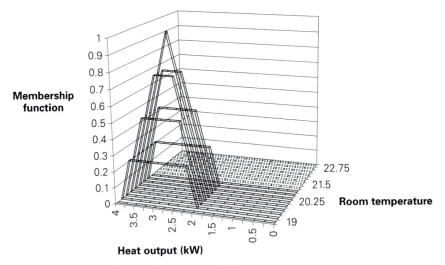

Fig. 5.29   Rule: if cool then heater high.

Then, from the min operation, the rule at 20.25°C is

$$\text{cool} \cap \text{high} = \left\{ \min\frac{0.75, 0}{4}, \min\frac{0.75, 0.25}{3.75}, \min\frac{0.75, 0.5}{3.5}, \dots \right\}$$

$$= \left\{ \frac{0}{4}, \frac{0.25}{3.75}, \frac{0.5}{3.5}, \frac{0.75}{3.25}, \frac{0.75}{3}, \frac{0.75}{2.75}, \frac{0.5}{2.5}, \frac{0.25}{2.25}, \frac{0}{2}, \dots \right\}$$

where the denominators refer to the heat output (kW) from the heater. The complete cool/high rule for all temperatures is shown in Fig. 5.29.

For the comfortable/medium rule at 20.25°C the min operation yields

$$\text{comfortable} \cap \text{medium}$$

$$= \left\{ \frac{0}{3}, \frac{0.25}{2.75}, \frac{0.25}{2.5}, \frac{0.25}{2.25}, \frac{0.25}{2}, \frac{0.25}{1.75}, \frac{0.25}{1.5}, \frac{0.25}{1.25}, \frac{0}{1.0}, \dots \right\}$$

Combining these two rules with the max operation gives

$$\left\{ \frac{0}{4}, \frac{0.25}{3.75}, \frac{0.5}{3.5}, \frac{0.75}{3.25}, \frac{0.75}{3}, \frac{0.75}{2.75}, \frac{0.5}{2.5}, \frac{0.25}{2.25}, \frac{0.25}{2}, \frac{0.25}{1.75}, \frac{0.25}{1.5}, \frac{0.25}{1.25}, \frac{0}{1}, \dots \right\}$$

This is shown in Fig. 5.27.

Matrices can be used to describe rules over all the temperatures considered. The matrix in Fig. 5.30 can be used to combine our cool/high and comfortable/medium rules. The boldface numbers indicate the temperatures and the heat outputs. They are shown graphically in Fig. 5.31.

	4	3.75	3.5	3.25	3	2.75	2.5	2.25	2	1.75	1.5	1.25	1	0.75	0.5	0.25	0
19	0	0	0	0	0	0	0	0	0	0	0	0	0	0	0	0	0
19.25	0	0.25	0.25	0.25	0.25	0.25	0.25	0.25	0	0	0	0	0	0	0	0	0
19.5	0	0.25	0.5	0.5	0.5	0.5	0.5	0.25	0	0	0	0	0	0	0	0	0
19.75	0	0.25	0.5	0.75	0.75	0.75	0.5	0.25	0	0	0	0	0	0	0	0	0
20	0	0.25	0.5	0.75	1	0.75	0.5	0.25	0	0	0	0	0	0	0	0	0
20.25	0	0.25	0.5	0.75	0.75	0.75	0.5	0.25	0.25	0.25	0.25	0.25	0	0	0	0	0
20.5	0	0.25	0.5	0.5	0.5	0.5	0.5	0.5	0.5	0.5	0.5	0.25	0	0	0	0	0
20.75	0	0.25	0.25	0.25	0.25	0.25	0.5	0.75	0.75	0.75	0.5	0.25	0	0	0	0	0
21	0	0	0	0	0	0.25	0.5	0.75	1	0.75	0.5	0.25	0	0	0	0	0
21.25	0	0	0	0	0	0.25	0.5	0.75	0.75	0.75	0.5	0.25	0	0	0	0	0
21.5	0	0	0	0	0	0.25	0.5	0.5	0.5	0.5	0.5	0.25	0	0	0	0	0
21.75	0	0	0	0	0	0.25	0.25	0.25	0.25	0.25	0.25	0.25	0	0	0	0	0
22	0	0	0	0	0	0	0	0	0	0	0	0	0	0	0	0	0
22.25	0	0	0	0	0	0	0	0	0	0	0	0	0	0	0	0	0
22.5	0	0	0	0	0	0	0	0	0	0	0	0	0	0	0	0	0
22.75	0	0	0	0	0	0	0	0	0	0	0	0	0	0	0	0	0
23	0	0	0	0	0	0	0	0	0	0	0	0	0	0	0	0	0

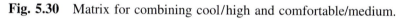

Fig. 5.30   Matrix for combining cool/high and comfortable/medium.

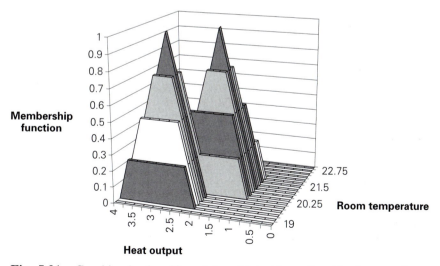

Fig. 5.31   Combined rules: if cool then high, if comfortable then medium.

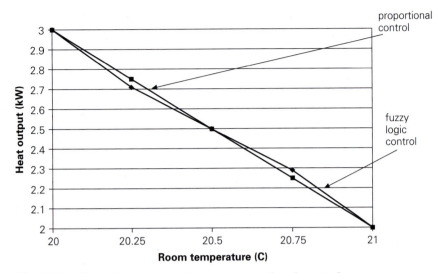

Fig. 5.32 Fuzzy logic control versus proportional control.

A controller needs a crisp, single-value output. The fuzzy output needs to be defuzzified. There are several ways to do this but a popular one is by using the **centroid** of the fuzzy output set. For the discrete values in our example, the centroid is given by

$$\frac{\sum\limits_{i=1}^{17} \mu(t_x, H_i)H_i}{\sum\limits_{i=1}^{17} \mu(t_x, H_i)}$$

for a given temperature t_x and the 17 discrete heat outputs H_i. In our case the temperature is 20.25°C and the heat outputs (kW) are 4, 3.75, . . . , 0.25, 0. The membership functions are given in the 20.25°C row of the rule matrix (Fig. 5.30). So at 20.25°C the defuzzified output is

$$\frac{(0 \times 4) + (0.25 \times 3.75) + (0.5 \times 3.5) + \ldots + (0.25 \times 1.5) + (0.25 \times 1.25) + (0 \times 1) + \ldots + (0 \times 0)}{0 + 0.25 + 0.5 + 0.75 + 0.75 + 0.75 + 0.5 + 0.25 + 0.25 + 0.25 + 0.25 + 0.25 + 0 + \ldots + 0}$$

$$= \frac{12.875}{4.75}$$

$$= 2.71 \text{ kW}$$

Figure 5.32 shows how the heater output varies with temperature and compares it with a proportional control schedule (Chapter 6). The two are very close as the five fuzzy sets for the temperature and heat output are very symmetrical. However, the fuzzy sets can be altered and a distinctly non-linear control surface formed to match the non-linear device it is controlling. And it is possible to add other rules which conventional linear controllers could not accept so easily.

References

[1] CIBSE (1986) CIBSE Guide, Volume A, Section A1, Environmental Criteria For Design, Chartered Institution of Building Services Engineers, London.

[2] CIBSE (1986) CIBSE Guide, Volume A, Section A2, Weather and solar data, Chartered Institution of Building Services Engineers, London.

[3] CIBSE (1986) CIBSE Guide, Volume A, Section A9, Estimation of plant capacity, Chartered Institution of Building Services Engineers, London.

[4] CIBSE (1986) CIBSE Guide, Volume A, Section A7, Internal heat gains, Chartered Institution of Building Services Engineers, London.

[5] Adams, S. and Holmes, M. (1977) *Determining Time Constants for Heating and Cooling Coils*, BSRIA Technical Note TN6/77, Building Services Research and Information Association, Bracknell.

[6] Marshall, S. A. (1984) *Introduction to Control Theory*, Macmillan, London.

[7] Nagrath, I. J. and Gopal, M. (1985) *Control Systems Engineering*, Wiley Eastern, New Delhi; new edition.

[8] Gille, Pelegrin and Decaulne (1959) *Feedback Control Systems*, McGraw-Hill, New York.

[9] CIBSE (1986) CIBSE Guide, Volume B, Section B4, Water service systems, Chartered Institution of Building Services Engineers, London.

[10] *Minimising the Risk of Legionnaires' Disease*, CIBSE Technical Memorandum 13, Chartered Institution of Building Services Engineers, London, 1987.

[11] Stoecker, W. F. and Stoecker, P. A. (1989) *Microcomputer Control of Thermal and Mechanical Systems*, Van Nostrand Reinhold, New York.

[12] DiStefano, J. J., Stubberud, A. R. and Williams, J. J. (1987) *Theory and Problems of Feedback and Control Systems*, Schaum's Outline series, McGraw-Hill, New York.

[13] Houpis, C. H. and Lamont, G. B. (1985) *Digital Control Systems Theory, Hardware, Software*, McGraw-Hill, New York.

[14] Ogata, K. (1987) *Discrete-time Control Systems*, Prentice-Hall, Englewood Cliffs NJ.

[15] Iserman, R. (1981) *Digital Control Systems*. Springer-Verlag, Berlin.

[16] Åstrom, K. J. and Witenmark, B. (1989) *Adaptive Control*. Addison-Wesley, Reading MA.

[17] Jang, R. J.-S. and Gulley, G. (1995) *Fuzzy Logic Toolbox for Use with MATLAB*, Math Works Inc., Natick MA.

[18] Mamdani, E. H. and Assilian, S. (1975) An experiment in linguistic synthesis with a fuzzy logic controller. *International Journal of Man–Machine Studies*, **7**(1).

[19] Zadeh, L. A. (1973) Outline of a new approach to the analysis of complex systems and decision processes. *IEEE Transactions on Systems, Man, and Cybernetics*, **3**(1).

[20] Zadeh, L. A. (1965) Fuzzy sets. *Information and Control*, **8**.

[21] Kosko, B. (1994) *Fuzzy Thinking*, HarperCollins, London.

[22] CIBSE (1999) CIBSE Guide A, Environmental Design, Chartered Institution of Building Services Engineers, London.

6
PID three-term direct digital control

Chapter 5 showed how on/off control either had the plant fully on or fully off. This could result in the controlled temperature overshooting the differential, due to the system and controller responses, especially at low loads. Figure 6.1 illustrates this for room temperature control with a thermostat, demonstrating that the mean room temperature varies about the thermostat setpoint according at the load.

A better form of control is to vary or modulate the plant output, so that as the temperature approaches the required value the output can be reduced to

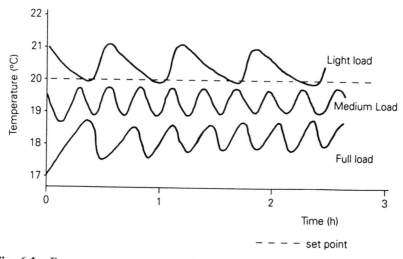

Fig. 6.1 Room temperature control with a thermostat.

stop any overshoot. This of course can be done only if the plant can be modulated, perhaps by using a modulating control valve or, in the case of a boiler, by having a modulating burner. All of this adds to the expense of the plant but produces better control.

The control used in BEMSs that can provide this modulating action is **proportional plus integral plus differential** (PID) control. As there are three distinct parts to PID control, it is also known as **three-term control**. Each term will now be examined.

6.1 Proportional control

A crude and simple example of proportional control is the ballcock valve in a cistern. As the water rises in the cistern, so the ball is raised and gradually closes the valve. When the cistern is empty, the valve is fully open. When the cistern is full, the valve is shut.

The water cylinder analogy we have used before can also illustrate proportional control. The two-port valve in Fig. 6.2 is a modulating valve. When the water level is at h_u, the valve is fully shut. When the water level is at h_1, the valve is fully open. The thermal analogy to the water height is the temperature. In Fig. 6.3 the flow of water into the cylinder is proportional to the height of the water as measured by the sensor, hence proportional control.

The difference between the upper and lower limits, h_u and h_1, is the **proportional band** (PB), also called the **throttling range**. It is similar to the differential for on/off control, and similar considerations have to be given to its setting. Too narrow and the control tends towards on/off control

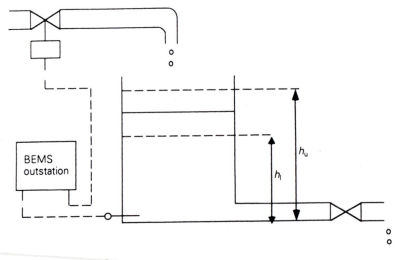

Fig. 6.2 Water analogy.

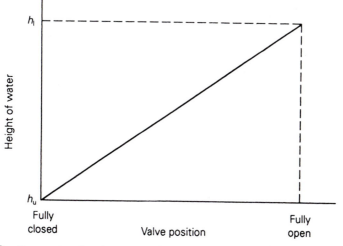

Fig. 6.3 Proportional valve control.

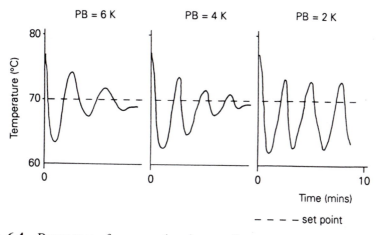

Fig. 6.4 Responses of a proportional controller.

and oscillation. Too wide and there is little control as the temperature offset can be large. These effects are shown in Fig. 6.4 for a heating valve when subjected to a step change increase in the heating load. The heating valve is a mixing valve (Chapter 8) with a water temperature sensor in the pipe, just downstream from the valve.

To set up a proportional controller, either in BEMS software or as a stand-alone proportional controller, the proportional band and **setpoint** are adjusted to the required settings. Traditionally the setpoint is the temperature (or in our analogy the height), in the middle of the proportional band (PB).

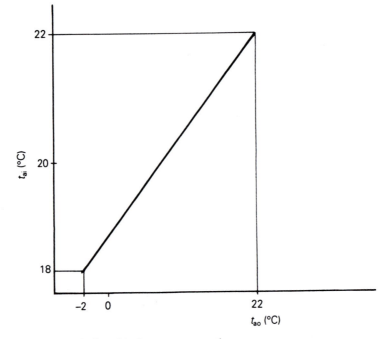

Fig. 6.5   The relationship between t_{ai} and t_{ao}.

Example 6.1
A proportional controller, set to $t_{set} = 20°C$, controls a room heater which can be modulated and at full output is rated at 4 kW. When the outside temperature is $t_{ao} = -2°C$, the heater is at full output and the room temperature is $t_{ai} = 18°C$. If the setpoint is in the middle of the proportional band, what is t_{ai} when t_{ao} is 4°C? Ignore thermal storage in the room structure.

Solution
The relationship between t_{ai} and t_{ao} is shown in Fig. 6.5. It is based on the two points given in the problem. One is where $t_{ao} = -2°C$ and $t_{ai} = 18°C$, and the other is derived from the fact that t_{set} is in the middle of the PB, so

$$PB = 2(t_{set} - t_1)$$
$$= 2(20 - 18)$$
$$= 4 \text{ K}$$

where t_1 is the lower control temperature of the PB (°C).
   This means that the upper control temperature is $t_u = 22°C$. So the second point of the graph is at t_{ai} with a value of 22°C when there is no heating, and no need for it, so t_{ao} must be 22°C as well. The equation for this line is

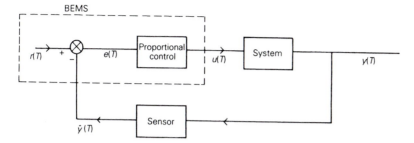

Fig. 6.6 Proportional control with a BEMS.

$$t_{ai} = \frac{1}{6}t_{ao} + 18.33 \qquad (6.1)$$

When $t_{ao} = 4°C$ then

$$t_{ai} = \frac{1}{6}4 + 18.33$$

$$t_{ai} = 19°C$$

In this example, t_{ai} is below the setpoint of 20°C. In fact, there is only one value of outside temperature at which the setpoint is achieved, and this is when $t_{ao} = 10°C$. The analogy with the water cylinder is that the valve on the outflow represents the heat loss to the outside. This valve can be identical to the modulating valve on the inflow, and the flow of water into the cylinder exactly matches the outflow. The water level will come to equilibrium, but for only one valve setting will the setpoint be achieved.

The difference between the setpoint and the measured value (either room temperature or water level in our examples) is the **error** or **offset** (Chapter 5). The block diagram for proportional control (Fig. 6.6) is very similar to the diagram for the on/off control. It is often assumed that the sensor introduces negligible error, so there is unity feedback of the measured value.

6.1.1 Proportional gain

The output signal, $u(T)$, from the proportional control block is simply a proportional relationship with $e(T)$:

$$u(T) = \frac{1}{PB}e(T) \qquad (6.2)$$

When a BEMS loop module is set up, or configured, the **proportional gain**, K_p, is sometimes used [11] instead of the proportional band (PB) or the

proportional control action [10, 18]. The relationship which is often used between K_p and PB is

$$K_p = \frac{100\%}{\text{PB}} = \frac{1}{\text{PB}} \; (K^{-1}) \tag{6.3}$$

The 100% represents the full range of output of the controlled device, e.g. a valve going from fully closed (0%) to fully open (100%).

Note that K_p has the dimension of inverse temperature, so the output, $u(T)$, is dimensionless. The proportional control equation (6.2) now becomes

$$u(T) = K_p \, e(T) \tag{6.4}$$

Where $u(T)$ is the signal from the loop to the driver, 0% to 100%. The driver then converts this signal to an electrical signal for the valve.

This example considers the proportional control from the steady state, time-independent situation. The dynamic, time-dependent approach is mathematically more complicated but it reveals more about the control action.

Example 6.2
Consider the dynamics of the 4 kW heater in the previous example, assuming that the room has a first-order response and the sensor introduces no delays or errors.

Solution
If t_{set} is the setpoint temperature and t_{ai} is the inside air temperature, then the error is

$$e = t_{set} - t_{ai}$$

and the heater output is

$$Q_{max} \frac{u(T)}{100} = Q_{max} \frac{K_p \, e(T)}{100} \text{ kW}$$

where Q_{max} is the maximum heater output, here 4 kW.

The differential equation describing the first-order room system and control is

$$\tau \frac{dt_{ai}}{dT} = \frac{Q_{max} K_p}{100H} (t_{set} - t_{ai}) - (t_{ai} - t_{ao}) \tag{6.5}$$

where H is the overall heat loss from the room (0.2 kW K^{-1}) and τ is the room time constant (s).

First-order equations such as equation (6.5) are discussed in greater detail in Chapter 9. The solution of equation (6.5) is

$$\left[\frac{K_p \Gamma t_{set} + t_{ao} - (1 + \Gamma K_p) t_{ai}(T)}{K_p \Gamma t_{set} + t_{ao} - (1 + \Gamma K_p) t_{ai}(0)} \right] = \exp\left[\frac{-T(1 + \Gamma K_p)}{\tau} \right] \tag{6.6}$$

where $\qquad \Gamma = \dfrac{Q_{max}}{100H} = 0.2 \text{ K}$

$t_{ai}(0)$ = inside temperature at time $T = 0$
$t_{ai}(T)$ = inside temperature at time T

The steady-state inside temperature is reached as T tends to infinity and the right-hand side of equation (6.6) tends to zero. This condition is fulfilled when

$$t_{ai} = \frac{K_p \Gamma t_{set} + t_{ao}}{1 + \Gamma K_p}$$

When $t_{ao} = 4°C$, as in the previous example, and with PB = 4 K ($K_p = 25\%$ K^{-1}) we have

$$t_{ai} = \frac{25 \times 0.2 \times 22 + 4}{1 + 0.2 \times 25}$$

$$= 19°C$$

which agrees with the previous example. Here the setpoint is 22°C, due to the definition of $u(T)$ and the error. The steady-state offset is given by

$$\text{offset} = t_{set} - t_{ai}$$

$$= t_{set} - \frac{K_p \Gamma t_{set} + t_{ao}}{1 + \Gamma K_p} \tag{6.7}$$

It is easier to examine equation (6.7) if t_{ao} is set to zero. As the gain is increased, so $K_p \Gamma / (1 + \Gamma K_p)$ tends to unity and the offset disappears. But notice that even if the gain is greater than 100% K^{-1}, the heater output cannot go above Q_{max} or 4 kW, so the output signal is effectively limited at a maximum at larger errors.

Another factor influencing the selection of the proportional gain is the effective change in τ. In equation (6.6) the value of τ is reduced by the factor $1/(1 + \Gamma K_p)$. A shorter value of τ speeds up the response but also increases the oscillation of the controlled element, as shown in Fig. 6.4 for a heating valve with a sensor close to it.

6.1.2   Loop modules and drivers

For a BEMS the software to implement proportional control action is with a driver and **loop controller**, or **loop module** (Fig. 6.7). The loop module is so called because this is where the feedback loop returns and where most of the control is carried out.

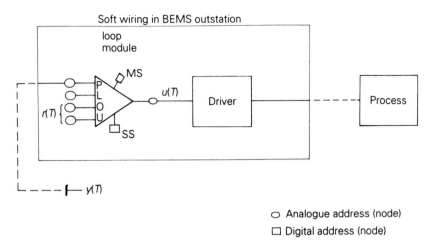

Fig. 6.7   BEMS loop module and driver.

The sensor sends the process variable signal, $y(T)$, to **node P**, an analogue input address of the loop module. Two setpoint analogue address points are shown: O for the setpoint temperature during the building's occupancy period and U for the setpoint temperature during the building's unoccupied period. For frost protection, U could be set to 10°C and O to 20°C. The module changes between O and U by changing the value of SS (setpoint selection) between 1 and 0; this is done using a **timing module**, which defines the occupancy times, or using a **logic module** (Chapter 14). Equation (6.2) is then used to derive the output from the loop module to the driver's analogue address. This is between 0% and 100%, where 100% corresponds to the bottom of the PB and 0% corresponds to its top, once the required temperature has been reached. Here the setpoint is at the top of the PB, so the error is positive when control is required, which simplifies calculation.

For testing the control of a system, perhaps during commissioning, the MS (manual selection) digital node bit is set to 1 and the analogue value in the analogue node (L) is sent to the loop output, regardless of the setpoints (O and U), the sensor input (P) or the control parameters.

The driver converts the 0% to 100% signal from the loop module to a voltage or current to drive the plant, e.g. a modulating valve. Two types of driver are used with loop modules for the proportional control of modulating valves: an **analogue output driver** and a **raise/lower driver** (Fig. 6.8). These drivers are similar to the **binary driver** in Chapter 5, except for the outputs.

The analogue driver converts the 0% to 100% signal into a continuous analogue signal to drive the valve. A hardware link in the example outstation

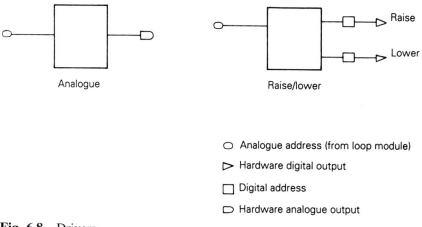

Analogue   Raise/lower

○ Analogue address (from loop module)

▷ Hardware digital output

□ Digital address

▭ Hardware analogue output

Fig. 6.8 Drivers.

of Chapter 2 can simply be changed to give either a voltage or current signal depending on the plant being driven. The 0% signal is for the valve fully closed, and the 100% signal is for the valve fully open. If the valve actuator or motor requires a different range of signal, then **offset** and **range** constants can be set in the driver's software. This is done in the configuration mode from the central station or a local portable PC (Chapter 2).

The raise/lower driver can be used to drive a valve with a **split-phase** motor. The raise digital output signal, when the raise bit is 1 and the lower bit is 0, drives the motor in a direction to open the valve. The lower signal, when the lower bit is 1 and the raise bit is 0, drives the motor in the reverse direction to close the valve. When both output bits are low, the valve is stationary. To position the valve, the **full drive time** of the valve (the time for the valve to go from fully open to fully closed) must be known, so that the driver drives the valve for the correct time to move the valve to the required position.

With valve wear and slippage, the full drive time may change with age. In this case a **position indicator potentiometer** can be used to provide feed-back to the driver of the valve's position. An alternative is **boundless control**. With this method, the 0% and 100% limits are reset when the valve reaches its fully closed and fully open positions and can move no further.

A **loop reschedule time** can be used to suppress the normal schedule service time, so that valves, actuators and relays are not constantly moved and worn each service time. This suppression takes the form of increasing the service time to the relevant output.

If sensors feeding the loops malfunction, there are **loop deviation alarms** and **loop failure reactions** to ensure the sensor malfunction does not result in control malfunction. A deviation alarm is raised when the loop error goes outside limits (set by the user at the control configuration stage). Once the relevant sensor feeding the loop develops a fault, the loop ignores its reading and defaults to a loop failure action. A default can take a number of modes. The first mode is for the loop to use the last within-limits sensor reading. A second mode is to take the present input value for the loop and disregard any faulty inputs, so the loop puts out a constant signal. The third mode is a variation on the second, in that the loop does not return to normal control until the operator has changed the loop to manual selection (MS) and back to automatic control. This form of manual reset safety button (as found on boiler thermostats) is used to draw the operator's attention to an important fault which has occurred.

With such default reactions, heating can be ensured, even when there is a fault with the BEMS. In one incident, recounted by a local authority energy manager, a BEMS outstation had been wrongly programmed to switch the heating off at the weekend. Unfortunately, the heating was for a block of flats. The BEMS operator at the civic centre had dialled up the block instead of a school. Emergency maintenance staff were called during the weekend by a councillor who lived in the now cold block of flats, and although they knew nothing of the BEMS system, the councillor demanded action. So the maintenance staff proceeded to take a crowbar to the outstation to open it. This put the outstation into failure mode, which then started the heating. All in the plant room were amazed!

6.2   Time proportional control

Although the controlled plant has to be capable of modulation, e.g. a boiler with a modulating burner or a modulating valve, on/off devices can be switched on for a certain time proportional to the error signal. This is **time proportional control**, implemented with a loop module connected to a time proportional driver (Fig. 6.9).

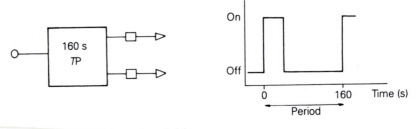

Fig. 6.9   A time proportional driver.

The time proportional (TP) driver keeps the plant on for a defined fraction of time. The time period, τ, is set by the operator when the TP driver is configured. The loop module output, denoted here as $x(\%)$, then keeps the plant on for a time $(x/100)\,\tau$. In Fig. 6.9 the output from the loop module is 25% and the time period is 160 s, so the plant is on for 40 s every 160 s.

6.3 Proportional plus integral (PI) control

There is always a persistent error, or offset, with proportional-only control, except at one point where the plant output provides the heat to maintain the setpoint temperature. This is when the heater, in the previous example, would be half on. With a BEMS loop, where the setpoint is at the top of the PB, the point of zero error would be when the setpoint temperature was achieved and the plant was off.

The persistent error of proportional control can be greatly reduced by increasing the control signal, $u(T)$, as the error persists with time. This is achieved by adding an integral term to the proportional term:

$$u(T) = K_p \left(e(T) + \frac{1}{T_i} \int_0^T e(T)\,\mathrm{d}T \right) \tag{6.8}$$

where T_i is the integral action time (s). The integral term may be considered as the 'memory' of the controller, looking at past errors. Sometimes integral action is defined using K_p/T_i, the integral control action constant [10, 18]. It is also defined in terms of the **integral gain**, K_i, or even the **reset rate** [11]. Another term for PI control is **proportional with automatic reset**.

The BEMS loop module can be configured to PI control by setting T_i. Typical settings for BEMS will be about 200 s. As T_i increases, the integral action reduces, and large values of T_i reduce the PI action to a P action. There is no point in making T_i less than the service schedule time, the time taken by the outstation microprocessor to service the other inputs and outputs before it gets back to the loop concerned. A typical value for this service time is 5 s.

Figure 6.10 shows the responses of a heating valve to changes in T_i, for a PI controller with K_p kept constant. The parameter T_i does indeed reduce the offset inherent in proportional-only control. But as T_i is reduced, the integral action becomes more dominant and the oscillations increase. Too much oscillation will reduce the life of the valve and actuator, so a compromise setting has to be used (see below).

With two control parameters in PI control, both interact with each other. A setting of one parameter cannot be assumed to produce stable operation unless the other parameter is also carefully set. This adds to the complexity of setting up a two-term, PI control loop.

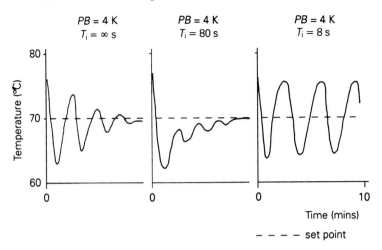

Fig. 6.10   Responses of a PI controller.

Example 6.3
Consider the use of a BEMS with PI control for (a) the room of Example 6.2
with the 4 kW heater, assuming that the room has a time constant of 20 min
and there is a distance velocity lag of 10 min from the heater to the sensor;
(b) a hot water heating valve with a response of less than 1 min and the
sensor close to it; and (c) a boiler with a non-modulating burner.

Solution
(a) From Example 6.2 the differential equation describing the room and the
heater's proportional control is

$$\tau \frac{dt_{ai}}{dT} = \frac{Q_{max} K_p}{100 H} (t_{set} - t_{ai}) - (t_{ai} - t_{ao}) \tag{6.5}$$

where H is the overall heat loss from the room (0.2 kW K^{-1}) and τ is the
room time constant (s). But now there is integral control as well, so the
heater output is

$$Q_{max} \frac{u(T)}{100} = \frac{Q_{max}}{100} K_p \left(e(T) + \frac{1}{T_i} \int e(T) dT \right) \tag{6.9}$$

where Q_{max} is the maximum heater output, 4 kW.
  With proportional-only control, equation (6.5) could be solved analytically; but equation (6.9) is more difficult. One solution would be to solve

equation (6.5) analytically for small time steps during which Q_{max} is regarded as constant; Q_{max} is then varied after each time step using equation (6.9). Another method would be to transpose equation (6.5) into a difference equation (Chapter 5 and Section 6.5) and solve it numerically [12].

Equations (6.5) and (6.9) will not be solved right now; instead the problem is examined from a more basic standpoint. Consider a step change either in the control or the load (the heat loss from the room). A step change shows the response of a system, as in the heating valve example of Fig. 6.10.

Suppose we take the control setpoint of 22°C from the previous example and introduce a step change to 20°C. The error drops from 3 K to 1 K. Then, if the control were proportional-only, the heating output would initially be reduced to 1 kW, after the 10 min distance–velocity lag, and the temperature would fall to its new steady-state value. This can be determined from the steady-state proportional control equation, as calculated in Example 6.1. With the new setpoint of 20°C, the steady-state equation is determined from the two points $t_{ai} = 20°C$, $t_{ao} = 20°C$, and $t_{ai} = 16°C$, $t_{ao} = -4°C$:

$$t_{ao} = 6t_{ai} - 100$$

With the outside temperature of 4°C, from Example 6.1, we have

$$t_{ai} = 17.3°C$$

(In Example 6.1 with the setpoint at 22°C, $t_{ai} = 19°C$.)

With $K_p = 25$, and $\Gamma = 0.2$, as before, then equation (6.6) shows that the time to reach its steady-state temperature is ∞, but to reach a temperature near to this, say 18°C, is

$$\ln\left[\frac{25 \times 0.2 \times 20 + 4 - (1 + 25 \times 0.2)18}{25 \times 0.2 \times 20 + 4 - (1 + 25 \times 0.2)19}\right] = \frac{-T(1 + 25 \times 0.2)}{20}$$

$$T = 3 \text{ min}$$

using equation (6.6) and taking logs of both sides.

With the distance–velocity lag of 10 min, this makes 13 min. If a larger control signal is used, so that the loop output signal is large enough to reduce the heater's output to zero, then the time for the temperature to fall to 17.3°C is given by equation (6.5) with Q_{max} set to zero. The differential equation can be solved to give

$$\ln\left[\frac{t_{ai}(T) - t_{ao}}{t_{ao}(0) - t_{ao}}\right] = \frac{-T}{\tau}$$

This solution is derived in detail in Chapter 9. Substituting the appropriate values shows that it take 2.4 min from the end of the distance–velocity lag to a temperature of 17.3°C. And 2.4 min is still quite some time.

The introduction of an integral term would reduce the offset, but it would introduce a large signal during the distance–velocity lag if T_i were not fairly large, effectively reducing the integral action. Even without the distance–velocity lag, the response of the system is still rather slow and it requires a large T_i. To determine the responses to various settings of K_p and T_i would take the operator some time, and this is needed in order to set up the PI controller (Section 6.7). In practice the distance–velocity lag of 10 min accounts only for the direct heat transfer from the heater to the sensor. But as will be seen in Chapter 9, there is also heat in the fabric of the building; this takes some hours to respond to the heater's output, which would make the integral term even more difficult to set up. Thus proportional-only control would be applicable in this case with the PB lowered to, say, 2 K to reduce the offset.

(b) For a hot water heating valve with a response of less than 1 min, and the sensor close to it, the response of the hot water in the pipe from the valve is going to be much faster. Here PI control is appropriate and easier to set up than in (a). Although there will be a distance–velocity lag between the valve and the sensor, this should be small as both are close together. The graphs in Figs 6.4 to 6.10 are from such a heating valve.

(c) Although the boiler is either on or off, it does not have a modulating burner, the implementation of on/off control in Chapter 5 used a PID loop as an error detector in its proportional-only mode. But if the integral term were also implemented, then at large loads it would be quite substantial as the temperature takes a long time to rise through the differential. The boiler is switched off when $u(T)$ is zero or low, but with an integral term it is often non-zero when the error has become zero itself (Fig. 6.11).

From calculus the integral operation in the integral term determines the area between the actual sensed temperature and the setpoint line. In Fig. 6.11 this is the area of the triangle formed by the sensed temperature and the upper differential temperature, and here it is also the setpoint. The error at the setpoint is zero, but the sensed temperature rises above the setpoint by an amount that gives the upper triangle an area K_p/T_i times that of the lower triangle. Due to the sampling process of the BEMS, the triangles will actually be stepped.

When the load is low, the water temperature in the boiler flow rises faster through the differential, so the integral term is not as large and the overlap of the differential is not as high. This is counter to the normal tendency for on/off overlap to be greater at low loads (Fig. 6.1) and smaller at high loads. With suitable selection of a setpoint, K_p and T_i, more accurate on/off control is possible.

With some BEMS, the PID loop output cannot go negative, so the integrated error is zero when the temperature is above the setpoint.

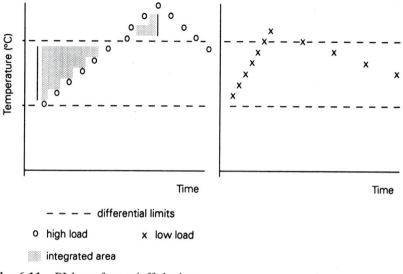

── ── differential limits

o high load   x low load

▒ integrated area

Fig. 6.11   PI loop for on/off device.

An ideal way to examine the control of a system before configuring a BEMS is to simulate the system on a computer, solving the differential equations of the various plant components. Hysteresis of the actuators should also be modelled. Shavit and Brandt have constructed a mathematical model of a heater battery and its controller, and they have examined various settings for K_p and T_i [13].

6.4   PID control

Full three-term control, PID control, has a differential term added to the PI equation (6.5) to give

$$u(T) = K_p\left(e(T) + \frac{1}{T_i}\int_0^T e(T)\,dT + T_d\,\frac{de(T)}{dT}\right) \qquad (6.10)$$

where T_d is the integral action time (s).

Whereas the integral term was the control's 'memory', the differential, or derivative, term 'looks' to the future by examining the slope, or rate of change, of the error. It is rather like an accelerator and brake. But this adds another parameter to be set up by the user, and it can add to the potential for instability (see later). As a result, when most BEMS outstations are configured, the differential term is invariably left out – T_d is given a default value of 0.

6.5   Discrete-time PID algorithms

So far the control equation (6.10) has been in its analogue form, but BEMS CPUs operate with digital signals. The sensor signal is sampled and converted at the ADC to provide a suitable signal for the CPU. Difference equations (Chapter 5) are more appropriate for this [6]. Recast as a difference equation, the PID algorithm becomes

$$u\big((k+1)T\big) = K_p\bigg(e\big((k+1)T\big) + \frac{T_{\text{sample}}}{T_i}\sum_{j=1}^{k} e(jT)$$

$$+ \frac{T_d\big(e((k+1)T) - e(kT)\big)}{T_{\text{sample}}}\bigg) \tag{6.11}$$

where T_{sample} is the sample time, which is the same as T but the subscript has been used to distinguish it more clearly from the integral and derivative times.

Equation (6.11) is often known as the **position** form of the PID, as opposed to the **velocity** form. The summation in equation (6.11) is the backward difference form of discrete integration, taking the integral during the last sample interval from k to $(k+1)$ as a rectangle of height $e(kT)$ and width T_{sample}. The backward form increases the stability at long sampling intervals such as those used in BEMS [7, 8]. Other forms of discrete integration, such as forward rectangular and trapezoidal summation [9] can give different forms.

A simpler form of equation (6.11) comes from the z-transform by regarding z^{-1} as the backward shift operator:

$$z^{-1}f(kT) = f\big((k-1)T\big)$$

Using rectangular integration over the sampling interval, T_{sample}, the integral of the function, $f(kT)$, termed $I(kT)$, can be approximated to

$$I(kT) = I\big((k-1)T\big) + T_{\text{sample}}f(kT)$$

or in terms of the z^{-1} operator:

$$(1 - z^{-1})\,I(kT) = T_{\text{sample}}f(kT)$$

Using this relationship and the z^{-1} operator for other difference relationships, equation (6.11) can be written as

$$U(z^{-1}) = K_p\bigg[1 + T_{\text{sample}}\frac{1}{T_i(1 - z^{-1})} + \frac{T_d}{T_{\text{sample}}}(1 - z^{-1})\bigg]E(z^{-1})$$

where $U(z^{-1})$ = z-transform of $u(kT)$
$E(z^{-1})$ = z-transform of $e(kT)$

Rearranging gives

$$(1 - z^{-1}) \, U(z^{-1}) = \{a_0 + a_1 z^{-1} + a_2 z^{-2}\} E(z^{-1})$$

where $a_0 = K_p \left(1 + \dfrac{T_{\text{sample}}}{T_i} + \dfrac{T_d}{T_{\text{sample}}} \right)$

$$a_1 = -K_p \left(1 + 2 \dfrac{T_d}{T_{\text{sample}}} \right)$$

$$a_2 = \dfrac{K_p \, T_d}{T_{\text{sample}}}$$

Changing this back to the form of a difference equation:

$$u(kT) = u((k - 1)T) + a_0 \, e(kT) + a_1 \, e((k - 1)T) + a_2 \, e((k - 2)T)$$

which is a much simpler equation than equation (6.11). Clarke discusses this equation in a good general paper on PID algorithms [14].

6.5.1   Noise, interference and derivative kick

Noise and interference can cause sudden changes in the sensor signal, which result in sudden error changes in the PID output. This is especially true in the derivative part of the PID control, where there are two sampled values, $e((k + 1)T)$ and $e(kT)$. If there is interference or noise to affect either of these values, especially if it causes a spike in e, then the rate of change measured by the derivative term will alter suddenly and dramatically. This immediately produces a large change in the control signal, u, due to the **derivative kick**. The following sampled value, unaffected by interference, will then go down again, producing a correspondingly large negative derivative control signal. A change in setpoint would also produce derivative kick and, to a lesser extent, **proportional kick** from the proportional term.

Avoidance of derivative kick from a setpoint change can be obtained by taking the derivative of the sensed variable, often the sensed temperature, instead of the error, i.e. the derivative term becomes

$$T_d \, \frac{y((k + 1)T) - y(kT)}{T_{\text{sample}}}$$

where $y(kT)$ is the system output sensed by the BEMS sensor. Proportional kick can similarly be avoided by this change, with only the integral term using the error.

But the noise and interference are still present, which especially disturb the derivative terms, and to a lesser extent, the other two terms as well. To overcome this, an averaging technique can be used to smooth out noisy signals. An average of four sensor signals is taken and for an unchanged setpoint the resulting average error is

$$e_{av}(k) = \frac{e(k) + e(k-1) + e(k-2) + e(k-3)}{4} \tag{6.12}$$

where $e_{av}(k)$ = average error at time kT
              $e(k)$ = latest error value at time kT
        $e(k-3)$ = earliest error considered, at time $(k-3)T$
The resulting derivative is

$$\frac{de_{av}}{dT} = \frac{e_{av}(k) - e_{av}(k-1)}{T_{sample}}$$

where $e_{av}(k-1)$ is the average error value at time $(k-1)T$.

Williams [16] recommends a four-point difference technique for the derivative term:

$$y^* = \frac{y(k) + y(k-1) + y(k-2) + y(k-3)}{4}$$

which is the same as equation (6.12) if the setpoint is not altered. But the derivative is calculated from

$$\frac{dy}{dT} = \left[\frac{y(k) - y^*}{1.5T_{sample}} + \frac{y(k-1) - y^*}{0.5T_{sample}} + \frac{y^* - y(k-2)}{0.5T_{sample}} + \frac{y^* - y(k-3)}{1.5T_{sample}} \right]$$

$$= \frac{1}{6T_{sample}} [y(k) - y(k-3) + 3y(k-1) - 3y(k-2)]$$

Filtering and spike, or **logical filters** [14], can also reduce problems with the digitized input signals for derivative control.

An **error deadband** around the setpoint stops small oscillations in the loop output, and small oscillations in the controlled plant, due to noise, when the controller has settled the system close to the setpoint. This is done by only changing the error value when it has significantly changed from the setpoint. This produces an **effective error**, e^* [15]:

$$e^* = 0 \text{ if } e \leqslant \text{deadband}$$

$$e^* = e - \text{deadband if } e > \text{deadband}$$

This applies if the setpoint is at the top of the proportional band, but if the BEMS has a setpoint in the proportional band with negative errors, then

$$e^* = 0 \text{ if } |e| \leqslant \text{deadband}$$

$$e^* = e - \text{deadband if } e < \text{deadband}$$

$$e^* = e + \text{deadband if } e > \text{deadband}$$

where $|e|$ is the modulus of the error.

In practice, most derivative control on BEMS is not implemented because of these problems, the consequent increase in set-up time with the third parameter, T_d, and the increased probability of unstable control.

6.5.2 Integral wind-up

It can be possible for a heating valve to be fully open and the controlled temperature still to be some way from the setpoint. Such an occurrence is during heat-up in the morning. The error will persist for some time and the integral term becomes enormously large; this is known as **integral wind-up**. It will cause overshoot when the setpoint is finally achieved (as happened with the on/off boiler control). Integral wind-up is typically prevented by limiting the integral term to a **saturation value** – the value which will produce a 100% output from the loop.

6.5.3 Velocity algorithm

In equation (6.11) the difference equation form of the PID control, the actuator and valve position can be exactly related to the output signal, $u((k+1)T)$. When u is zero the valve is shut, and when it is 100% the valve is fully open. If this is not precisely the case (i.e. the valve is fully open at 80%) then a position offset constant can simply be added to equation (6.11). Equation (6.11) is therefore called the **position algorithm** and it is transmitted to the valve actuator via a DAC as an analogue signal.

An alternative PID algorithm is the **velocity** or **incremental algorithm**, derived from the difference between successive values of the position algorithm in equation (6.11). Here it is given without the derivative term:

$$D^*(k+1) = u((k+1)T) - u(kT)$$

$$= K_p \left(e((k+1)T) - e(kT) + \frac{T_{sample}}{T_i} e(kT) \right) \qquad (6.13)$$

With this algorithm the change or increment in the control signal is used to control a device directly. Equation (6.13) can also be used with valves controlled by **stepper motors**, where a pulse moves the shaft through a certain rotation or step [17]. Similar actuator devices not requiring feedback of their position can also make use of this algorithm.

If the algorithm is used for proportional-only control, there may be drifting as there is no reference to a setpoint in equation (6.13), except in the last integral term. The setpoint disappears in the proportional term:

$$e((k+1)T) - e(kT) = t_{set} - t((k+1)T) - t_{set} - t(kT)$$

$$= t((k+1)T) - t(kT)$$

where $t(kT)$ is the sensed temperature at time kT and t_{set} is the setpoint temperature.

Advantages of the velocity algorithm are that integral wind-up is avoided. Since there is no integral summation term, the velocity integral term is limited to the single error term. If the setpoint is suddenly changed, there is no bump in the controller, as occurs with the position algorithm.

6.5.4   Supervisory jacket software

The **supervisory jacket** is the software to ensure as far as possible that the PID loop control in the BEMS is protected from noisy signals, and that integral wind-up and derivative kick are minimized. Within the supervisory jacket there are default values for T_i and T_d in case the user does not set them during the configuration. It also contains the relevant configuration user interface for setting up a loop control. Having the supervisory jacket means that a BEMS program for PID loop control extends beyond the code to execute the PID equation. This applies to other BEMS control programs too, and it partly explains why the cost of development of a complete BEMS's software can run to over a million pounds.

6.6   Stability

We can generalize the feedback control system of Fig. 6.6, with the proportional control and the system having transfer function $G(s)$ and the sensor feedback loop having transfer function $H(s)$. Written in terms of Laplace transforms, the relationship between the error, $e(T)$, and the setpoint, $r(T)$, is given by

$$E(s) = R(s) - G(s)H(s)Y(s)$$

$$Y(s) = \frac{G(s)\ R(s)}{1 + H(s)G(s)}$$

The expression $G(s)/(1 + H(s)G(s))$ is the **closed-loop transfer function** and $H(s)G(s)$ is the **open-loop transfer function**.

If the feedback through the sensor is accurate, the sensor has little thermal mass and the lags are short, then there is **unity feedback** and the closed-loop transfer function becomes $G(s)/(1 + G(s))$. To cover many different types of transfer function, this can be written as

$$\frac{b_0 + b_1 s + b_2 s^2 + \ldots + b_m s^m}{a_0 + a_1 s + a_2 s^2 + \ldots + a_n s^n}$$

where the bs and as are constants. This can be simplified to

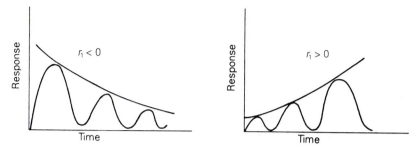

Fig. 6.12 Transient response e^{r_1T}: $r_1 < 0$ is stable, $r_1 > 0$ is unstable.

$$\frac{C_1}{s - r_1} + \frac{C_2}{s - r_2} + \frac{C_3}{s - r_3} + \ldots + \frac{C_n}{s - r_n}$$

where the Cs are constants and the rs are the roots of the **characteristic equation**

$$a_0 + a_1s + a_2s^2 + \ldots + a_ns^n = 0$$

From Table 5.2 we know that the transform of $C_1/(s - r_1)$ is $C_1 \exp(r_1T)$ or $C_1 e^{r_1T}$. These roots of the characteristic equation determine the transient response of the system and its control. If a root, say, r_1, is negative, then e^{r_1T} decreases with time and the transient response reaches a steady, stable value (Fig. 6.12). If the root is positive, the transient response increases and the system cannot achieve a stable state. In practice the system's response will not increase indefinitely but will oscillate, or **hunt**.

If the root is complex, i.e. $r_1 = \sigma + j\mu$ then the transform yields $e^{\sigma T}e^{j\mu T}$, and

$$e^{j\mu T} = \cos \mu T + j \sin \mu T$$

which is a steadily oscillating function.

These observations lead to the Nyquist criterion [1] that the system and its control are unstable if the denominator of the transfer function has any roots that are positive [2–4].

6.7 Configuring a PID loop for stability

For simple plant, one can derive a straightforward differential equation to determine whether the system will be stable for various settings. But for most practical heating plant, the differential equations and transfer functions will be complex. One can get an idea about the response of a system by looking at a BEMS graph of the system when it is turned on, or when the setpoint is changed – the response to a step function. There are also various software programs for analysing the response of systems and determining the transfer functions.

Table 6.1   Settings for closed-loop ultimate cycling

P control	$K_p = 0.5K_p^*$		
PI control	$K_p = 0.45K_p^*$	$T_i = 0.8\tau_p$	
PID control	$K_p = 0.6K_p^*$	$T_i = 0.5\tau_p$	$T_d = 0.125\tau_p$

When commissioning a plant for a building, there is a limit on how much time can be spent in trying to determine the transfer functions for all its separate items. Zeigler and Nichols [5] have determined empirical rules for two commissioning procedures: closed-loop ultimate cycling and open-loop process reaction curves.

6.7.1   Closed-loop ultimate cycling

Closed-loop ultimate cycling is the most common method for BEMS configuration of loop modules. The loop is set for proportional control only; T_i is set to ∞ or a very large value, and T_d is set to 0. The proportional gain, K_p, is initially set to a small value and progressively increased until the system starts to oscillate or hunt. The gain where hunting begins is called K_p^*. It is also necessary to measure the period of oscillation, τ_p. Recommended empirical settings are given in Table 6.1.

6.7.2   Open-loop process reaction curves

Process reaction curves are not as useful for configuring a loop as they entail disconnecting the loop module from the control element (e.g. the valve) to obtain an open loop (Fig. 6.13). A step input signal is then applied to the control element and the response of the system is measured by the BEMS. The response generally takes the form of an *n*th order system (Fig. 6.14).

Such an *n*th order system is approximately described mathematically by the transfer function

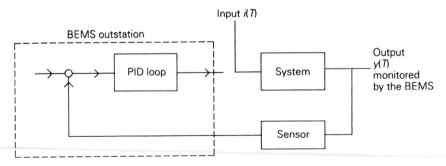

Fig. 6.13   Opened loop.

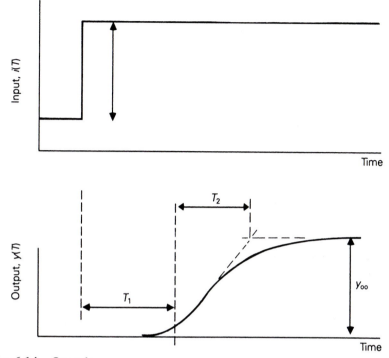

Fig. 6.14 Open-loop response.

Table 6.2 Settings for open-loop process reaction curves

P control	$K_p = \dfrac{iT_2}{y_\infty T_1}$		
PI control	$K_p = 0.9\dfrac{iT_2}{y_\infty T_1}$	$T_i = 3.3T_1$	
PID control	$K_p = 1.2\dfrac{iT_2}{y_\infty T_1}$	$T_i = 2T_1$	$T_d = 0.5T_1$

$$\frac{y(T)}{i(T)} = \frac{Ke^{-sT}(T_1)}{1 + sT_2}$$

where T_1 = dead time (s)
T_2 = effective time constant (s)
$y(T)$ = output (°C)
$i(T)$ = input (°C)
K = gain

Figure 6.14 shows that the dominant slope of the curve is y_∞/T_2, where y_∞ is the output the system tends to as T tends to infinity. Table 6.2 shows the empirical settings for stable control of the system.

References

[1] Nyquist, H. (1932) Regeneration theory. *Bell Systems Technical Journal*, January.

[2] DiStefano, J. J., Stubberud, A. R. and Williams, I. J. (1987) *Feedback and Control Systems*, Schaum's Outline series, McGraw-Hill, New York.

[3] Healy, M. (1975) *Principles of Automatic Control*, Hodder and Stoughton, London.

[4] Morris, N. M. (1983) *Control Engineering*, 3rd edn, McGraw-Hill, London.

[5] Zeigler, J. G. and Nichols, N. B. (1942) Optimum settings for automatic controllers. *Transactions of ASME*, **64**.

[6] Spiegel, M. R. (1971) *Calculus of Finite Differences and Difference Equations*, Schaum's Outline series, McGraw-Hill, New York.

[7] Bristol, E. H. (1977) Designing and programming control algorithms for DDC systems. *Control Engineering*, January.

[8] Stoecker, W. F. and Stoecker, P. A. (1989) *Microcomputer Control of Thermal and Mechanical Systems*, Van Nostrand Reinhold, New York.

[9] Ogata, K. (1987) *Discrete-time Control Systems*, Prentice-Hall, New York.

[10] CIBSE (1985) Automatic Controls and their implications for systems design, *Applications Manual*, CIBSE, London.

[11] ASHRAE (1999) ASHRAE Handbook, Heating, ventilating, and air-conditioning applications. American Society of Heating, Refrigerating and Air-Conditioning Engineers, Atlanta.

[12] Chapra, S. C. and Canale, R. P. (1987) *Numerical Methods for Engineers with personal computer applications*, McGraw-Hill, New York.

[13] Shavit, G. and Brandt, S. G. (1982) The dynamic performance of a discharge air-temperature system with a P-I controller. *ASHRAE Trans.*, **88**, Pt 2.

[14] Clarke, D. W. (1984) PID algorithms and their computer implementation, *Transactions of the Institute of Measurement and Control*, **6**(6).

[15] Nesler, C. G. and Stoecker, W. F. (1984) Selecting the proportional and integral constants in the direct digital control of a discharge air temperature. *ASHRAE Trans.*, **90**, Pt 2.

[16] Williams, T. J. (1984) *The Use of Digital Computers in Process Control*, Instrument Society of America, North Carolina.

[17] Chesmond, C. J. (1986) *Control System Technology*, Edward Arnold, London.

[18] CIBSE (2000) CIBSE Guide, Automatic Control Systems, Chartered Institution of Building Services Engineers, London.

7
Building heat loss and heating

Having dealt with the basic elements of control, on/off and PID, it is necessary to appreciate the heating system and its sizing to further our development of its control. The main element of designing a heating system is to
determine the heat loss from the building when the inside is maintained at a
comfortable temperature. As was briefly mentioned in Chapter 4 on sensors
and their responses, an index of comfort is not the inside air temperature, t_{ai},
but the **dry resultant temperature**, or resultant temperature, t_{res}:

$$t_{res} = \frac{t_r + t_{ai}\sqrt{(10v)}}{1\sqrt{(10v)}} \tag{7.1}$$

where t_{ai} = inside air temperature (°C)
  t_r = mean radiant temperature (°C)
  v = inside air speed (m s^{-1})
In typical interiors v is low, of the order of 0.1 m s^{-1}, so

$$t_{res} = \tfrac{1}{2}t_r + \tfrac{1}{2}t_{ai} \tag{7.2}$$

Typically most design centres on a t_{res} of about 20°C. The CIBSE Guide [1]
gives the recommended design values for t_{res} [24]. The guide has been updated with summer and winter values. Table 7.1 shows the old and new
recommended values.

People's assessment of comfort is also influenced by outside temperatures, and the temperature change experienced upon entering a building [2].
(Further details of comfort can be found in the literature [3, 4].)

Many designers still refer to, and use in calculation, air temperature, t_{ai}, as
it simplifies calculations. The error introduced by using t_{ai} instead of t_{res} is
often not large because when a building is in equilibrium, t_r is reasonably

Table 7.1   Recommended dry resultant temperatures

Type of building	t_{res}	winter t_{res}	summer t_{res}
Bank	20	19–21	21–23
Canteen	20	22–24	24–25
Church	18	19–21	22–24
Factory			
Sedentary work	18	19–21	21–23
Light work	16	16–19	–
Heavy work	13	11–14	–
Residence			
Living room	21	22–23	23–25
Bedroom	18	17–19	23–25
Offices	20	21–23	22–24
Schools	18	19–21	21–23
Shops	18	19–21	21–23

close to t_{ai}, the exact difference depending on the type of heating system. This is considered in a later example.

7.1   Heat loss

The heat loss from a building is made up of losses through the building fabric and by infiltration and ventilation. The losses occur due to the processes of **heat transfer**. Consider the heat transfer in a room heated by a radiator, as shown in Fig. 7.1. The single panel, or single column, radiator transfers heat to the room air by **convection**, and to the room surfaces by **radiation**. For a single panel radiator, the split between radiation and convection is equal [5].

The air in the room is warmed by the radiator and then transfers some of its heat to the room surfaces. The internal room surfaces are warmer than the external wall and window, so there is transfer of heat from the warm surfaces to the colder external surfaces by radiation. Heat is then transferred by **conduction** through the external wall and window to their outside surfaces. At the outside surface of the wall and window, heat is transferred and lost to the air and surroundings by convection and radiation.

7.1.1   Conduction

Conduction is the process of heat transfer through a substance such as a wall. It is described by Fourier's equation [6]:

$$\rho C_p \frac{dt}{dT} = k \left[\frac{\partial^2 t}{\partial x^2} + \frac{\partial^2 t}{\partial y^2} + \frac{\partial^2 t}{\partial z^2} \right] \qquad (7.3)$$

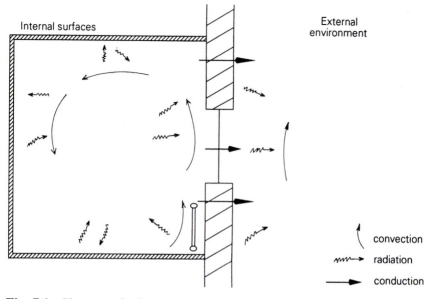

Fig. 7.1   Heat transfer in a room.

where T = time (s)
   t = temperature (°C)
   k = thermal conductivity (W m^{-1} K^{-1})
   ρ = density (kg m^{-3})
   C_p = specific heat capacity (kJ kg^{-1} K^{-1})
   x, y, z = coordinates in space or dimensions (m)

The above equation is a highly complex one, indeed there is a large book solely devoted to solutions of it [7]. For the fabric of buildings, the equation can be reduced to one dimension, x, ignoring the small errors that would arise at the building corners and other points where the heat flow is not uniformly perpendicular to the surfaces.

For traditional building heat loss calculations the time element is surprisingly ignored, which reduces Fourier's equation to a **steady-state equation**, i.e. the temperature does not vary with time:

$$0 = k \left[\frac{d^2 t}{d x^2} \right] \tag{7.4}$$

By integration and application of boundary conditions this leads to

$$Q = -kA \frac{dt}{dx} \tag{7.5}$$

Table 7.2 Conductivities of materials

Material	Conductivity ($W\,m^{-1}\,K^{-1}$)
Copper	388
Aluminium	202
Mild steel	45
Concrete	1.4
Water	0.6
Air	0.026

where Q is the rate of heat flow (W) and A is the cross-sectional area (m^{-2}). The minus sign in the above equation is due to the convention that Q is positive as x increases but t decreases along the x-axis. This can be written as

$$Q = kA\frac{(t_{hot} - t_{cold})}{l} \tag{7.6}$$

where t_{hot} = temperature at the hot end (°C)
t_{cold} = temperature at the cold end (°C)
l = the length of the conductor (m)
Some textbooks refer to heat flux, q, and this is

$$q = Q/A \quad (W\,m^{-2})$$

Typical conductivities of materials are given in Table 7.2. More conductivities are contained in the CIBSE Guide [8, 24].

We can compare the flow of heat to the flow of electricity through a resistance, as was considered earlier in the water/electrical/thermal analogies of Chapter 4 on sensors and their responses. Ohm's law for electricity flow gives

$$V = IR \tag{7.7}$$

where V = potential difference (or voltage) (V)
I = current (A)
R = resistance (Ω)
Comparing this with equation (7.6), the heat flow, Q, being analogous to I and temperature difference being analogous to V, then the thermal resistance is

$$R_{th} = \frac{1}{kA} \tag{7.8}$$

This makes equation (7.6) become

$$(t_{hot} - t_{cold}) = QR_{th} \tag{7.9}$$

where R_{th} is the thermal resistance ($K\,W^{-1}$).

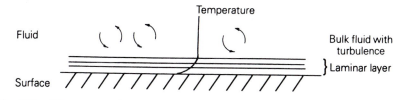

Fig. 7.2 The convection process.

The North Americans use thermal resistance as defined above [9], but it is worth noting that the thermal resistance referred to in the CIBSE Guide [8, 24] is

$$R^*_{th} = \frac{1}{k} \tag{7.10}$$

where R^*_{th} is the resistance (m^2 K W^{-1}).

Equation (7.10) substituted into equation (7.6) gives

$$\frac{QR_{th}}{A} = qR_{th} = t_{hot} - t_{cold}$$

7.1.2 Convection

Heat transfer by convection takes place in fluids where the heated or cooled fluid particles move, taking their heat or 'coolth' with them. There are two types of convection, **natural** and **forced**.

Natural convection takes place when the fluid flow is due only to fluid temperature differences, consequently changes in fluid density. It therefore primarily takes place in the vertical plane around hot and cold bodies such as radiators and room surfaces.

Forced convection occurs when the fluid flow is enhanced by mechanical means such as a pump or fan. With both natural and forced convection, there is a stationary laminar film of fluid close to the solid surface, as shown in Fig. 7.2.

The heat transfer is initially through this layer by conduction (the film does not move), the genuine convection taking place in the bulk of the fluid. The thickness of the film and hence the resistance to heat transfer is dependent on the fluid velocity. Hence with forced convection there is more heat transfer than natural convection, consequently the heat transfer coefficient for forced convection is larger than for natural convection. Both processes are described by the following general equation, often referred to as Newton's law of cooling:

$$Q = h_c A (t_{surf} - t_{fluid}) \tag{7.11}$$

Table 7.3   CIBSE convective heat transfer coefficients

Direction of heat flow	$h_c(W\ m^{-2}\ K^{-1})$
Horizontal	3.0
Upward	4.3
Downward	1.5
Average	3.0

where h_c = heat transfer coefficient (W m^{-2} K^{-2})
   A = heat transfer area (m^2)
   t_{surf} = surface temperature (°C)
   t_{fluid} = bulk fluid temperature (°C)

Equation (7.11) is deceptively simple as h_c is very complex. It is usually obtained from empirical data where it is often quoted in the dimensionless **Nusselt number**, Nu:

$$Nu = \frac{h_c l}{k} \tag{7.12}$$

where l is a characteristic linear dimension (m).

As the heat transfer equations are often complex and long, they are simplified by using the dimensionless **Reynolds number**, Re; **Prandtl number**, Pr; and the **Grashof number**, Gr. These dimensionless numbers are defined in reference [6], and values are given in the literature [9, 10].

The general form of the heat transfer equation for natural convection is

$$Nu = Cf\{Gr^x Pr^y\} \tag{7.13}$$

where f signifies 'a function of', x, y are indices and C is a constant.

The general form of the heat transfer equation for forced convection is

$$Nu = Kf\{Re^v Pr^w\} \tag{7.14}$$

where f signifies 'a function of', v, w are indices and K is a constant.

Table A3.4 of the CIBSE Guide [8, 24] gives values of h_c for still air conditions (defined as the air speed at the surface being less than 0.1 m s^{-1}). These values are shown in Table 7.3.

The values in Table 7.3 are for specific conditions of air speed and they are only valid over a small range of room temperatures [11]. This then reduces the complex convection equation to a linearized form like the conduction equation and Ohm's law:

$$Q = h_c A(t_{surf} - t_{ai}) \tag{7.15}$$

7.1.3   Radiation

Radiation is the transfer of heat through space by electromagnetic waves which are similar to light, except that light waves have a higher frequency

Table 7.4   Radiation factors

	Area	$F_e F_a$
Large parallel surfaces of area A_1 and A_2	A_1 or A_2	$\dfrac{\varepsilon_1 \, \varepsilon_2}{\varepsilon_1 + \varepsilon_2 - \varepsilon_1 \, \varepsilon_2}$
Small area A_1 within an enclosed area	A_1	ε_1
Small surfaces well separated, so that amount of radiation reradiated back to emitter is small	A_1	$\varepsilon_1 \, \varepsilon_2 \, F_a$

and shorter wavelength. The thermal radiation from a black body is given by the Stefan–Boltzman law:

$$Q = \sigma A \; T^{*4} \tag{7.16}$$

where σ = Stefan–Boltzman constant (5.67×10^{-8} W m^{-2} K^{-4})
   A = heat transfer area (m^2)
   T^* = absolute temperature (K)
For an actual material, the radiation will be more like that from a grey body:

$$Q = \sigma \; \varepsilon \; A \; T^{*4} \tag{7.17}$$

where ε is the emissivity of the surface, which for most building materials is 0.9.

The radiation exchanged between two surfaces, 1 and 2 is

$$Q_{12} = A_1 \sigma \; F_a \; F_e \; (T_1^{*4} - T_2^{*4}) \tag{7.18}$$

where F_a = a form factor determined by the relative geometries of the two surfaces
   F_e = a factor taking into account the emissivities and absorptivities of the surfaces
   A_1 = surface area (m^2) of surface 1 (Table 7.4).

For further details on these factors see the CIBSE Guide [10]. For normal rooms, $F_a \, F_e$ can be taken as 0.87, giving an accuracy of within 10% to equations (7.18) and (7.12).

It is convenient to linearize the radiation heat transfer equation to

$$Q = A \varepsilon h_r \; (t_1 - t_2) \tag{7.19}$$

and it can be shown (equation (7.10)) that by rearranging equation (7.18) into

$$Q_{12} = A_1 \sigma \; F_a \; F_e \; (T_1^* - T_2^*)(T_1^* + T_2^*)(T_1^{*2} - T_2^{*2}) \tag{7.18}$$

then at normal ambient temperatures $(T_1^* - T_2^*)$ dominates, and the radiation heat transfer coefficient, h_r, is given by

$$h_r = \sigma(T_1^* + T_2^*)(T_1^{*2} + T_2^{*2}) \tag{7.20}$$

Here is another approximation that reinforces this [18]:

Table 7.5　Values of h_r from CIBSE Guide, Section A3

Surface temperature (°C)	$h_r(W\ m^{-2}\ K^{-1})$
−10	4.1
0	4.6
10	5.1
20	5.7

Table 7.6　Values of h_r from CIBSE Guide, Section C3 [10]

$t_1 = 20°C$ when t_2 is	$h_r(W\ m^{-2}\ K^{-1})$
−10°C	4.9
0	5.2
10	5.4
20	5.7
25	5.9
30	6.0

$$(T_1^* + T_2^*)(T_1^{*2} + T_2^{*2}) \approx 4T_{av}^{*2}$$

where T_{av}^* is the average temperature of the building surfaces radiating heat.

Table 7.5 gives values of h_r quoted in the CIBSE Guide, Section A3 [8, 24]. Although not stated there, one infers that the temperatures of the transmitting and receiving surfaces are similar, so that the h_r values are only valid for surfaces up to 10°C different. Volume C, Section C3 gives more details for values of h_r. And Table 7.6 gives some of the values when one of the surfaces is kept at 20°C.

Example 7.1

How much heat transfer in a warm room is made up of radiation and convection? The room is cubic and 3 m in height. One wall is external facing, the other surfaces are internal. The emissivity of the surfaces is 0.9.

Solution

If we consider a room with one external wall whose inside surface temperature is 10°C (280 K), with the inside air at 20°C and the other surfaces at 18°C (288 K), the heat transfer will be primarily from the warm walls and the warm air to the cold external wall.

The radiation that the external wall receives from the other warmer walls can be derived by first calculating h_r:

$$h_r = \sigma\ (T_1^* + T_2^*)(T_1^{*2} + T_2^{*2})$$
$$= 5.67 \times 10^{-8}(280 + 288)(280^2 + 288^2)$$
$$= 5.2\ W\ m^{-2}\ K^{-1} \tag{7.20}$$

Assuming that the cold wall is a small area surrounded by the warmer walls, so that $F_a F_e = \varepsilon$, and that $\varepsilon = 0.9$, then the heat received from the warmer surfaces is

$$Q_{rad} = 5.2 \times 0.9 \times 9 \times (288 - 280)$$
$$= 337\ W$$

For convective heat transfer from the warm air to the cold wall, taking $3.0\ W\ m^{-2}\ K^{-1}$ for the horizontal heat transfer coefficient from Table 7.3, we obtain

$$Q_{conv} = 3.0 \times 9 \times (20 - 10)$$
$$= 270\ W$$

Radiation is the larger heat transfer process. It is interesting to calculate the dry resultant temperature of the room. The mean radiant temperature is approximately the mean surface temperature:

$$t_r \approx \frac{(5 \times 18) + (1 \times 10)}{6}$$
$$\approx 16.7°C$$

So
$$t_{res} = \tfrac{1}{2} t_{ai} + \tfrac{1}{2} t_r,$$
$$= 18.4°C$$

The BEMS sensor will be primarily monitoring the air temperature.

7.1.4 Surface resistance

The linearized radiation and convection heat transfer processes may be combined into a single heat transfer process for building surfaces. This gives rise to a combined resistance to heat flow at the surface, the **surface resistance**, R_s. In the CIBSE Guide [8] R_s is defined as

$$R_s = \frac{1}{Eh_r + h_c}\ (m^2\ K\ W^{-1}) \qquad (7.21)$$

where E = emissivity factor $\Theta \varepsilon_1 \varepsilon_2$
Θ = CIBSE A3 notation for form or shape factor, F_a
$\varepsilon_1, \varepsilon_2$ = emissivities of surfaces
h_r = radiative heat transfer coefficient ($W\ m^{-2}\ K^{-1}$)
h_c = convective heat transfer coefficient ($W\ m^{-2}\ K^{-1}$)

For outside surfaces the outside surface resistance, R_{so}, is evaluated with the form factor, F_a or Θ, as unity. The CIBSE Guide [8, 24] quotes values of R_{so} for sheltered, normal and severe exposures of the surfaces. These are necessary due to the influence of wind speed on the convection component of the

Table 7.7   Values of R_{so} (m² K W⁻¹)

	Sheltered	*Normal*	*Severe*
High emissivity ($\varepsilon = 0.9$)	0.08	0.06	0.03
Low emissivity ($\varepsilon = 0.05$)	0.11	0.07	0.03

Table 7.8   Values of R_{si} (m² K W⁻¹)

Element	*Heat Flow*	*High-emissivity surface*
Wall	Horizontal	0.12
Ceiling, floor or roof	Upward	0.10
Ceiling and floor	Downward	0.14

heat loss. Normal exposure corresponds to a wind speed at roof level of 3 m s⁻¹. Table 7.7 gives some values of R_{so} for a wall.

Most building materials have high emissivities and correspond to the $\varepsilon = 0.9$, but polished metal surfaces as in some wall claddings have low emissivities. Table 7.8 gives some values for the inside surface resistance, R_{si}.

The derivation of R_{si} is related to the inside environmental temperature, t_{ei}. This temperature was introduced by CIBSE to account for the fact that room surface temperatures are different to (often lower than) the room air temperature. Therefore the environmental temperature is employed to combine the mean room surface temperature and the air temperature:

$$t_{ei} = \tfrac{1}{3}t_{ai} + \tfrac{2}{3}t_{m} \tag{7.22}$$

where t_{ai} is the room air temperature (°C), t_{m} is the mean temperature of the room surfaces, and

$$t_{m} = \frac{\Sigma(At_{s})}{\Sigma A}$$

where A is the surface area (m²) and t_{s} is the surface temperature (°C).

The derivation of t_{ei} is given in the CIBSE Guide [13, 24]. So the heat exchange at an inside room surface and the rest of the room and air is

$$Q = A_{s}\frac{(6Eh_{r} + h_{c})}{5}[t_{ei} - t_{si}] \tag{7.23}$$

where Q = heat flow (W)
  A_{s} = area of room surface considered (m²)
  t_{si} = surface temperature (°C)

This gives

$$R_{si} = \left[\frac{6Eh_r + h_c}{5} \right]^{-1}$$

which is slightly different to the resistance defined in equation (7.21). This is due to the use of t_{ei} in equation (7.22) [8, 24].

It is often convenient to generalize heat flow equations for heat transfer from emitters and within rooms. The CIBSE Guide [10] gives a generalized heat transfer equation which will be the basis of emitter heat transfer equations used later on:

$$Q = C (T_s^* - T_a^*)^n \qquad (7.24)$$

where C = a constant, $0.64 \leqslant C \leqslant 1.4$
T_s^* = absolute temperature of the surface (K)
T_a^* = absolute temperature of the air (K)
n = an index, either 1.33 or 1.25

7.2   Temperature relationships

A number of temperatures such as t_{ei}, t_{ai} and t_m were mentioned in Section 7.1 and t_{res}, t_r were introduced in the discussion on comfort at the beginning of the chapter. These temperatures can be related to each other and it is necessary to relate them in heat loss and comfort calculations.

At the centre of a cubical room, and also approximately for most other rooms, the mean temperature of the room surfaces, t_m, effectively equals the mean radiant temperature, t_r, referred to earlier in equation (7.2) for the dry resultant temperature:

$$t_m \approx t_r$$

This was the assumption made in the previous example in working out t_{res}.

CIBSE also refers to t_{res}, the dry resultant temperature, as t_c at the centre of the room and it is often t_c that is referred to as the design criterion:

$$t_c = \tfrac{1}{2} t_{ai} + \tfrac{1}{2} t_m$$

CIBSE uses factors F_1 and F_2 to relate the various design temperatures:

$$F_1 = \frac{t_{ei} - t_{ao}}{t_c - t_{ao}}$$

$$F_2 = \frac{t_{ai} - t_{ao}}{t_c - t_{ao}}$$

Tables of F_1 and F_2 (F_{1cu}, F_{2cu} in [24]) are given in the Guide [14, 24] for various types of emitter and their variations in radiant and convective heat output. For example, a single panel radiator, with 50% radiant and 50% convective output has higher values of F_1 but lower values of F_2 than a fan convection heater with 100% convective heat output. This means that t_{ei} is higher for the radiator, and t_{ai} is higher for the convector.

In well-insulated buildings with convective heating

$$t_{ei} \approx t_{ai}$$

Because t_{ei} and t_{ai} can be close in some cases, it leads many designers to be lazy and to simply use t_{ai} as a design criterion for comfort. In fact in other countries, such as the United States, t_{ai} is used still, and it is only recently that CIBSE in the United Kingdom has adopted t_{ei}.

7.3   Heat loss resistance network

Now that the internal and external surface resistances have been derived to complement the conduction resistance, the overall resistance network of a wall can be derived. Consider the example room at the beginning of this chapter. The resistance network for the external wall is shown in Fig. 7.3 and the resistances are as follows:

R_{sic} = inside surface convection resistance (m² K W⁻¹)
R_{sir} = inside surface radiation resistance (m² K W⁻¹)
R_{soc} = outside surface convection resistance (m² K W⁻¹)
R_{sor} = outside surface radiation resistance (m² K W⁻¹)
R_c = conduction resistance of wall = $1/k$ (m² K W⁻¹)
t_{si} = inside surface temperature (°C)
t_{so} = outside surface temperature (°C)

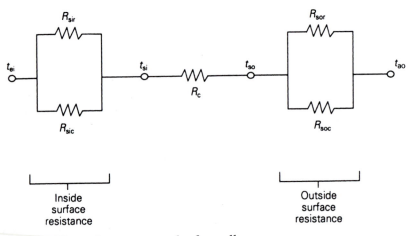

Fig. 7.3   The resistance network of a wall.

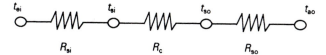

Fig. 7.4 The simplified resistance network.

As the radiation and convection resistances are in parallel, just as in electrical resistance calculations we can write

$$\frac{1}{R_{si}} = \frac{1}{R_{sic}} + \frac{1}{R_{sir}} \qquad (7.25)$$

or

$$R_{si} = \frac{R_{sir} R_{sic}}{R_{sic} + R_{sir}}$$

where R_{si} is the inside surface resistance ($m^2\ K\ W^{-1}$).

Comparing equation (7.25) with the inverse of the CIBSE surface resistance derived from equation (7.24),

$$R_{si}^{-1} = \frac{6Eh_r + h_c}{5}$$

it can be calculated that

$$R_{sir} = \frac{5}{6Eh_r} \quad \text{and} \quad R_{sic} = \frac{1}{h_c}$$

With the surface resistances combined, the network is as shown in Fig. 7.4.

The overall, equivalent resistance of the whole wall, R_{eq}, is

$$R_{eq} = R_{si} + R_c + R_{so}$$

7.4   *U*-value or overall thermal transmittance

Buildings have a number of walls and windows as well as ceilings and floors, and each of these elements has its own equivalent resistance. To calculate the equivalent resistance of this network of resistors, it is necessary to sum six or more resistors in parallel. The network for the four external walls of a simple one-room building is shown in Fig. 7.5.

The equivalent resistance of this simple room is

$$\frac{1}{R_1} + \frac{1}{R_2} + \frac{1}{R_3} + \frac{1}{R_4} \qquad (7.26)$$

This is rather tedious so the reciprocal of the resistances, the transmittances (or *U*-values), are used in heat loss calculations. The CIBSE Guide [8, 24]

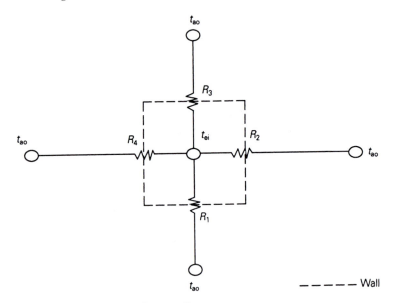

Fig. 7.5 The network for four walls.

quotes *U*-values for common building elements. Many walls and building elements consist of a composite of layers and the derivation of the *U*-value for such a composite is

$$U = \frac{1}{R_{si} + R_1 + R_2 + \ldots + R_a + R_{so}} \qquad (7.27)$$

where U = overall thermal transmittance (W m^{-2} K^{-1})
   R_{si} = inside surface resistance (m^2 K W^{-1})
   R_1 = conduction resistance of element 1 of fabric (m^2 K W^{-1})
   R_2 = conduction resistance of element 2 of fabric (m^2 K W^{-1})
   R_a = airspace resistance, typically 0.18 m^2 K W^{-1}
   R_{so} = outside surface resistance
The airspace is included for cavity walls.
   Using *U*-values, equation (7.26) is simplified to

$$U_1 + U_2 + U_3 + U_4$$

The heat loss through the fabric of a building, Q_{fab}, is therefore given by

$$Q_{fab} = (t_{ei} - t_{ao}) \sum_i U_i A_i \qquad (7.28)$$

7.5   Infiltration and ventilation loss

Heat is not only lost through the building fabric, but also through warm air leaking out of the building and cold air leaking in, or infiltrating. Deliberate leakage, by window opening for natural ventilation or mechanical ventilation with a fan system, also results in heat loss. The infiltration and ventilation loss, Q_v, is given by

$$Q_v = \rho C_p G \, (t_{ai} - t_{ao}) \qquad (7.29)$$

where ρ = density of air (kg m^{-3})
   C_p = specific heat capacity of air (kJ kg^{-1} K^{-1})
   G = infiltration and ventilation of air (m^3 s^{-1})
Putting in standard values for density of 1.2 kg m^{-3}, and for C_p of 1.02 kJ kg^{-1} K^{-1} and putting in the room volume, V, gives

$$Q_v = \tfrac{1}{3} NV \, (t_{ai} - t_{ao}) \qquad (7.30)$$

where V is the volume of the building (m^3) and N is the number of air changes per hour (h^{-1}).

Determination of N is difficult, although techniques are being developed and much research is being conducted in this area [15, 16]. In heating system design, empirical values of N are taken from the CIBSE Guide [16, 24]. Typically most values of N are centred around 1.0.

The $NV/3$ term can be considered as an infiltration heat resistance, R_v:

$$R_v = \frac{3}{NV} \qquad (7.31)$$

7.6   Total building heat loss

The total heat loss from a building due to fabric heat losses and infiltration and ventilation heat losses is

$$
\begin{aligned}
Q &= Q_{fab} + Q_v \\
&= \Sigma \, (UA)(t_{ei} - t_{ao}) + (NV/3)(t_{ai} - t_{ao}) \qquad (7.32)
\end{aligned}
$$

Using the factors F_1 and F_2, a common temperature difference may be used for both loss terms:

$$Q = \{F_1 \, \Sigma \, (UA) + F_2 NV/3\}(t_c - t_{ao}) \qquad (7.33)$$

Note that traditionally there is no consideration of internal heat gains, such as the heat from the lights, equipment, the occupants and solar radiation (7.17), in heat loss calculations. If these gains were included, they would offset the heat losses and the heating system would be smaller. As these gains mostly occur during occupancy, the smaller heating system might not be able to bring the building up to temperature before the occupancy period during colder weather. So ignoring the internal heat gains acts as a safety factor on the design.

It may, however, be necessary to consider the influence of internal gains on the temperature of the building, and the CIBSE Guide [13, 24] details the equations for this. Solar radiation is an interesting case. Under steady-state conditions any solar radiation entering a room is not stored in the fabric, but absorbed and then partly transmitted through the room surface and partly retransmitted back into the room to the environmental point (where t_{ei} is measured). The steady-state surface factor, $\bar{F}_s$, determines the heat retransmitted to the environmental point:

$$\bar{F}_s = \frac{1/U - R_{si}}{1/U}$$

$$= 1 - UR_{si} \qquad (7.34)$$

If the room surface receiving the radiation is internal and separates two rooms at equal temperatures, then $\bar{F}_s$ is unity.

Many heat emitters are either fixed to walls or embedded in room surfaces. Not all of the heat is emitted into the room but some is lost directly to the surface. This is termed **back loss**, Q_{bl}, and for a wall-mounted radiator it is

$$Q_{bl} = A_R U(t_R - t_{ei}) \qquad (7.35)$$

where A_R = area of radiator (m^2)

$\quad U$ = U-value of surface that radiator is on (W m^2 K^{-1})

$\quad t_R$ = mean radiator temperature (°C)

For a radiator heating system, Q_{bl} must be added to the building heat loss to size the system.

For embedded heating systems, such as ceiling panels or underfloor heating, a steady-state **back loss factor**, $\bar{F}_{bl}$, is used and is given by

$$\bar{F}_{bl} = 1 - UR_s \qquad (7.36)$$

where R_s is the resistance between the surface and the environmental point (°C).

The conductance (inverse resistance) network linking all the temperatures in a room, t_{ei}, t_{ai}, t_c and t_{ao}, is shown in Fig. 7.6.

Point A in Fig. 7.6 is the **inside air point**; E is the **environmental point**; C is the **dry resultant point**; and O is the **outside air point**. The conductances $h_{ac} \Sigma A$ and $h_{ec} \Sigma A$ are derived in the CIBSE Guide in the appendix to Section A5 [13, 24], where it is calculated that

$$h_{ac} = 6 \text{ W m}^{-2} \text{ K}^{-1}$$

$$h_{ec} = 18 \text{ W m}^{-2} \text{ K}^{-1}$$

So the heat flow from the dry resultant point to the inside air point is

$$Q_{ae} = h_{ac} \Sigma A(t_c - t_{ai})$$

and likewise from the environmental point to the outside air point, it is

$$Q_{fab} = \Sigma (UA)(t_{ei} - t_{ao})$$

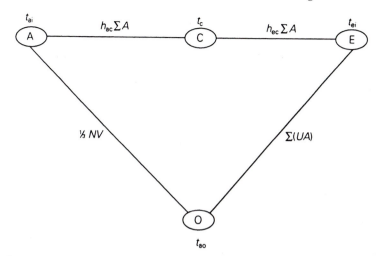

Fig. 7.6   Conductance network relating temperatures.

Example 7.2

Compare the heat loss from a factory heated by radiant heaters (90% radiant, 10% convective) to the heat loss when heated by fan convective heaters (100% convective). The total inside surface area of the factory is 450 m² and $\Sigma (UA) = 470$ W K^{-1}, $NV/3 = 280$ W K^{-1}. Primarily sedentary work is done in the factory, so the design $t_c = 19°C$ and the design $t_{ao} = -1°C$. Section A9 of the CIBSE Guide [14, 24] gives tables of F_1 and F_2 for different heating systems and values of $\Sigma (UA)/\Sigma A$ and $NV/3 \Sigma A$. For the fan convective heaters in this factory, the tables give $F_1 = 0.95$ and $F_2 = 1.17$. For the radiant heaters, $F_1 = 1.06$ and $F_2 = 0.83$.

Solution

Sometimes t_{ai} is wrongly used as the design criterion, and also in fabric heat loss calculations instead of t_{ei}. This example shows why it is an error.

The heat loss is given by equation (7.33):

$$Q = \{F_1 \Sigma (UA) + F_2 NV/3\} (t_c - t_{ao}) \qquad (7.33)$$

For the radiant heating, this is

$$Q = \{(1.06 \times 470) + (280 \times 0.83)\} (19 - [-1])$$

$$= \{498.2 + 232.4\} (20)$$

$$= 14.6 \text{ kW}$$

The air temperature, t_{ai}, is calculated from

$$F_2 = \frac{t_{ai} - t_{ao}}{t_c - t_{ao}}$$

so

$$0.83 = \frac{t_{ai} - [-1]}{19 - [-1]}$$

$$t_{ai} = (0.83 \times 20) - 1$$

$$= 15.6°C$$

For the fan convective heaters, the heat loss is

$$Q = \{(0.95 \times 470) + (280 \times 1.17)\} (19 - [-1])$$

$$= \{446.5 + 327.6\} (20)$$

$$= 15.5 \text{ kW}$$

The air temperature, t_{ai}, is given by

$$F_2 = \frac{t_{ai} - t_{ao}}{t_c - t_{ao}}$$

so

$$1.17 = \frac{t_{ai} - [-1]}{19 - [-1]}$$

$$t_{ai} = (1.17 \times 20) - 1$$

$$= 22.4°C$$

For the radiant heaters, the heat loss (14.6 kW) is less than for the fan convectors (15.5 kW), but the radiant heaters produce a lower air temperature (15.6°C) compared to the fan convectors' t_{ai} (22.4°C).

If the heat loss had been calculated wrongly on a design t_{ai} of 19°C, then the heat loss would have been

$$Q = \{\Sigma (UA) + (1/3)NV\} (t_{ai} - t_{ao})$$

$$= \{470 + 280\} (19 - [-1])$$

$$= 15 \text{ kW}$$

Care is needed in selecting sensors for the control of radiant heating systems as most sensors respond primarily to air temperature, which (as shown here) can be significantly different from comfort temperature, t_c.

7.7   Heating system design

Having determined the steady-state heat loss from the building, the heating system is sized to match it, but in practice the system often surpasses it by about 25%. This excess acts as a safety factor.

The heat output from the boiler system to match the building heat loss is given by

$$\dot{m} C_p (t_f - t_r) \qquad (7.37)$$

where $\dot{m}$ = mass flow rate of water (kg s^{-1})
$\quad C_p$ = specific heat capacity of water (kJ kg^{-1} K^{-1})
$\quad t_f$ = flow water temperature from the boiler system (°C)
$\quad t_r$ = return water temperature to the boiler system (°C)

The boiler system then delivers hot water to the heat emitters. These then heat the building and their output is given by

$$B(t_m - t_{ai})^n \qquad (7.38)$$

where t_m = the emitter mean water temperature (°C)
$\quad = (t_f + t_r)/2$
$\quad B$ = a constant relating to heat transmission from the emitter and the emitter size
$\quad n$ = 1.3 for radiators
$\quad\quad$ = 1.5 for natural convectors
$\quad\quad$ = 1.0 for fan convectors

Note that equation (7.38) is a form of equation (7.24), for heat transfer involving both radiation and convection. A single panel radiator, for instance, gives out its heat in the form of 50% radiation and 50% convection.

Equations (7.37) and (7.38) can be equated to the building heat loss to form the **heat balance equation** (which can be derived from the conservation of energy):

$$p_1 \dot{m} C_p (t_f - t_r) = p_2 B (t_m - t_{ai})^n$$
$$= \{F_1 \Sigma (UA) + F_2 NV/3\} (t_c - t_{ao}) \qquad (7.39)$$

or

$$p_1 \dot{m} C_p (t_f - t_r) = p_2 B (t_m - t_{ai})^n$$
$$= \frac{1}{F_2} \{F_1 \Sigma (UA) + F_2 NV/3\} (t_{ai} - t_{ao}) \qquad (7.40)$$

where p_1 is the boiler system plant size ratio related to the initial design condition heat loss:

$$\{F_1 \Sigma (UA) + F_2 NV/3\} (t_c - t_{ao})_{des} \qquad (7.41)$$

where $(t_c - t_{ao})_{des}$ is the design condition temperature difference.

The values of t_{ai} in equation (7.40) and t_c in equation (7.39) will differ from the design values when p_1 is non-zero.

Similarly, p_2 = emitter plant size ratio again related to the initial design condition heat loss

$n = 1.5$ for natural convectors
$\quad = 1.3$ for radiators
$\quad = 1.0$ for fan convectors

In the design process one would aim to make $p_1 = p_2$ but the available sizes of boilers and emitters from manufacturers will undoubtedly make p_1 and p_2 slightly different.

Having sized the emitters and the boiler system, the final design task is to size the pipework and the pump to ensure that an adequate flow of hot water gets to all the emitters.

7.8   Intermittent heating

So far, the heat loss from a building has been considered as a steady-state process, i.e. there is no time element and temperatures do not change with time. This is a reasonable assumption for the design process, especially if the heating system were to be run 24 h a day. But this is a rarity now and heating systems are switched off outside the occupancy period to save energy. However, if the heating is switched off for say 8 h a day, the saving in energy is not $8/24 = 33\%$ but much less. This is due to the building fabric storing energy and acting as a storage heater. The building has thermal capacitance, to store heat energy, as well as thermal resistance.

7.8.1   Admittance

When the temperatures change with time, the electrical analogy of heat flow used earlier to describe heat conduction can be used again, but with an alternating voltage source. Also the thermal capacitance of the fabric of the building has to be included as a capacitor or capacitors. So the electrical analogy for the steady-state resistor analogy in Fig. 7.4 becomes an alternating resistor capacitor network for varying temperatures (Fig. 7.7).

The n capacitors represent the fabric capacity, and the conduction resistance, R_c, has been split into $(n + 1)$ resistances, where

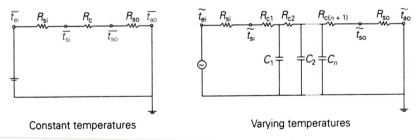

Constant temperatures   Varying temperatures

Fig. 7.7   Constant and varying temperature networks.

$$R_{c1} = R_{c2} = \ldots = R_{c(n+1)} = \frac{R_c}{n+1} \qquad (7.42)$$

for a homogeneous wall. The value of n depends on the accuracy required and the number of elements, if it is a multicomponent wall. For a multi-component wall, the resistors and capacitors will not necessarily be equal. One model of a building, based on a resistor–capacitor network, uses one capacitor for the fabric with reasonable accuracy [19].

The varying temperatures are denoted in the CIBSE Guide as $\tilde{t}$, and the constant temperatures as $\bar{t}$. The period of the alternating temperature (analogous to the alternating voltage) is 24 h. So $\bar{t}$ is the mean temperature over 24 h, and $\tilde{t}$ is the variation about this mean. CIBSE refers to $\tilde{t}$ as the 'cyclic variation' in $\bar{t}$ [13], where it is the amplitude of the temperature variation about the mean [21]:

$$t_{ei}(T) = \bar{t}_{ei} + \tilde{t}_{ei} \qquad (7.43)$$

Although the outside temperature is shown as $\tilde{t}_{ao}$, in fact it is 'earthed' and does not vary from the mean value, $\bar{t}_{ao}$.

As with alternating electrical circuits, the waveform of the temperatures and heat flows is assumed to be a sine wave:

$$Q = \tilde{Q} \sin(\omega T) \qquad (7.44)$$

$$t = \tilde{t} \sin(\omega T - \Phi) \qquad (7.45)$$

where $\tilde{Q}$ = amplitude of the heat flow (W)
$\tilde{t}$ = amplitude of the temperature (°C)
T = time (s)
ω = frequency (rad s^{-1})
   = $2\pi/T_{period}$
   = $2\pi/(24 \times 3600)$ for 24 h
T_{period} = time period of the cycle (s)
   = 24×3600 for 24 h
Φ = temperature phase lag behind the heat flow (rad)

This assumption of a sine wave allows Fourier's equation (7.3) to be solved exactly, although the actual daily variations in reality may not be exact sine waves at all [20].

For an interior source of alternating heat flow, the fabric has an impedance of its 'RC circuit', relating the interior temperature to the interior heat flow, $\tilde{Q}_i$:

$$Z_{ii} = \tilde{t}_{ei}/\tilde{Q}_i \qquad (7.46)$$

and another impedance, Z_{io}, relating the exterior temperature to the inside heat flow:

Table 7.9   Overall resistances and impedances

Element	Overall resistance $(m^2 \ K \ W^{-1})$	Modulus of impedance $\mid Z_{ii} \mid (m^2 \ K \ W^{-1})$
Cast concrete floor with carpet or wood-block finish	2.5	0.3
Cast concrete floor with PVC tiles or bare screed	2.5	0.17
Single glazing	0.18	0.18

$$Z_{io} = \tilde{t}_{ao}/\tilde{Q}_i \qquad (7.47)$$

If the alternating heat source is placed on the exterior side of the wall, with the interior environmental point earthed, then there are two more impedances:

$$Z_{oo} = \tilde{t}_{ao}/\tilde{Q}_{oo} \quad \text{and} \quad Z_{oi} = \tilde{t}_{ei}/\tilde{Q}_{oo}$$

Z_{oo} differs from Z_{ii} because the alternating heat source has R_{so} facing it when it is outside. R_{si} faces an internal heat source. The surface resistance is important in determining the impedance. A carpet on a concrete floor increases the impedance by about twice that of the bare concrete floor [22]. In Table 7.9 the impedance of the glazing is the same as its resistance due to its having negligible thermal capacitance.

As with steady-state resistances of building fabric, it is easier to manipulate the reciprocal impedances, **admittances**. The admittance value in the CIBSE Guide refers to the admittance derived from Z_{ii}, here given the symbol Y_{ii}. Note that Z_{ii} is a complex number, as is electrical impedance:

$$Z_{ii} = R + \frac{1}{j\omega C} \qquad (7.48)$$

$$Y_{ii} = \frac{1}{Z_{ii}}$$

where R = resistance of the wall and surface resistance $(m^2 \ K \ W^{-1})$
$\quad\quad C$ = thermal capacitance of the wall $(J \ m^{-2} \ K^{-1})$
$\quad\quad j = \sqrt{(-1)}$.
Similarly, Y_{ii} is also a complex number, but to simplify matters CIBSE quote values of the modulus of Y_{ii}, which is denoted as Y:

$$Y = \mid Y_{ii} \mid$$

$$= \left| \frac{j\omega C}{1 + j\omega CR} \right| = \left| \frac{j\omega C + \omega^2 C^2 R}{1 + \omega^2 C^2 R^2} \right|$$

$$= \left[\frac{\omega^2 C^2 + \omega^4 C^4 R^2}{(1 + \omega^2 C^2 R^2)^2} \right]^{0.5} = \left[\frac{\omega^2 C^2}{(1 + \omega^2 C^2 R^2)} \right]^{0.5} \qquad (7.49)$$

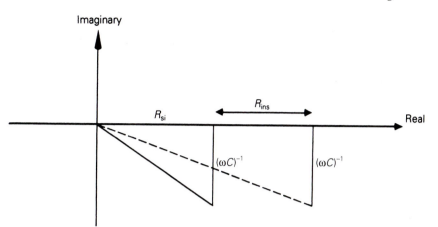

Fig. 7.8 Influence of insulation.

Bassett and Pritchard [11] show that the equation for deriving Z_{ii} is

$$Z_{ii} = \frac{R_{so} \cosh \sqrt{(j\omega RC)} + \sqrt{(R/j\omega C)} \sinh \sqrt{(j\omega RC)}}{\cosh \sqrt{(j\omega RC)} + R_{so} \sqrt{(j\omega C/R)} \sinh \sqrt{(j\omega RC)}} + R_{si} \qquad (7.50)$$

where R = total wall conduction resistance (m² K W⁻¹)

C = thermal capacitance of the wall (J m⁻² K⁻¹)

The sinh and cosh functions can be put in terms of exponential functions, in terms of sin and cos [8], or in terms of a series [11] to enable calculation in a computer program without the facility of imaginary numbers.

Notice the potential influence of the inside surface resistance, R_{si}. If insulation is put on the inside of the wall, the insulation resistance, R_{ins}, will be added to R_{si} (as would be the resistance of the carpet on the floor in Table 7.9). Both the carpet and insulation have high resistances but very small thermal capacities, so they are considered as resistances alone. This will increase the overall resistance of the fabric element, increasing its impedance (decreasing its admittance) and increasing its time constant (Fig. 7.8). Z_{ii} is the equivalent impedance of the resistor and capacitor of the fabric element. The time constant of the fabric element, τ, is RC (Chapter 4); this means the element will be slower at soaking up heat and will have a lower admittance and capacity to store heat.

Section A3 of the CIBSE Guide [8, 24] uses matrix notation to relate internal temperatures and heat flows to their external equivalents. For a homogeneous fabric element the matrix equation (omitting the ~ over the temperatures and heat flows) is

$$\begin{bmatrix} t_i \\ q_i \end{bmatrix} = \begin{bmatrix} M_1 & M_2 \\ M_3 & M_4 \end{bmatrix} \begin{bmatrix} t_o \\ q_o \end{bmatrix} = \begin{bmatrix} \{(M_1 \times t_o) + (M_2 \times q_o)\} \\ \{(M_3 \times t_o) + (M_4 \times q_o)\} \end{bmatrix} \qquad (7.51)$$

where t_i = internal temperature
t_o = outside temperature
q_i = internal heat flux (W m^{-2})
q_o = external heat flux (W m^{-2})
M = a matrix element

Included in the matrix elements are the internal and external resistances; the external resistances are for an external wall. Internal walls contribute to the heat storage of the building, so they are included in the total building admittance, unlike the steady-state ΣU value, which is primarily for external fabric:

$$\begin{bmatrix} M_1 & M_2 \\ M_3 & M_1 \end{bmatrix} = \begin{bmatrix} 1 & R_{so} \\ 0 & 1 \end{bmatrix}\begin{bmatrix} m_1 & m_2 \\ m_3 & m_1 \end{bmatrix}\begin{bmatrix} 1 & R_{si} \\ 0 & 1 \end{bmatrix}$$

where $m_1 = \cosh(p + jp)$

$$m_2 = \frac{1\sinh(p + jp)}{k(p + jp)}$$

$$m_3 = \frac{k(p + jp)\sinh(p + jp)}{1}$$

$$p = \left[\frac{\omega l^2 \rho C_p}{2k}\right]^{0.5}$$

and where l = thickness of the element (m)
ρ = density of element (kg m^{-3})
ω = frequency (rad s^{-1})
k = conductivity (W m^{-1} K^{-1})

From this the admittance is given by

$$Y_{ii} = \frac{M_1}{M_2} \tag{7.52}$$

This can be derived from the matrix in equation (7.51) with $t_o = 0$:

$$t_i = M_1 t_0 + M_2 q_0$$

$$= M_2 q_0$$

$$q_i = M_3 t_0 + M_1 q_0$$

$$= M_1 q_0$$

from which

$$q_i = \frac{M_1}{M_2} t_i$$

$$= Y_{ii} t_i$$

Example 7.3

Calculate the admittance for a 220 mm brick wall with the following data:

Density	$\rho = 1700 \text{ kg m}^{-3}$
Conductivity	$k = 0.84 \text{ W m}^{-1} \text{ K}^{-1}$
Specific heat capacity	$C_p = 800 \text{ J kg}^{-1} \text{ K}^{-1}$
Inside surface resistance	$R_{si} = 0.12 \text{ m}^2 \text{ K W}^{-1}$
Outside surface resistance	$R_{so} = 0.06 \text{ m}^2 \text{ K W}^{-1}$

Solution

This example comes from the CIBSE Guide [8, 24]. We will solve it by the matrix method but first we will use equation (7.50):

$$Z_{ii} = \frac{R_{so} \cosh \sqrt{(j\omega RC)} + \sqrt{(R/j\omega C)} \sinh \sqrt{(j\omega RC)}}{\cosh \sqrt{(j\omega RC)} + R_{so} \sqrt{(j\omega CR)} \sinh \sqrt{(j\omega RC)}} + R_{si} \qquad (7.50)$$

where R is the total wall conduction resistance (m² K W⁻¹) and C is the thermal capacitance of the wall (J m⁻² K⁻¹).

Putting $j\omega C = S$, the susceptance, we have

$$S = 21.8j$$

and

$$R = \frac{0.22}{0.84} = 0.26 \text{ m}^2 \text{ K W}^{-1}$$

Using the expansions

$$\cosh \sqrt{(RS)} = 1 + \frac{RS}{2} + \frac{(RS)^2}{24} + \frac{(RS)^3}{720}$$

$$\sqrt{(R/S)} \sinh \sqrt{(RS)} = R \left\{ 1 + \frac{RS}{6} + \frac{(RS)^2}{120} + \frac{(RS)^3}{5040} \right\}$$

$$\sqrt{(j\omega C/R)} \sinh \sqrt{(j\omega RS)} = S \left\{ 1 + \frac{RS}{6} + \frac{(RS)^2}{120} + \frac{(RS)^3}{5040} \right\}$$

Hence

$$Z_{ii} = \frac{0.06 \,(-0.34 + 2.59j) + (0.19 + 0.24j)}{(-0.34 + 2.59j) + 0.06 \,(-19.8 + 16j)}$$

$$= 0.077 - 0.08j$$

Adding on the internal resistance, the total impedance is

$$Z_{ii} + R_{si} = 0.197 - 0.08j$$

Inverting this yields

$$Y_{ii} = \frac{0.197 + 0.08j}{(0.197 - 0.08j)\,(0.197 + 0.08j)}$$

$$= 4.34 + 1.76j$$

The CIBSE admittance is

$$|Y_{ii}| = \sqrt{(4.34^2 + 1.76^2)}$$

$$= 4.68 \text{ W m}^{-2}\text{ K}^{-1}$$

In comparison, the CIBSE matrix method gives

$$p = 1.688$$

and using the expressions

$$\cosh(p + jp) = \tfrac{1}{2}\{(e^p + e^{-p})\cos p + j(e^p - e^{-p})\sin p\}$$

$$\sinh(p + jp) = \tfrac{1}{2}\{(e^p + e^{-p})\cos p + j(e^p - e^{-p})\sin p\}$$

which gives

$$\begin{bmatrix} M_1 & M_2 \\ M_3 & M_1 \end{bmatrix} = \begin{bmatrix} 1 & 0.12 \\ 0 & 1 \end{bmatrix} \begin{bmatrix} m_1 & m_2 \\ m_3 & m_1 \end{bmatrix} \begin{bmatrix} 1 & 0.06 \\ 0 & 1 \end{bmatrix}$$

where $m_1 = -0.33 + 2.59j$
$m_2 = 0.19 + 0.24j$
$m_3 = -0.19 + 15.9j$
so the overall matrix is

$$\begin{bmatrix} M_1 & M_2 \\ M_3 & M_1 \end{bmatrix} = \begin{bmatrix} (-2.72 + 4.5j) & (-0.013 + 0.821j) \\ (-19.9 + 15.9j) & (-1.52 + 3.54j) \end{bmatrix}$$

from which

$$Y_{ii} = \frac{(-1.52 + 3.54j)}{(-0.013 + 0.821j)}$$

$$= 4.35 + 1.78j$$

which is very similar to the non-matrix method.

Thankfully there are tables of admittances given in the CIBSE Guide [8, 24], and these tedious calculations rarely have to be done to derive the values.

The admittance calculated here relates to the swing in internal temperature due to internal swings in the internal heat flux, probably due to the heating or gains. The influence of outside swings in temperature and heat flux is related to Y_{oi} or, in CIBSE terms, a **decrement factor** and **phase lag** [8, 24], relating the amplitude and phase of the internal sine wave to the external sine wave producing it.

7.8.2   CIBSE response factor

Chapter 5 considered the need for control and quoted the design conditions for a heavyweight building and a lightweight building. But 'heavyweight' and 'lightweight' were not defined in detail. They refer to the thermal capacity of the building. CIBSE [13, 22] uses the **response factor**, f_r, to define the thermal weight of a building:

$$f_r = \frac{\Sigma\,(YA) + NV/3}{\Sigma\,(UA) + NV/3} \tag{7.53}$$

where $\Sigma\,(YA)$ = sum of the internal and external element products of admittances and areas (W K^{-1})

$\Sigma\,(UA)$ = sum of the external element products of U-values and areas (W K^{-1})

N = air change rate, number of air changes per hour

V = volume of building (m^3)

The denominator approximates to the steady-state heat loss as in equation (7.33):

$$Q = \{F_1\,\Sigma\,(UA) + F_2 NV/3\}\,(t_c - t_{ao}) \tag{7.33}$$

but without F_1 and F_2.

Correspondingly, the numerator of equation (7.53) gives the swing in the heat gain about the mean value:

$$\tilde{Q} = \left(\Sigma\,(YA) + \tfrac{1}{3}NV\right)\tilde{t}_{ei}$$

The greater the thermal capacity, the greater the numerator of equation (7.53) and the larger the value of f_r. CIBSE defines [23] the weight of a building as in Table 7.10. From the response factor, the rate of energy supply for an intermittently heated building [22] is

$$Q = \frac{\{\Sigma\,(UA) + NV/3\}\,(t_{im} - t_{om})}{\eta_H} \tag{7.54}$$

Table 7.10   Thermal weight in terms of the response factor

Nominal building classification	Response factor
Heavyweight	$f_r \geq 6$
Lightweight	$f_r \leq 4$

where Q = rate of energy supply (W)
   t_{im} = mean inside temperature over heating season (°C)
   t_{om} = mean outside temperature over heating season (°C)
   η_H = heating system efficiency
and

$$(t_{im} - t_{om}) = \frac{Hf_r(t_{im} - t_{om})}{Hf_r + (24 - H)} \tag{7.55}$$

where H is the average daily heating period, including the preheating period (h), and t_r is the required internal temperature (°C).

As the building weight increases, f_r increases and equation (7.55) shows that $(t_{im} - t_{om})$ increases. From equation (7.54) the rate of energy supply must also increase. Hence lightweight buildings should be more efficient than heavyweight buildings. An alternative determination of energy supply is given in Chapter 14 and the influence of thermal weight is considered in Chapter 9.

The CIBSE response factor should not be confused with the ASHRAE response factors, which are the transfer functions (Z-transfer functions), of building elements [9].

7.8.3   Sizing an intermittent heating system

For intermittent heating, CIBSE determines that the heating system size should be $F_3 Q$ where

$$F_3 = 1.2 \frac{(24 - n)(f_r - 1)}{24 + n(f_r - 1)} + 1$$

$$Q = \{F_1 \Sigma (UA) + F_2 NV/3\} (t_c - t_{ao}) \tag{7.33}$$

and n is the daily plant operation time (h). For an 8 h plant operation period this yields $F_3 = 1.8$, which seems rather large. An alternative expression is used in Section A5 of the CIBSE Guide [13], but it is rather complex. Really this is a difficult problem for the admittance method as it is dealing with heat input from a heating system, which is more like a step function than a diurnal sine wave. A simpler and more appropriate method, considering the system as a first-order system, is used in Chapter 9 to examine heating system size.

References

[1]   CIBSE (1988, 1999) CIBSE Guide, Volume A, Section A1, Environmental criteria for design, Chartered Institution of Building Services Engineers, London.

[2] Humphreys, M. A. (1978) Outdoor temperatures and comfort indoors. *Building Research and Practice*, March–April.

[3] Fanger, P. O. (1972) *Thermal Comfort*, McGraw-Hill, New York.

[4] Humphreys, M. A. (1976) Field studies of thermal comfort. *Building Services Engineer*, **44**, April.

[5] CIBSE (1988, 1999) CIBSE Guide, Volume A, Section A9, Estimation of plant capacity, Chartered Institution of Building Services Engineers, London.

[6] Holman, J. P. (1990) *Heat Transfer*, McGraw-Hill, New York.

[7] Carslaw, H. S. and Jaeger, J. C. (1959) *Conduction of heat in solids*, Oxford University Press, Oxford.

[8] CIBSE (1988, 1999) CIBSE Guide, Volume A, Section A3, Thermal properties of building structures, Chartered Institution of Building Services Engineers, London.

[9] ASHRAE (1997) ASHRAE Handbook Fundamentals, American Society of Heating, Refrigerating and Air Conditioning Engineers, Atlanta.

[10] CIBSE (1988) CIBSE Guide, Volume C, Reference Data Section C3, Heat transfer, Chartered Institution of Building Services Engineers, London.

[11] Bassett, C. R. and Pritchard, M. D. W. (1968) *Environmental Physics Heating*, Longman, Harlow.

[12] Shaw, E. W. (1970) *Heating and Hot Water Services*, Crosby Lockwood, London.

[13] CIBSE (1988, 1999) CIBSE Guide, Volume A, Section A5, Thermal response, Chartered Institution of Building Services Engineers, London.

[14] CIBSE (1988, 1999) CIBSE Guide, Volume A, Section A9, Estimation of plant capacity, Chartered Institution of Building Services Engineers, London.

[15] Liddament, M. W. (1986) *Air Infiltration Calculation Techniques – an Applications Guide*, Air Infiltration and Ventilation Centre, Warwick.

[16] CIBSE (1986, 1999) CIBSE Guide, Volume A, Section A4, Air infiltration, Chartered Institution of Building Services Engineers, London.

[17] CIBSE (1986, 1999) CIBSE Guide, Volume A, Section A7, Internal heat gains, Chartered Institution of Building Services Engineers, London.

[18] O'Callaghan, P. W. (1978) *Building for Energy Conservation*, Pergamon, Oxford.

[19] Crabb, J. A., Murdoch, N. and Penman, J. M. (1987) *A Simplified Thermal Response Model*, Building Services Engineering Research and Technology 8, Chartered Institution of Building Services Engineers, London.

[20] Levermore, G. J. (1989) Which program? *Chartered Institution of Building Services Engineers Journal* (London), **11**(3).

[21] Fisk, D. J. (1981) *Thermal Control of Buildings*, Applied Science Publishers, London.

[22] CIBSE (1999) CIBSE Building Energy Code 1, Energy demands and targets for heated and ventilated buildings, Chartered Institution of Building Services Engineers, London.

[23] CIBSE (1988, 1999) CIBSE Guide, Volume A, Section A8, Summer-time temperatures, Chartered Institution of Building Services Engineers, London.

[24] CIBSE (1999) CIBSE Guide A, Environmental design, Chartered Institution of Building Services Engineers, London.

8
Compensation

Although PID loop control has been discussed in Chapter 4, a modified version is used in most BEMSs to control the heating. This modified PID control is a **compensator loop controller**. It enables the heating system output to be varied in relation to the outside temperature. A whole building's heating, or its zones, can be controlled by compensator loops, without the need for a myriad of room thermostats, wiring and valves. If windows are opened, the heating output is not increased and wasted. Hence compensation control is widely used, but there are complications and this chapter will explore them.

8.1 Fundamentals

Before we consider in detail the compensator loop controller, it is worth considering a simple heating system consisting of a boiler pump and emitter (Fig. 8.1). The boiler and pump are switched on and off by a BEMS outstation receiving signals from a room temperature sensor. This would be the type of system for a domestic installation. To determine the average steady-state temperatures in the system, use the balance equation from Chapter 7:

$$p_1 \dot{m} C_p(t_f - t_r) = p_2 B(t_m - t_{ai})^n$$

$$= \frac{1}{F_2} \{F_1 \Sigma (UA) + F_2 NV/3\} (t_{ai} - t_{ao}) \qquad (7.40)$$

Assuming $p_1 = 1.0$, $p_2 = 1.0$ and

$$A = \frac{1}{F_2} \{F_1 \Sigma (UA') + F_2 NV/3\}$$

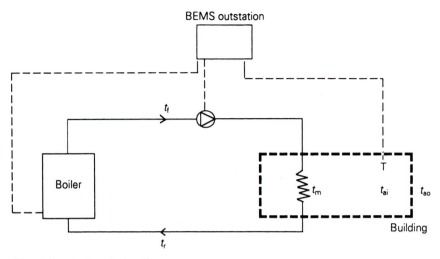

Fig. 8.1 A simple heating system.

then where A' = fabric area (m²) and

$$\dot{m}C_p(t_f - t_r) = B(t_m - t_{ai})^n = A(t_{ai} - t_{ao})$$

Putting in the design conditions, simplified for easy calculation:

$$t_f = 80°C, \quad t_r = 70°C, \quad t_{ai} = 20°C$$

$$t_{ao} = 0°C, \quad n = 1.3 \text{ for a radiator system}$$

and the building heat loss of 100 kW gives

$$\dot{m}C_p = 10 \text{ kW K}^{-1}$$

$$A = 5 \text{ kW K}^{-1}$$

$$B = 0.546$$

It is interesting to consider how the system temperatures change as the heat load changes. Consider when $t_{ao} = 10°C$. Assuming that the heating system can be controlled to maintain the room at 20°C, the heat load has now changed to 50 kW:

$$5(20 - 10) = 50 \text{ kW}$$

The boiler output would need to match this, so

$$10(80 - 75) = 50 \text{ kW}$$

However, the radiator is now producing

$$0.546(77.5 - 20)^{1.3} = 106 \text{ kW}$$

The dynamics of the heating system have not been considered here, but obviously t_m, the mean water temperature in the radiator, is too high and has to be reduced to get the correct 50 kW for the emitter. Consequently, t_f has to be reduced. The required value of t_m can be found by solving the heat balance equation

$$0.546(t_m - 20)^{1.3} = 5(20 - 10)$$

giving

$$t_m = 52.3°C$$

and so

$$t_f = 54.8°C \quad \text{and} \quad t_r = 49.8°C$$

These values from the balance equation are the average, steady-state values. If the boiler cycles then the values are the average values during the cycle periods.

So far, the control of the heating system has not been considered. A simple control would be to have the outstation controlling the pump only; this is the control in most domestic heating systems. When the pump turns off, the boiler will heat the water in itself and the surrounding pipework, until it is switched off by its own thermostat at 80°C. So t_f will remain near to 80°C, especially during the pump off time. In order to obtain $t_m = 52°C$, the return water temperature, t_r would need a value well below 52°C, near to 24°C if t_f was around 80°C. This would mean large amplitude cycling, and with the delays and dead times in the system, it would lead to poor control. If t_m could not be kept near to 52°C, then the room temperature, t_{ai}, would rise. This is a strong possibility due to the cycling of the system and the potential power of the heating (Fig. 8.2).

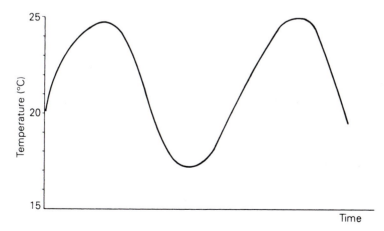

Fig. 8.2   The effect of simple heating control.

The load will affect the overshoot (Chapter 4). If the boiler is controlled by the BEMS, as well as the pump, there is a better chance of keeping the boiler flow temperature down.

8.2   The compensator circuit

The best way of achieving a lower temperature in the flow water is by using a three-port mixing valve (Fig. 8.3). Here the three-port valve reduces t_f, the flow water temperature, to the appropriate level according to the heating load, by reducing the water flow from the boiler while increasing the flow of return water to be mixed.

The value of t_f is set in proportion to the outside temperature, t_{ao}, sensed by the temperature sensor sited on an external, north-facing wall.

A typical schedule for a BEMS compensator loop is shown in Fig. 8.4. The equation for this schedule is

$$t_f = 80 - 2t_{ao} \tag{8.1}$$

The signal from the t_{ao} sensor is used to determine t_f, which is monitored by a water temperature sensor just after the valve. The valve is controlled by the signal from the t_f sensor. Hence the t_f sensor and the valve are known as the submaster control, receiving its control setpoint from the master t_{ao} sensor. North Americans call this a reset controller. Control of the valve is by a PI control loop whose setpoint is determined, or reset, by the t_{ao} signal.

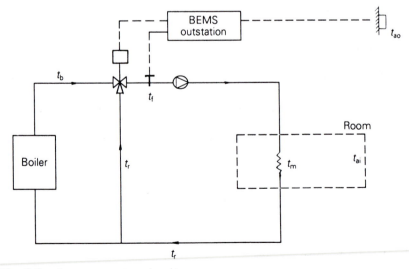

Fig. 8.3   A compensator circuit.

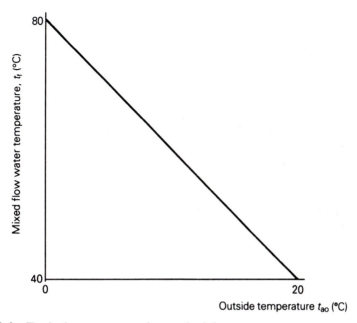

Fig. 8.4 Typical compensator loop schedule.

Compensator control enables the heating system to be adjusted in relation to the coldness of the weather. It obviates the need for many thermostats and valves to be installed, at considerable expense in medium and large buildings.

This is compensator control, and most BEMS outstations contain compensator loops to perform it.

8.3 BEMS compensator loop

The configuration for a compensator is shown in Fig. 8.5. Sensor S1, a type-1 sensor or flow water sensor, sends back a signal of t_f to loop 1, input P. Sensor S2, a type-3 sensor, sends a signal of t_{ao} to the function module, F1, input G. This function module calculates the schedule of t_f against t_{ao}, from the values set for E, H and F. The output from F1 is given by the function equation

$$D = (E \times G) + (F \times H)$$

For a straight-line schedule with the equation

$$t_f = -2t_{ao} + 80$$

the function module is configured to give

$$D = (-2 \times G) + (1 \times 80)$$

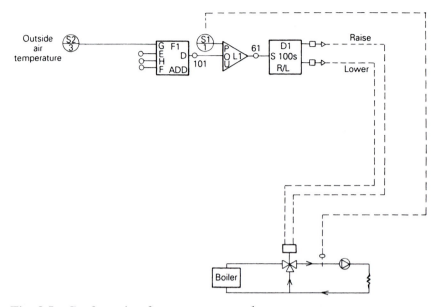

Fig. 8.5   Configuration for a compensator loop.

where D = output to L1
G = input from S2 (t_{ao})
F = 80
H = 1

Output D from the function module, F1, is then the input, via an analogue node, 101, to the PID loop, L1. This input from F1 is the occupation setpoint, O, for loop L1 and is the required value for t_f. Loop L1 then sends a control signal, via analogue node 61, to the raise/lower driver, D1. In turn, D1 sends a digital signal to a raise relay or to a lower relay, which moves the valve actuator or motor to open or close the valve.

This motor may well be a split-phase motor which moves in one direction when receiving an in-phase signal from the raise relay, and moves in the other direction when receiving an antiphase signal. This will move the valve either forward for more recirculation or back for less recirculation. The '100 s' inside the raise/lower box is the time to drive the valve from fully open to fully shut.

8.4   Alternative compensator loop control

Compensation can be carried out without the need for a three-port mixing valve, by operating straight on to the boiler control. For a single boiler it can either switch off the boiler at a flow water temperature related to the outside

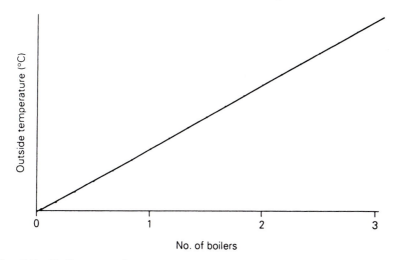

Fig. 8.6   Boiler control.

temperature, or it can control the boiler by a time proportional signal, again related to t_{ao}. Neither of these methods is as satisfactory as using a three-port valve, because boiler cycling can interfere with the control.

For multiple boilers, compensator control can be used to sequence the number of boilers required. The control signal for this is simply a proportional control (Fig. 8.6).

8.5   Inside air temperature variation

The compensator loop commonly used in most BEMSs simply regulates the heating system in relation to the outside temperature, t_{ao}. There is no feedback of room temperature, t_{ai}. Hence t_{ai} can vary slightly, according to the compensator schedule used.

Example 8.1
Consider a heating system, with radiators, and the following design conditions:

$$t_{ao} = 0°C, \; t_{ai} = 20°C, \text{ design heat loss} = 100 \text{ kW}$$

$$t_f = 80°C, \; t_r = 70°C$$

$$t_f = 80 - 2t_{ao} \quad \text{(compensator loop schedule)}$$

What is t_{ai} when $t_{ao} = 10°C$?

Solution
From the design conditions, the constants in the heat balance equation (7.40) are

$$\dot{m}C_p = 10, \quad B = 0.546, \quad A = 5$$

So at $t_{ao} = 10°C$ and $t_f = 60°C$,

$$10(60 - t_r) = 0.546(t_m - t_{ai})^{1.3} = 5(t_{ai} - 10)$$

Putting this in terms of t_m and t_{ai} only,

$$20(60 - t_m) = 0.546(t_m - t_{ai})^{1.3} = 5(t_{ai} - 10)$$

From the boiler output and the heat loss parts of this equation, t_{ai} may be found in terms of t_m:

$$20(60 - t_m) = 5(t_{ai} - 10)$$

and we get

$$250 - 4t_m = t_{ai} \tag{8.2}$$

Substituting into the balance equation:

$$0.546(t_m - 250 + 4t_m)^{1.3} = 5(250 - 4t_m - 10)$$
$$0.546(5t_m - 250)^{1.3} = 5(240 - 4t_m) \tag{8.3}$$

There are two ways of solving this. The first is by guessing and adjusting until both sides are equal. There are limits to guide such guesses; however, the left-hand side, the emitter output, cannot be negative, so

$$5t_m - 250 \geqslant 0$$
$$t_m \geqslant 50°C$$

Similarly, on the right-hand side we have

$$240 - 4t_m \geqslant 0$$
$$t_m \leqslant 60°C$$

So a suitable guess would be $t_m = 55°C$:

left-hand side = 35.9 kW   right-hand side = 100 kW

Therefore t_m needs to be increased. Eventually t_m will be determined.

A quicker but more complicated way is by Newton's method of approximation. Using this method the roots of a function of x, $f(x) = 0$, are found where $f(x) = 0$ is a generalized function of x. In our case

$$f(x) \equiv f(t_m) = 0.546(5t_m - 250)^{1.3} - 5(240 - 4t_m)$$

This method is readily applicable to a computer program. The principle of the method can be explained geometrically (Fig. 8.7).

Newton's method starts by making an educated guess as to the root of the function, t_{m0}. The true, unknown root is shown as t_{mr}. A tangent at t_{m0} cuts

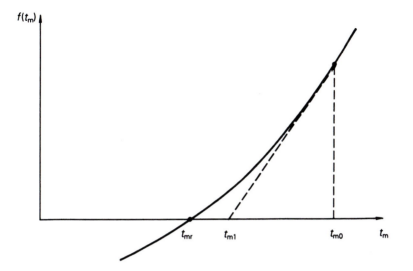

Fig. 8.7   Newton's method for finding the root of $f(t_m) = 0$.

the t_m axis nearer to t_{mr}, at t_{m1}. Repeating the process gets closer to the true root, t_{mr}. From the construction of the tangent it can be seen that

$$\frac{df(t_{m0})}{dt_m} = \frac{f(t_{m0})}{t_{m0} - t_m}$$

Rearranging gives

$$t_{m1} = t_{m0} - \frac{f(t_{m0})}{f'(t_{m0})}$$

where $f'(t_{m0}) = \dfrac{df(t_{m0})}{dt_m}$

Instead of differentiating $f(t_m)$, a difference approximation can be made, as shown in (Fig. 8.8):

$$f'(t_{m0}) = \frac{f(t_{m0} + k) - f(t_{m0})}{k}$$

where k is a small increment in t_m; this yields

$$t_{m1} = t_{m0} - \frac{kf(t_{m0})}{f(t_{m0} + k) - f(t_{m0})}$$

Let us take the same value of t_m as in our guess method, $t_{m0} = 55°C$, with $k = 1°C$. This yields

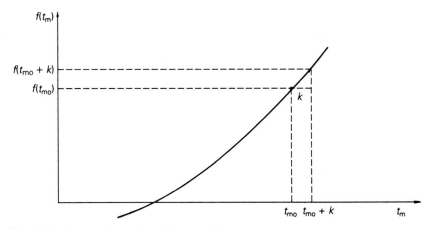

Fig. 8.8   Newton's method by small increment.

$$t_{m1} = 55 - \frac{f(t_{m0})}{\{0.546(5 \times 56 - 250)^{1.3} - 5(240 - 4 \times 56)\} - f(t_{m0})}$$

where $f(t_{m0}) = 1 \times \{0.546[5 \times 55 - 250]^{1.3} - 5(240 - 4 \times 55)\}$. Further calculation gives

$$t_{m1} = 55 - \frac{\{0.546[65.66] - 100\}}{\{0.546(83.23) - 80\} - \{0.546[65.66] - 100\}}$$

which works out to

$$t_{m1} = 57.168°C$$

This can be checked by putting 57.2°C into equation (8.3):

left-hand side = 57.26 kW   right-hand side = 56.64 kW

These results are close together, but if greater accuracy is required then t_{m1} could be used in Newton's method as the guessed value to produce t_{m2}. And t_{ai} can be found by substituting t_{m1} into equation (8.4):

$$t_{ai} = 250 - 4t_m$$

$$= 21.3°C$$

The room temperature has risen above its design value by 1.3°C.

With straight-line schedules such as in Example 8.1, compensator loops invariably allow the room temperature to vary slightly as the outside temperature varies. If the schedule is dropped to say $t_f = 80 - 3t_{ao}$, to stop the room temperature rising, it then falls below the design temperature of 20°C (Fig. 8.9).

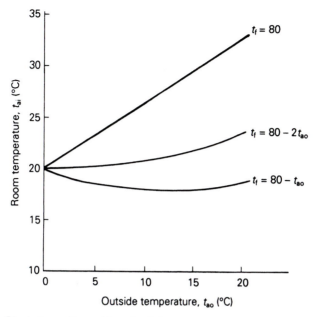

Fig. 8.9 Variation of t_{ai} with schedule.

8.6 Variation of heating system output

It is interesting to note that the heating system under compensator loop control will be putting out heat until the room temperature, t_{ai}, equals the emitter mean water temperature, t_m. In other words, until the heat emitter output is zero, then

$$B(t_m - t_{ai})^{1.3} = 0$$

The other terms in the balance equation, the boiler output and the building heat loss, will also have to be zero for there to be no heat output. This zero output occurs when

$$t_f = t_r = t_m = t_{ao} = t_{ai}$$

So for the schedule $t_f = 80 - 2t_{ao}$, the zero output temperature is when

$$t_f = 80 - 2t_f$$

$$t_f = 26.67°C$$

Changing the schedule to $t_f = 80 - 3t_{ao}$ produces a zero output temperature of

$$t_f = 20°C$$

The heating system output can be derived from the heat balance equation. If the heating system output is defined in terms of the heat load factor, Φ, where

$$\Phi = \frac{\text{heat output}}{\text{maximum heat output}}$$

then for the boiler output we have

$$\Phi = \frac{t_f - t_m}{t_{f\,des} - t_{m\,des}}$$

$$\Phi = \frac{0.5\,\delta t_b}{0.5\,\delta t_{b\,des}} = \frac{\delta t_b}{\delta t_{b\,des}} \qquad (8.4)$$

where $t_{f\,des}$ = design mixed flow water temperature (°C)
$t_{m\,des}$ = design return water temperature (°C)
$\delta t_b = t_f - t_r$
$\delta t_{b\,des} = t_{f\,des} - t_{r\,des}$

Parameter t_m is used in the upper equation to maintain consistency with the emitter output term in the balance equation. Similarly, for the heat emitter and the building heat loss the outputs are respectively

$$\Phi^{1/1.3} = \Phi^{0.769} = \frac{\delta t_e}{\delta t_{e\,des}} \qquad (8.5)$$

and

$$\Phi = \frac{\delta t_a}{\delta t_{a\,des}} \qquad (8.6)$$

where $\delta t_e = t_m - t_{ai}$
$\delta t_{e\,des} = t_{m\,des} - t_{ai\,des}$
$\delta t_a = t_{ai} - t_{ao}$
$\delta t_{a\,des} = t_{ai\,des} - t_{ao\,des}$

Adding together equations (8.4), (8.5) and (8.6) yields

$$\delta t_a + \delta t_e + 0.5\,\delta t_b = \Phi\,\delta t_{a\,des} + \Phi^{0.769}\delta t_{e\,des} + 0.5\Phi\,\delta t_{b\,des}$$

$$t_f - t_{ao} = \Phi\,\delta t_{a\,des} + \Phi^{0.769}\delta t_{e\,des} + 0.5\Phi\,\delta t_{b\,des} \qquad (8.7)$$

Substituting the compensator loop schedule equation into this gives the relationship between Φ and t_{ao}. Using the schedule and the design values in Example 8.1,

$$80 - 3t_{ao} = 25\Phi - 55\Phi^{0.769} \qquad (8.8)$$

A graph of this equation, which has a shallow curve, is shown in Fig. 8.10.

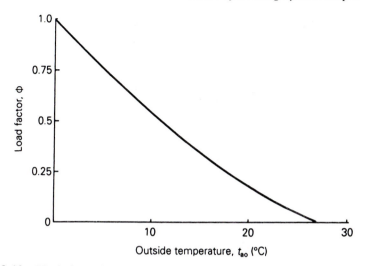

Fig. 8.10 Variation of load factor with t_{ao}.

Equation (8.8) can be used to check the solution to Example 8.1. There $t_{ao} = 10°C$ and the load was 57.26 kW (from the left-hand side of the equation), or 56.64 kW (from the right-hand side of the equation). Taking 56.64 kW gives

$$\Phi = 0.566$$

Putting this value into equation (8.8) gives

$$t_{ao} = 10.1°C$$

close to the given value.

To ensure that the BEMS, with its compensator loop, does not allow the heating system to continue heating in mild weather, a function module can be used to stop the heating beyond certain values of t_{ao}. With a schedule of $t_f = 80 - 2t_{ao}$, the zero output temperature is $t_{ao} = 26.67°C$, as we have seen. Unless the heating is manually switched off, there will be heating during the summer. The addition of a **comparator function module** and a **gate function module** (Fig. 8.11) closes the three-port valve to the boiler when $t_{ao} \geqslant 15°C$.

The comparator's function is given by

$$D = 1 \quad \text{when } E < F$$

$$D = 0 \quad \text{when } E > F$$

So with F set at 15°C then $D = 0$ when E, the t_{ao} value, is above 15°C. The gate module's function is

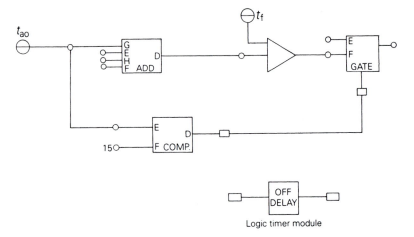

Fig. 8.11   How to stop compensated heating in mild weather.

$$D = F \quad \text{when } B = 1$$

$$D = E \quad \text{when } B = 0$$

The value of E can be set to 0, so the signal to the driver closes the three-port valve to the boiler as if there was no error signal from the mixed flow temperature sensor and loop.

To ensure there is no rapid switching of the valve when t_{ao} hovers around 15°C, a **logic timer module** (Fig. 8.11) can be inserted between the comparator and the gate. This timer module introduces a delay (which can be up to 9 h) of say 15 min before the 0, or off, signal is transmitted to the gate. If at the end of the 15 min delay t_{ao} has dropped below 15°C, the output signal to the gate remains 1, or on.

If this heating shutdown were required from an older stand-alone compensator controller, then a separate external thermostat would have to be wired up to switch off the heating. With the BEMS it is simply a reconfiguration, using the existing outside temperature sensor, so no wiring and separate thermostat are required.

8.7   The ideal schedule

The ideal schedule to ensure that t_{ai} stays constant as the outside temperature varies is given by solving the balance equation (7.40), for constant t_{ai}:

$$2\dot{m}C_p(t_f - t_m) = B(t_m - t_{ai})^n = A(t_{ai} - t_{ao})$$

where n is the index for emitter output (for radiators $n = 1.3$). This equation leads to the schedule

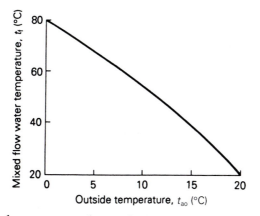

Fig. 8.12 Ideal compensator loop schedule.

$$t_f = \frac{A}{2\dot{m}C_p}(t_{ai} - t_{ao}) + t_m$$

and t_m can be equated to terms in t_{ao} and t_{ai} from the balance equation:

$$t_m = [(A/B)(t_{ai} - t_{ao})]^{1/n} + t_{ai}$$

hence

$$t_f = \frac{A}{2\dot{m}C_p}(t_{ai} - t_{ao}) + \left[\frac{A}{B}(t_{ai} - t_{ao})\right]^{1/n} + t_{ai} \qquad (8.9)$$

This schedule is shown in Fig. 8.12; notice how it is curved, especially at the higher values of t_{ao}.

This schedule requires knowledge of the constants in the balance equation: $\dot{m}C_p$, B, A and t_{ai}. The constants may not be readily available, and t_{ai} needs to be measured, entailing the use of an inside sensor. But a number of compensator loops are curved to reflect the non-linear output of heat emitters.

When occupants feel cold in a building, they often touch the heat emitters to determine whether the heating is working. Unless the radiator is very hot, they sometimes incorrectly conclude that the heating system is not working. This could apply in spring and autumn periods when a BEMS compensator loop will maintain the emitter mean temperature at a reduced level due to its schedule. If the schedule reduces the mean temperature too steeply with respect to t_{ao}, there is a greater risk of the occupants feeling warm radiators rather than hot radiators and complaining of the heating not working. And with steep schedules, the room temperature will be reduced, as Fig. 8.9 shows. This is why many compensator loop schedules have a value of t_f around 40°C when t_{ao} is near 20°C.

8.8   Plant size implications

We have considered the heat balance equation with the various elements equally sized, i.e. $p_1 = 1.0$ and $p_2 = 1.0$. But the balance equation can also be used to examine the variations in t_{ai} when there are different plant sizes.

Example 8.2
Consider a radiator heating system controlled by a BEMS compensator loop with a three-port mixing valve. The design conditions are the same as in Example 8.2. How does t_{ai} vary when $t_{ao} = 10°C$. Consider two cases: (a) the boiler is 100% oversized and (b) the emitters are 10% oversized. The schedule is now $t_f = 80 - 3t_{ao}$, whereas it was $t_f = 80 - 2t_{ao}$ in Example 8.1.

Solution
Before answering the main question it will help in later comparison to determine what t_{ai} would be if everything were the same size. The compensator schedule makes $t_f = 50°C$ and calculation reveals that

$$t_m = 47.85°C \quad \text{and} \quad t_{ai} = 18.6°C$$

As a check, the emitter output is

$$0.546(47.85 - 18.6)^{1.3} = 43.5 \text{ kW}$$

and the heat loss is

$$5(18.6 - 10) = 43 \text{ kW}$$

Returning to part (a), the boiler is 100% oversized, so $p_1 = 2.0$ and $p_2 = 1.0$. The oversizing assumes that the water flow rate has been increased to allow a 10 K drop across the boiler on full load. The balance equation becomes

$$2 \times 2\dot{m}C_p(t_f - t_m) = B(t_m - t_{ai})^{1.3} = A(t_{ai} - t_{ao})$$

so

$$40(t_f - t_m) = 0.546(t_m - t_{ai})^{1.3} = 5(t_{ai} - t_{ao})$$
$$40(50 - t_m) = 0.546(t_m - t_{ai})^{1.3} = 5(t_{ai} - 10)$$

Following through Newton's method with $t_{mo} = 47$, and $k = 2$ yields

$$t_m = 48.9°C \quad \text{and} \quad t_{ai} = 19.0°C$$

This is above t_{ai} for $p_1 = 1.0$, where it was shown earlier that $t_{ai} = 18.6°C$ (lower than the design value of 20°C).
(b) Now the emitters are 10% oversized, hence $p_2 = 1.1$ and $p_1 = 1.0$. The balance equation becomes

$$20(t_f - t_m) = 1.1 \times 0.546(t_m - t_{ai})^{1.3} = 5(t_{ai} - t_{ao})$$

so

$$20(50 - t_m) = 0.6552(t_m - t_{ai})^{1.3} = 5(t_{ai} - 10)$$

Analysis yields

$$t_m = 47.68°C \quad \text{and} \quad t_{ai} = 19.3°C$$

Comparing this with (a) and Example 8.1, it can be seen that a 10% increase in emitter size produces a greater increase in room temperature than 100% increase in boiler size! The emitter size has a considerable influence on the room temperature, more so than the boiler size. An increase in room temperature would result in a corresponding increase in energy consumption, A δt_{ai}, where δt_{ai} is the increase in room temperature.

These influences of boiler and emitter size can be examined in more detail by partial differentiation of equation (8.9), where t_f is kept constant and the plant size ratios are included:

$$t_f = \frac{A}{2\dot{m}p_1 C_p}(t_{ai} - t_{ao}) + \left[\frac{A}{p_2 B}(t_{ai} - t_{ao})\right]^{1/n} + t_{ai} \tag{8.9}$$

Now consider Example 8.2, with $A = 5$, $\dot{m}C_p = 10$, $B = 0.546$, $n = 1.3$ for radiators, but to see the influence of boiler size, leave p_1 as a variable:

$$t_f = \frac{(t_{ai} - t_{ao})}{4p_1} + \left[\frac{9.16}{p_2}(t_{ai} - t_{ao})\right]^{0.769} + t_{ai}$$

For assessing boiler size, $p_2 = 1.0$ and differentiating (partially) with constant t_f:

$$0 = -\frac{(t_{ai} - t_{ao})}{4p_1^2}\partial p_1 + \frac{\partial t_{ai}}{4p_1} + 0.769 \times 9.16\, \partial t_{ai}\,[9.16(t_{ai} - t_{ao})]^{-0.231} + \partial t_{ai}$$

$$\frac{(t_{ai} - t_{ao})}{4p_1^2}\partial p_1 = \partial t_{ai}\left[\frac{1}{4p_1} + 7.05\,[9.16(t_{ai} - t_{ao})]^{-0.231} + 1\right] \tag{8.10}$$

For Example 8.2 with t_{ai} varying around 18.6°C and t_{ao} at 10°C:

$$8.6\frac{\partial p_1}{4p_1^2} = \partial t_{ai}\left[\frac{1}{4p_1} + 7.05\,[79.1]^{-0.231} + 1\right]$$

$$= \partial t_{ai}\left[\frac{1}{4p_1} + 3.57\right]$$

For a boiler 100% oversized (somewhat stretching the calculus), $p_1 = 1.0$, $\partial p_1 = 1.0$, then

$$\partial t_{ai} = 0.57°C$$

So the new temperature with the oversized boiler will be 19.2°C, reasonably close to the answer of 19°C in Example 8.2. But this is not much of an increase, considering the increase in boiler size. Equation (8.10) can be used to show that, over the various values of t_{ao}, the increase in temperature will be small and that the increase in power, $A \partial t_{ai}$, will also be small [1].

For the emitter size, the same exercise can be conducted with p_1 set to unity:

$$t_f = \frac{(t_{ai} - t_{ao})}{4} + \left[\frac{9.16}{p_2} (t_{ai} - t_{ao}) \right]^{0.769} + t_{ai}$$

Rearranging this before differentiation:

$$p_2 \{t_f - 1.25t_{ai} + 0.25t_{ao}\}^{1.3} = 9.16(t_{ai} - t_{ao})$$

Differentiating this yields

$$1.3p_2\{\}^{0.3} + \partial p_2\{\}^{1.3} = 9.16 \, \partial t_{ai}$$

where $\{\} = \{t_f - 1.25t_{ai} + 0.25t_{ao}\}$

For Example 8.2 with t_{ai} varying around 18.6°C and t_{ao} at 10°C, and with the emitters 10% oversized, i.e. $\partial p_2 = 0.1$ and $p_2 = 1.0$, we have

$$\partial t_{ai} = 0.6°C$$

making $t_{ai} = 19.2°C$, close to the value of 19.3°C obtained in Example 8.2. This is a much larger increase than with the oversized boiler, even though the emitters are only 10% oversized.

8.9   Internal gains

So far we have not considered the heat gains in the building due to occupants, equipment and solar radiation through windows. If these gains are sufficient to raise the temperature of the room by t_g, there is a heat gain term, At_g, to subtract from the heat loss from the building term in the heat balance equation. So the heat balance equation becomes

$$\dot{m}C_p(t_f - t_r) = B(t_m - t_{ai})^n = A(t_{ai} - t_{ao}) - At_g$$

The best way to examine this is using an example.

Example 8.3
Consider a BEMS compensator loop with a schedule

$$t_f = 80 - 2t_{ao}$$

The heating system has radiators and there are heat gains giving $t_g = 2°C$. What is the room temperature when $t_{ao} = 10°C$?

Table 8.1 Utilization of internal heat gains

Emitter	Index, p	Heat gain utilized (%)
Radiator	1.3	30
Natural convector	1.5	32
Fan convector	1.0	25

Solution

This is similar to Example 8.1 but now there are heat gains. In Example 8.1 $t_{ai} = 21.3°C$ and $t_m = 57.2°C$. Here t_g will raise the room temperature, which in turn reduces the radiator heat output. The balance equation is therefore

$$2 \times 10(60 - t_m) = 0.546(t_m - t_{ai})^{1.3} = 5(t_{ai} - 12)$$

hence

$$4(60 - t_m) = t_{ai} - 12$$

or

$$t_{ai} = 252 - 4t_m$$

Substituting into the emitter part of the balance equation:

$$0.546(t_m - 252 + 4t_m)^{1.3} = 5(252 - 4t_m - 12)$$

or

$$0.546(t_m - 252)^{1.3} = 5(240 - 4t_m)$$

Solving this by Newton's method:

$$t_m = 57.3°C \quad \text{so} \quad t_{ai} = 22.8°C$$

Without any heat gains, i.e. $t_g = 0°C$, then t_{ai} would be 21.3°C, as in Example 8.1. It is interesting that t_{ai} has not simply risen by $t_g = 2°C$ to 23.3°C. The radiator system under BEMS compensator loop control has actually used some of the gains. This is due to the inherent feedback of the radiator, reacting to the gains and raising its return water temperature, t_r, as a consequence.

It has been shown [1] that by differentiating equation (8.9), modified to account for t_g, in a similar way to analysing the influence of boiler size and emitter size, that a radiator system will use 30% of the heat gains. If $t_g = 2°C$ the room temperature rise will be 1.4°C, close to the numerical results obtained above. Table 8.1 shows the utilization of gains for fan convector emitters, natural convector emitters and radiators.

8.10 Inside temperature and cascade control

Although between 25% and 32% of the heat gains are utilized by the emitters, the remaining 75% and 68% of the heat gains are still wasted, besides which the compensator loop still allows heating in mild weather. To overcome these deficiencies, the loop can be controlled on internal temperature, t_{ai}, rather than external temperature, t_{ao}. Example 8.4 illustrates such an application.

Example 8.4
The control of a radiator heating system with a BEMS compensator loop is changed from outdoor control to indoor control, using an internal sensor. The outdoor schedule was $t_f = 80 - 2t_{ao}$. The indoor schedule maintains t_f at 80°C when t_{ai} is 19°C, and when t_{ai} is 21°C it keeps t_f at 40°C. What is t_{ai} when t_{ao} is 10°C? Assume the same design conditions as in previous examples.

Solution
The equation for the new schedule is

$$t_f = 460 - 20t_{ai}$$

and the balance equation is

$$20(t_f - t_m) = 0.546(t_m - t_{ai})^{1.3} = 5(t_{ai} - 10)$$

Substituting t_f in the schedule into the balance equation:

$$20(460 - 20t_{ai} - t_m) = 0.546(t_m - t_{ai})^{1.3} = 5(t_{ai} - 10)$$

Equating the boiler output and building heat loss part of this equation:

$$462.5 - 20.25\, t_{ai} = t_m$$

Substituting for t_m in the emitter part of the balance equation and equating it to the heat loss:

$$0.546(462.5 - 21.25\, t_{ai})^{1.3} = 5(t_{ai} - 10)$$

Solving this numerically, by Newton's method:

$$t_{ai} = 20.24°C$$

The loop controlled from outside would have allowed the temperature, t_{ai}, to rise to 21.3°C, as in Example 8.1. So the internally controlled compensator loop can maintain a closer control of room temperature.

The closer control can also be seen from the lower zero output temperature for inside control, derived from the schedule

$$t_f = 460 - 20\, t_{ai}$$

The zero output temperature occurs when $t_f = t_{ai}$:

$$t_{ai} = 460 - 20\, t_{ai}$$

$$t_{ai} = 21.9°C$$

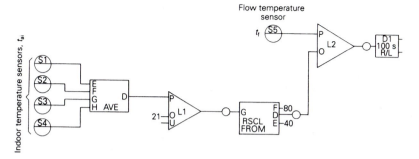

Fig. 8.13 Configuration for indoor compensator loop control.

With the loop controlled from outside, the zero output temperature was $t_{ai} = 26.7°C$.

Significant savings have been reported from changing BEMS compensator loop control from outdoor to indoor [2]. However, a suitable representative position must be found for the inside sensor. This is not easy; the average temperature from a number of sensors placed around the building gives better results but it costs more. If a window is opened near one of the sensors, the heating is increased to compensate for the loss!

The configuration for indoor control could be similar to that for outdoor control, namely using an adding function element or module. But if the inside temperature, t_{ai}, is well below its setpoint, perhaps on a cold day, or if someone opens a window, then ridiculously high values of t_f will be set. For instance, with the above indoor schedule suppose $t_{ai} = 15°C$, then

$$t_f = 460 - 20 \times 15$$

$$= 160°C$$

A wider indoor schedule could be used, but a better way is to feed the average inside temperature into a PID loop whose output becomes the setpoint for the t_f PID loop, as used in the outdoor compensator loop. Hence for the indoor temperature control there are two PID loops, one cascaded into the next; this is **cascade control** (Fig. 8.13).

The AVE module averages the four inside sensor signals. The RSCL FROM module rescales the 0% to 100% output from loop 1 to give the indoor schedule (40°C to 80°C) as the setpoint for loop 2. For the above indoor schedule, the rescale values are 80 and 40. Notice that when t_{ai} is low, say 15°C, then with the setpoint of loop 1 at 21°C and a gain of 50 (proportional band = 2°C), the error is 6°C. But the loop cannot put out a signal greater than 100%, corresponding to the three-port valve being fully open to the boiler.

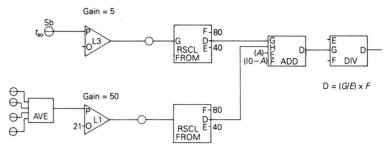

Fig. 8.14 Configuration for internal and external compensator loop control.

A compromise is to use an inside sensor and an outside sensor and combine their signals. The schedule for this, using the above inside and outside schedules, is

$$t_f = A(80 - 2t_{ao}) + (1 - A)\{460 - 20t_{ai}\}$$

where A is the authority of the sensors, $A \leq 1$.

When $A = 0.5$ both sensors have equal authority and contribute equally to determining the setpoint for t_f. There should be limits on t_{ai}, and to a lesser extent on t_{ao}. So in the configuration for combined inside and outside control (Fig. 8.14), a PID loop (L3) has been used for the outside schedule.

Loop 3 has a gain of 5 (proportional band = 20°C), and the RSCL FROM module provides the upper and lower limits of t_f on the schedule. For instance, the upper limit is 80°C when there is a 100% signal from L3 which corresponds to an error of 20°C, i.e. $t_{ao} = 0°C$.

The ADD module combines the indoor and outdoor schedules, and here the authority can be applied:

$$D = [A \times G] + [(10 - A) \times H]$$

Here A has to be an integer, and $A \leq 10$. The DIV module brings the signal back to a value between 80 and 40 for the setpoint of loop 2, controlling t_f. The DIV module equation is

$$D = (G/E) \times F$$

$$= (G/10) \times 1$$

The user still has to decide the authority to use for this combined control. Potentially better control means more decisions for the operator.

8.11 Three-port valve

An essential element of compensator control is the three-port valve. This regulates the water recirculated from the return pipe, mixing it with hot water from the boiler at temperature t_b (Fig. 8.15).

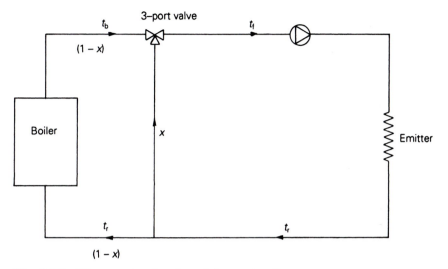

Fig. 8.15 Three-port valve for mixing water flows.

A fraction x of return water is recirculated to be mixed at the valve with a fraction $(1 - x)$ of boiler water. If differences in the specific heat capacity of water, C_p, between t_b and t_r are ignored, and they are very small, then at the valve a heat balance, similar to equation (7.40), gives the relationship between x and the average temperatures:

$$xmC_p(t_r - t_0) + (1 - x)mC_p(t_b - t_0) = mC_p(t_r - t_0)$$

$$xt_r + (1 - x)t_b = t_f$$

where t_0 is a reference temperature, say 0°C.
   Rearranging for x gives

$$x = \frac{t_b - t_f}{t_b - t_r}$$

Consider the heating system on half-load, as in Example 8.1. There $t_f = 60°C$ and $t_r = 54.4°C$, so

$$x = \frac{80 - 60}{80 - 54.4}$$

$$= 0.78$$

Some 78% of the return water is recirculated, and only 22% comes from the boiler. This shows that the valve action is not proportional to the load but should have a non-linear characteristic. This is due to the non-linear heat

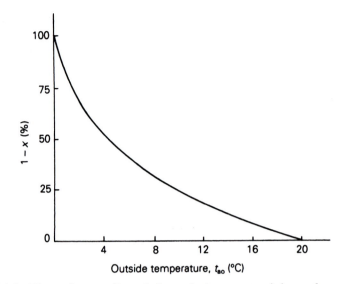

Fig. 8.16   Flow of water $(1 - x)$ through three-port mixing valve.

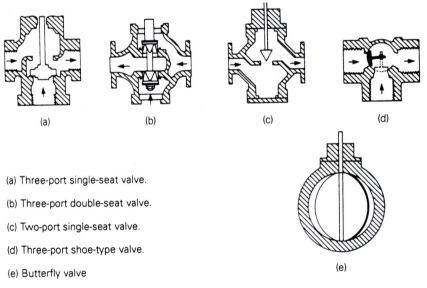

(a) Three-port single-seat valve.

(b) Three-port double-seat valve.

(c) Two-port single-seat valve.

(d) Three-port shoe-type valve.

(e) Butterfly valve

Fig. 8.17   Different types of valve.

emitter. The ideal valve characteristic is shown in Fig. 8.16. Some common two- and three-port valves are shown in Fig. 8.17.

To appreciate three-port valves it is easier to begin with a two-port valve. For a two-port valve to control a non-linear heat emitter, an **equal percentage**,

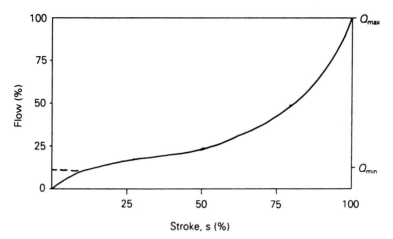

Fig. 8.18   Equal percentage valve characteristic.

or **logarithmic**, flow characteristic is required under constant pressure conditions (Fig. 8.18).

A two-port valve produces an equal percentage increase in opening area for equal increments of valve movement, or **stroke**, s. Mathematically this is expressed as

$$s = K \ln\left[\frac{Q}{Q_0}\right]$$

where s = valve stem position, or stroke, as a percentage of the maximum (100% at valve fully open)

Q = flow as a percentage of maximum flow, Q_{max}, when the valve is fully open (m^3 s^{-1})

Q_0 = flow of the valve at minimum controllable flow, as a percentage of maximum flow, Q_{max}

K = constant

Manufacturers sometimes use this relationship:

$$n = \ln\left[\frac{Q_{max}}{Q_{min}}\right]$$

where Q_{max} = maximum flow at valve full open (m^3 s^{-1})

Q_{min} = minimum controllable flow (m^3 s^{-1})

n = index, typically 4

This is the logarithmic value of another valve parameter, its **rangeability**, defined as the ratio Q_{max}/Q_{min}. For most valves this is 50.

For a three-port mixing valve, as used in compensator control, the port controlling the flow of water from the boiler circuit has a shape similar to the logarithmic two-port valve. The port controlling the recirculation water on a three-port mixing valve has a linear characteristic [6, 7].

The most useful parameter of a valve is the **capacity index**, C_v in imperial units, K_v in continental units and A_v in SI units. The SI units are the flow rate through the valve ($m^3 s^{-1}$) when a pressure of 1 Pa is dropped across the valve.

Manufacturers usually quote K_v values, which also approximately equal C_v values. The K_v value is the flow rate ($m^3 h^{-1}$) when a pressure drop of 10^2 kPa (1 bar) exists across the valve. The equation relating K_v to the flow and pressure is

$$K_v = \frac{Q_h}{\Delta P^{0.5}}$$

where ΔP is the pressure drop (Pa or kPa), and Q_h is the volume flow rate ($m^3 h^{-1}$).

D'Arcy's equation [3–5] uses a resistance to relate the volume flow rate through a pipe to the pressure drop across it:

$$\Delta P = RQ^2$$

where Q is volume flow rate ($m^3 s^{-1}$).

Using D'Arcy's equation, it can be seen that the resistance is related to the capacity index:

$$K_v = \frac{3600}{\sqrt{R}}$$

The factor of 3600 is to convert Q_h, measured per hour, to D'Arcy's Q, measured per second.

The authority, N, is

$$N = \frac{\Delta P_v}{\Delta P_v + \Delta P_s}$$

where ΔP_s is the pressure drop across the whole system, apart from the valve (Pa), and ΔP_v is the pressure drop across the valve, or relevant port of a three-port valve (Pa); see Fig. 8.19.

In terms of resistance, the authority is

$$N = \frac{R_v}{R_v + R_s}$$

where R_s is the resistance of the whole system, apart from the valve (Pa $s^{0.5}$ $m^{-1.5}$) and R_v is the resistance across the valve or the relevant port of a three-port valve (Pa $s^{0.5}$ $m^{1.5}$).

It is recommended that N is not less than 0.3 for a mixing valve to provide good control [10], which means the resistance across the valve is $0.3/0.7 = 43\%$ of the system it controls.

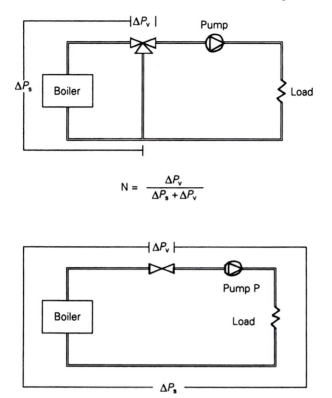

$$N = \frac{\Delta P_v}{\Delta P_s + \Delta P_v}$$

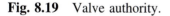

$$N = \frac{\Delta P_v}{\Delta P_s + \Delta P_v}$$

$$= \frac{\Delta P_v}{P}$$

Fig. 8.19   Valve authority.

Example 8.5
For an equal percentage valve, it is stated that there is a 4% reduction in the initial flow for a 1% movement of the stroke, or movement of the valve stem. What is the mathematical equation and its constants if the index is $n = 4$?

Solution
If the index is $n = 4$, then

$$4 = \ln\left[\frac{100}{Q_{min}}\right]$$

where Q_{max} has been put in as a percentage, giving

$$Q_{min} = 1.83\%$$

so the valve has a rangeability of 54.6.

From the valve being fully open, the stroke is reduced by 1%:

$$99 = K \ln\left[\frac{96}{1.83}\right]$$

$$= 3.96\ K$$

$$K = 25$$

Further details on valves can be found in the literature [6–10]. Suffice it to say that good valve selection is required for efficient compensator loop control, no matter how good is the BEMS.

References

[1]   Levermore, G. J. (1986) A simple model of a heating system with a compensator control. *BSERT*, **7**(2).

[2]   ETSU (1986) *Monitoring of a Microprocessor-based Energy Management System. A Demonstration Project at Various Tenanted Office Buildings*, ETSU Final Report ED/105/127 1986, Harwell.

[3]   Hansen, E. G. (1985) *Hydronic System Design and Operation: A Guide to Heating and Cooling with Water*, McGraw-Hill, New York.

[4]   CIBSE (1988) CIBSE Guide, Volume C, Section C4, Flow of fluids in pipes and ducts, Chartered Institution of Building Services Engineers, London.

[5]   ASHRAE (1997) ASHRAE Handbook Fundamentals, American Society of Heating, Refrigerating and Air Conditioning Engineers, Atlanta, 1989.

[6]   Wolsey, W. H. (1975) *Basic Principles of Automatic Controls with Special Reference to Heating and Air Conditioning Systems*, Hutchinson Educational, London.

[7]   Wolsey, W. H. (1971) A theory of 3-way valves with some practical conclusions. *Journal of the Institute of Heating and Ventilating Engineers*, **39**, May.

[8]   CIBSE (1986) CIBSE Guide, Volume B, Section B11, Automatic controls, Chartered Institution of Building Services Engineers, London.

[9]   CIBSE (1986) CIBSE Guide, Volume B, Section B16, Miscellaneous equipment, Chartered Institution of Building Services Engineers, London.

[10] CIBSE (1985) CIBSE Applications Manual, Automatic controls, and their implications for systems design. Chartered Institution of Building Services Engineers, London.

9
Optimizer control

A standard control loop on almost all BEMSs is the **optimizer loop**. This switches on the heating plant at the latest possible time to get the building warm by the start of occupancy. It can also be used with air-conditioning systems with both heating and cooling. In terms of the notation in Fig. 9.1, the optimizer determines the shortest **preheat time**, $T_3 - T_2$, to get the occupied space up to the design temperature, t_d. Temperature t_p is the inside temperature at which the optimizer switches on the heating and starts the preheat, which occurs at the time denoted by T_2. The outside temperature is t_{ao}.

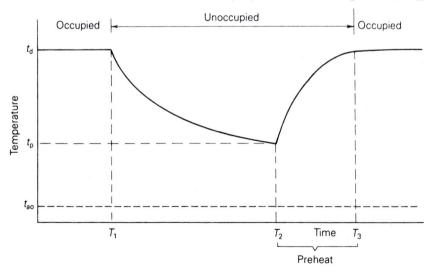

Fig. 9.1 Inside temperature during intermittent heating.

9.1   Optimizer development

In the United Kingdom, initial work on optimizers was carried out by the then Department of Environment's Property Services Agency (PSA), using early electromechanical optimizers based on time switches. When they were installed, considerable savings were reported [1]. However, the savings were assessed by comparison with a compensator control with a **night setback setting**. Night setback is the reduction of the occupancy temperature by a few degrees, with the heating ticking over to maintain it. It is not comparable to the modern practice of intermittent heating, where the system is switched off overnight. So the considerable savings found by the PSA would be lower when compared with intermittent heating.

The PSA work on electromechanical optimizers initiated the development of optimizers into microprocessor-based controllers. The modern BEMSs now incorporate these optimizers into their own outstations, using similar algorithms for their optimizer loops.

It is quite revealing to examine these electromechanical optimizers to appreciate optimum start control, even though they are now dated. One type of electromechanical optimizer has an electrical resistor placed inside a room thermostat. During the night a voltage would be applied to the resistor, which would heat the thermostat and raise its temperature by an amount Δt:

$$\Delta t = \frac{V^2}{R(UA)_{t/s}} \qquad (9.1)$$

where V = voltage applied (V)
　　　　　R = resistance of resistor (Ω)
　　　$(UA)_{t/s}$ = heat loss from the thermostat (W K^{-1})

With the additional heat, the thermostat has an apparent temperature, t_{app}, of

$$t_{app} = t_{ai} + \Delta t \qquad (9.2)$$

Suppose the thermostat were to be set at 20°C and δt were 10°C, then t_{app} would be 10°C above room air temperature, t_{ai}. The thermostat would close when t_{app} dropped below the setpoint, t_{set}, fulfilling the condition

$$t_{app} \leq t_{set} \qquad (9.3)$$

In the present case this would be

$$t_{app} \leq 20 \quad \text{or} \quad t_{ai} \leq 10$$

This could be interpreted as the setpoint of the thermostat, t_{set}, being reduced to an effective setpoint, $t_{eff\ set}$:

$$t_{eff\ set} = t_{set} - \Delta t$$

and the heating would come on when

$$t_{ai} \leq t_{eff\ set} \qquad (9.4)$$

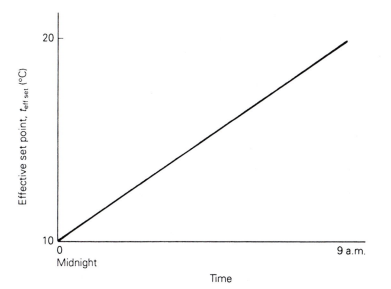

Fig. 9.2 Change in effective setpoint due to the optimizer.

In our example this would give

$$t_{\text{eff set}} = 20 - 10$$

$$= 10°C$$

so from equation (9.4) the optimizer would therefore close and bring on the heating if t_{ai} fell below 10°C.

For optimum start control, a time switch would be set to engage a little potentiometer (a variable resistor) at a set time. The time switch would then turn the potentiometer and reduce the current to the room thermostat's resistor. This would happen over a number of hours before the start of occupancy. By doing this the optimizer would be reducing Δt, hence raising the effective setpoint of the thermostat over a period of time. As Fig. 9.2 shows, the lower the room temperature, t_{ai}, the lower $t_{\text{eff set}}$ would have to be to bring on the heating, so the earlier the heating would be switched on. Figure 9.2 relates to the data above, with the heating coming on under optimum start control at the earliest possible time of midnight. The start of occupancy is taken as 9 am.

9.2 Self-adapting optimizers

For the electromechanical optimizer in Section 9.1 the preheat time varies linearly with $t_{\text{eff set}}$ (Fig. 9.2). This means that the preheat time also varies with the room temperature at the start of preheat, t_{p}, which leads to the preheat curve of Fig. 9.1 being approximated to the straight line in Fig. 9.3.

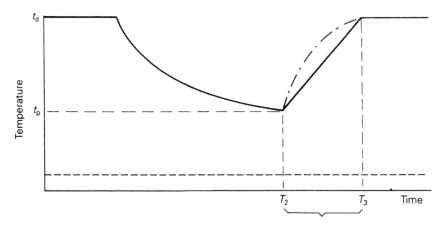

Fig. 9.3   Preheat response approximated to a straight line.

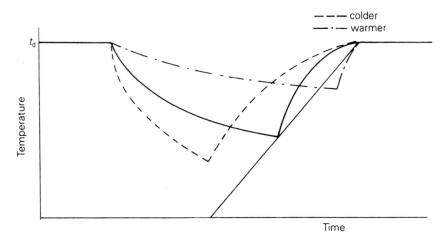

Fig. 9.4   Variation in preheat curves with weather.

Unfortunately, as the cooling of the building changes with the weather, so does the preheat curve (Fig. 9.4). A new straight line is needed for each preheat curve, otherwise the optimizer will be very inaccurate.

For the electromechanical optimizer in Section 9.1, the preheat line cannot easily be changed, as the potentiometer is turned by the time switch for a fixed period of time. Only the setpoint of the room thermostat, t_{set}, can be changed, which simply moves the line up or down (Fig. 9.5).

Flexible movement of the line was achieved by the later optimizers. They were based on microprocessors and employed **self-adaptation** to tune the line to the response of the building and heating system. Self-adaptation is quite straightforward, provided the building and heating system response

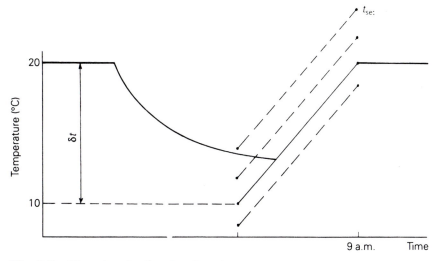

Fig. 9.5 Changing the line by changing t_{set}.

during preheat is still regarded as linear, like Fig. 9.3, rather than the actual situation in Fig. 9.1.

If the slope of the preheat line, S, is given by

$$S = \frac{t_d - t_p}{T_3 - T_2} \tag{9.5}$$

then a self-adapting algorithm is

$$S_{i+1} = wS_e + (1 - w)S_i \tag{9.6}$$

where S_i = slope used for previous day, day i (K h^{-1})
S_{i+1} = adapted slope to be used for current day's preheat, day $i + 1$ (K h^{-1})
S_e = slope that would have produced an exactly correct preheat time for the previous day
w = weighting, or forgetting, factor ($0 \leqslant w \leqslant 1.0$)

S_e can be determined from the overshoot or undershoot of the temperature at the occupancy time, T_3 (Fig. 9.6):

$$S_e = \frac{t_d + \Delta t_{over/under} - t_p}{T_3 - T_2} \tag{9.7}$$

Initially, when the optimizer has just been installed, the slope has to be guessed. Gradually equation (9.6) will correct it to a closer approximation of the true slope. The larger the value of w, the quicker this process will be. When $w = 1.0$ then $S_{i+1} = S_e$.

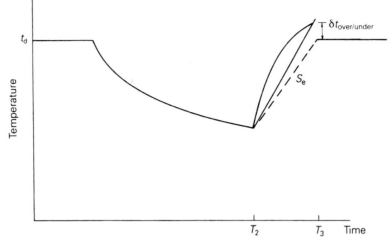

Fig. 9.6   Determining the best slope, S_e.

But if there is a brief disturbance to the heating system (e.g. one boiler in a multiple-boiler installation failing to fire, the slope will be drastically altered. If the boiler failed to fire, i.e. the burner went to **lockout**, because its burner control unit sensed a malfunction, the burner will most probably have its manual reset button pressed once its locked out burner is discovered, and it will again become operational. So the slope will have to change again dramatically. It would have been better if w had been set to a lower value, say less than 0.5.

Invariably the advice given by BEMS manufacturers is to start with w high and then to reduce it once the building response has been approximated. Some adaptation will always be needed with a straight-line preheat algorithm as the true preheat line is curved.

9.3   Non-linear algorithms

Birtles and John [2] examined a building's preheating process and derived an empirical curve of preheat time $(T_3 - T_2)$ against preheat temperature difference $(t_p - t_d)$, as shown in Fig. 9.7. The equation for this curve is

$$\ln(T_3 - T_2) = A_{BJ} (t_p - t_d) + B_{BJ} \qquad (9.8)$$

where A_{BJ} is a constant associated with the thermal weight of the building, and B_{BJ} is a constant associated with the time between switching on the heating and the interior starting to heat up.

Only the inside temperature is considered in equation (9.8) in determining the preheat time. This accords with other findings that inside air temperature

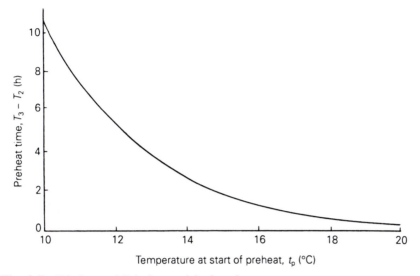

Fig. 9.7   Birtles and John's empirical preheat curve.

correlates well with the preheat time, whereas there is little correlation with outside air temperature [9, 3].

However, Birtles and John [2] later modified their algorithm to include outside air temperature as the initial algorithm was not as accurate when the outside air temperature was low and the plant size was on the small side. Resulting from this, equation (9.8) is modified to

$$\ln(T_3 - T_2) = A_{BJ}\,(t_p - t_d) + B_{BJ} + C_{BJ}\,t_{ao} \tag{9.9}$$

where C_{BJ} is a constant, less than A, so that the inside temperature is still the dominant term at most moderate temperatures.

Equation (9.9) is the basis of an optimizer algorithm called BRESTART, and its self-adaptation is by recursive least squares. The least-squares method is discussed in Chapter 14 and its use for parameter identification is dealt with in the literature [4].

9.4   An optimizer model

Equation (9.9) has been found from observation, i.e. it is empirical. But it is useful to understand its basis, and some insight can be obtained by applying simple theory.

Analytical derivations of preheat times for walls have been developed from Dufton's equation [5–7]. As quoted in reference [6], Dufton's equation is

$$\Theta_{si} - \Theta_{so} = pQ\left[\frac{5T}{4k\rho C_p}\right]^{0.5} \tag{9.10}$$

where T = duration of heating, or preheat time (s)
   Θ_{si} = inside surface temperature (°C)
   Θ_{so} = outside surface temperature (°C)
   p = plant size ratio (plant size to steady-state heat loss)
   Q = steady-state heat loss (kW)
   k = thermal conductivity of fabric (W m^{-1} K^{-1})
   ρ = density (kg m^{-3})
   C_p = specific heat capacity (kJ kg^{-1} K^{-1})

From this equation it can be seen that the preheat time, T, is reduced as the heating plant size is increased. McLaughlin, McLean and Bonthron [6] discuss an empirical way of determining the average value of ρC_p for a building, so that the preheat time can be estimated.

A more informative approach is to consider the heating system from the heat balance equation discussed in previous chapters. Here, though, there are additional terms for storage of heat in the fabric of the building and the inside air. These are the two terms added to the heat loss term:

$$p_1 \dot{m} C_p (t_f - t_r) = p_2 B (t_m - t_{ai})^n$$

$$= \frac{H_{des}}{F_2} (t_{ai} - t_{ao}) + (mC_p)_{fab} \frac{dt_{fab}}{dT} + (mC_p)_{air} \frac{dt_{ai}}{dT} \qquad (9.11)$$

where $(mC_p)_{fab}$ = thermal capacity of the building's fabric, both internal
   and external (kJ K^{-1})
   t_{fab} = mean temperature of the building's fabric, both internal and external (°C)
   $(mC_p)_{air}$ = thermal capacity of the air inside the building (kJ K^{-1})
   t_{ai} = temperature of the air inside the building (°C)
   H_{des} = $\{F_1 \Sigma (UA) + F_2 NV/3\}$ (kW K^{-1})
$H_{des}(t_c - t_{ao})_{des}$ = design heat loss = Q_{des} (kW)
   $(t_c - t_{ao})_{des}$ = design temperature difference (K)

The other terms have been defined in earlier chapters. This equation is an oversimplification of a complex situation, involving radiative, convective and conductive heat transfer processes. For a more detailed exposition on dynamic thermal models the reader is referred to Clarke's book [8] and Fisk's book [9]. However, Crabb, Murdoch and Penman [10] with their Excalibur model, which uses only two time constants – a short time constant for the air and lightweight fabric, and a longer time constant for the main fabric elements – have achieved satisfactory results. Excalibur is similar to equation (9.11), except that Excalibur's steady-state heat loss term more correctly relates to the fabric temperature rather than the overall steady-state heat loss. So equation (9.11) is used and later simplified, to demonstrate the underlying principles of the building dynamics, rather than as a rigorous model.

Before much analysis can be carried out on equation (9.11), and indeed before it can be used for a simple optimizer algorithm, some simplifying assumptions need to be made. The first is that the boilers are at maximum output during the preheat period and this dictates the heating system output. This is a reasonable assumption, except when p_1 is much larger than p_2 and the emitters will limit the heat output. Such a case will be considered later.

So the heat output from the heating system is

$$pQ_{des}$$

where $$p = p_1$$
$$= \text{boiler system plant size ratio}$$
$$p_1 \dot{m} C_p(t_f - t_r) = p_1 Q_{des}$$
$$Q_{des} = H_{des}(t_c - t_{ao})_{des}$$
$$= \text{design heat loss (kW)}$$

Equation (9.11) becomes

$$pQ_{des} = \frac{H_{des}}{F_2}(t_{ai} - t_{ao}) + (mC_p)_{fab}\frac{dt_{fab}}{dT} + (mC_p)_{air}\frac{dt_{ai}}{dT} \qquad (9.12)$$

The second assumption is that the mean temperature of the fabric, t_{fab}, can be related linearly to the inside air temperature, t_{ai}, and the outside air temperature, t_{ao}. This unfortunately eliminates the time delay for the heat to diffuse through the fabric, but Fig. 9.8 shows that this is not a gross distortion,

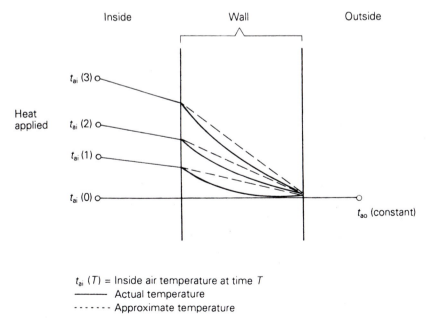

$t_{ai}(T)$ = Inside air temperature at time T
——— Actual temperature
······ Approximate temperature

Fig. 9.8 Temperature development in the wall.

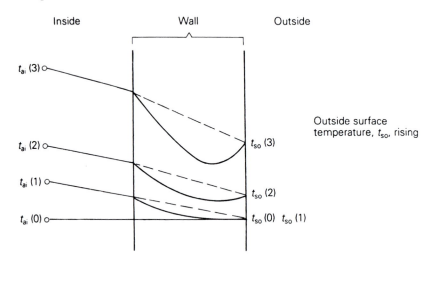

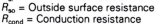

Fig. 9.9   Temperature development when the outside surface heats up.

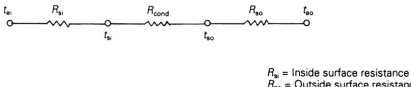

R_{si} = Inside surface resistance
R_{so} = Outside surface resistance
R_{cond} = Conduction resistance

Fig. 9.10   Steady-state resistor network for the wall.

although if the outside surface of the structure receives an appreciable amount of heat from the sun, as in summer, then the assumption leads to distortion (Fig. 9.9).

Using the steady-state resistor network of a solid wall (Fig. 9.10) we have

$$t_{fab} = \frac{t_{si} + t_{so}}{2} \tag{9.13}$$

From the resistor network it can be shown that

$$t_{si} + t_{so} = (t_{ei} - t_{ao})\{1 - (R_{si} - R_{so})U\} + 2t_{ao} \tag{9.14}$$

Relating t_{ei} to t_{ai} by the temperature ratios given in the CIBSE Guide [12, 17]:

$$F_1 = \frac{t_{ei} - t_{ao}}{t_c - t_{ao}}$$

and

$$F_2 = \frac{t_{ai} - t_{ao}}{t_c - t_{ao}}$$

So equation (9.14) becomes

$$t_{si} + t_{so} = \frac{F_1}{F_2}(t_{ai} - t_{ao})\{1 - (R_{si} - R_{so})U\} + 2t_{ao} \tag{9.15}$$

and the mean fabric temperature, t_{fab}, is

$$t_{fab} = \frac{F_1}{2F_2}(t_{ai} - t_{ao})\{1 - (R_{si} - R_{so})U\} + t_{ao} \tag{9.16}$$

Assuming the outside temperature is constant, then equation (9.16) can be substituted into equation (9.11) to give

$$pQ_{des} = \frac{H_{des}}{F_2}(t_{ai} - t_{ao}) + (m^*C_p)_{fab}\frac{dt_{ai}}{dT} + (mC_p)_{air}\frac{dt_{ai}}{dT} \tag{9.17}$$

where m^* is the effective thermal mass (kg) and is given by

$$m^* = m\frac{F_1}{2F_2}\{1 - (R_{si} - R_{so})U\}$$

Notice that this effective thermal mass is approximately half the actual mass of the solid wall, which would consequently lead to a time constant of approximately

$$\tau \approx \frac{mC_p}{2U}$$

almost half the normal time constant. This arises because the fabric is now associated with the room air temperature, which is almost twice the fabric temperature. So instead of the fabric time constant being the time to rise up to within $0.63t_{fab}$, it is now effectively the time to rise to within $0.315t_{ai}$, with $t_{ai} \approx 2t_{fab}$.

For a multilayered wall, each element would have to be considered with its mean temperature to calculate the effective thermal mass of the wall. This is dealt with in a later example.

More important, the internal fabric of the building (the partition walls, intermediate floors and ceilings) will be capable of storing a considerable

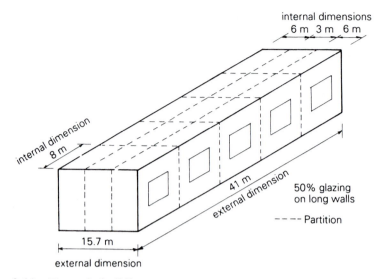

Fig. 9.11 Example building.

amount of heat. Mean temperatures of the internal fabric will be close to the internal temperature of the building, and so quite close to t_{ai}. In the optimizer model, the effective thermal mass of the internal fabric is assumed to be equal to the actual mass, with a mean temperature of t_{ai}.

With the inclusion of the internal and external fabric of the building for storing heat, as well as the air, the total effective thermal mass of the building can be grouped into one summed term, Σm^*C_p, so that equation (9.17) can be expressed as

$$pQ_{des} = \frac{H_{des}}{F_2}(t_{ai} - t_{ao}) + \Sigma m^*C_p\frac{dt_{ai}}{dT} \tag{9.18}$$

In comparison with the fabric, the air stores very little heat. This is demonstrated in Example 9.1.

Example 9.1
Calculate the heat stored in the office building shown in Fig. 9.11 under steady-state design conditions of $t_{ao} = 0°C$ and $t_{ai} = 20°C$. The building is single storey, with individual offices and a central corridor running along its long axis. Figure 9.11 shows the external and internal dimensions of the building, as may be ascertained from the following details:

Walls: 105 mm brick, 50 mm air gap, 100 mm concrete block
Roof: 100 mm concrete block
Partitions: 100 mm block
Window: single-glazed with 6 mm glass, each long wall
            having 50% glazing

For the wall, $R_{si} = 0.12$ m^2 K W^{-1} $R_{so} = 0.06$ m^2 K W^{-1}
For the roof, $R_{si} = 0.10$ m^2 K W^{-1} $R_{so} = 0.04$ m^2 K W^{-1}
For the air gap, $R_{air} = 0.20$ m^2 K W^{-1}

	brick	*block*	*window*
Conductivity (W m^{-1} K^{-1})	0.8	0.25	1
Density (kg m^{-3})	1700	750	2500
C_p (J kg^{-1} K^{-1})	800	800	910

Solution

This building also featured in Section 4.8.1. The steady-state average temperatures of the external fabric elements can be calculated from the resistance network between the inside and outside temperatures. It is assumed that the partitions are at the inside temperature of 20°C. The floor slab has been ignored. Mean fabric temperatures are as follows.

Element	Mean temperature (°C)
Brick	2.4
Block	13.0
Window	6.8
Roof	8.9
Partitions	20.0

Now the heat stored in the fabric elements, raising them from 0°C to their mean temperatures, can be calculated from

$$m C_p (t_{fabric} - t_{ao\ des})$$

where t_{fabric} is the mean fabric temperature (°C).

The heat stored in the fabric elements and the air in the rooms (not the wall air gap) is given in Table 9.1. It is assumed in the above calculation for

Table 9.1 Heat stored in example building

Element	Heat stored	
	MJ	*Fraction of total (%)*
Window	13	1
Air	59	4
Brick	100	8
Block	229	17
Roof	340	25
Partitions	614	45
Total	1355	100

the air that the corridor is maintained at the same temperature as the rooms, with $t_{ai} = 20°C$.

From the above, it can be seen that the windows and the air have little heat content compared to the rest of the elements. If the differential equation for the heat stored and heat lost from a fabric element is considered, then a time constant for the element can be derived. Considering equation (9.17) for one external element of fabric alone (with the effective thermal mass and using the air temperature) then the cooling of the element, without heating, is

$$0 = \frac{H_{des}}{F_2}(t_{ai} - t_{ao}) + m^*C_p \frac{dt_{ai}}{dT}$$

This is a first-order differential equation with a time constant

$$\tau = \frac{m^*C_p F_2}{H_{des}}$$

Considering the element on its own, the surface resistances are ignored, we have

$$\tau = \frac{mC_p}{2UA} = \frac{\rho/C_p}{2U}$$

where ρ is the density (kg m^{-3}) and l is the thickness of the element (m). The U-value here is without any surface resistances. For this external element, the effective thermal mass is half the actual mass. This is because the effective thermal mass relates to the warm surface temperature of the element, which has twice the temperature rise over the outside surface temperature as the mean fabric temperature.

For internal fabric elements of the building, the mean fabric temperature is taken as the mean inside temperature, so the effective thermal mass is equal to the actual mass, giving the time constant as

$$\tau = \frac{(mC_p)}{UA}$$

The time constants for the building in this example are given below, calculated on the heat stored method.

Element	Time constant τ (h)
Brick	5.8
Block	5.1
Window	0.7
Roof	4.5
Partitions	6.7
Air (1 air change per hour)	1.0

For the whole building, it is very tempting to simply add the time constants together, but this is only possible if the internal resistance of the elements is negligible, and the element internal temperature is essentially uniform [11]. The whole building time constant is found from the following equation, developed from equation (9.17), except that the heat stored, being much less than the heat in the fabric, has been omitted and all the fabric effective thermal capacities have been summated:

$$pQ_{des} = \frac{H_{des}}{F_2}(t_{ai} - t_{ao}) + \sum m^*C_p \frac{dt_{ai}}{dT} \qquad (9.18)$$

giving

$$\tau = \frac{F_2 \sum m^*C_p}{H_{des}}$$

The term $\sum m^*C_p$ is the effective thermal capacity for the whole building, and it can be derived from the design steady-state conditions. From equation (9.18) the heat stored in the building is simply the last storage term integrated from the time when all the fabric is at $t_{ao\ des}$ until $t_{ai\ des}$ is achieved, i.e. the heat stored is

$$(t_{ai\ des} - t_{ao\ des}) \sum m^*C_p$$

This was earlier determined as 1355 MJ (which includes the small amount of storage in the air), so $\sum m^*C_p = 67.8$ MJ K^{-1}. For this building, with a single-panel radiator heating system, $F_1 \approx 1$ and $F_2 \approx 1$. Hence $H_{des} = 3.2$ kW K^{-1}, so

$$\tau = \frac{67.8 \times 10^6}{3.2 \times 10^3} = 2.12 \times 10^4 \text{ s}$$

$$= 5.9 \text{ h}$$

This is a surprisingly short time, but there are two reasons for this. The first is that there is a lot of glazing which increases the heat loss without adding to the thermal storage. Over half the losses are through the windows and air (Table 9.2).

Table 9.2 Heat loss in example building

Element	Heat loss	
	kW	*Fraction of total (%)*
Window	17.1	27.7
Air	16.0	25.0
Brick and block wall	6.5	10.1
Roof	23.8	37.2
Total	64.0	100

Second, the block has a low density, reducing the storage both in the external and internal walls. If the external wall were solid brick of 255 mm thickness, then ignoring surface resistances the time constant would be

$$\tau \approx \frac{pC_p l^2}{2k}$$

$$= \frac{1700 \times 800 \times (0.255)^2}{2 \times 0.8}$$

$$= 15.4 \text{ h}$$

a considerable increase on 5.2 h for the 105 mm brick leaf. Notice that τ increases as the square of thickness, l.

These time constants have been derived from simple theory, but the BEMS user can simply log the inside temperature of a room and during the heating's off period examine a plot of the cooling curve to determine the dominant long-term time constant. The early rapid drop in temperature will be due to cooling of the air, but the later cooling will be due to the more massive fabric. Normalizing the axes and making a log-linear graph will make any first-order responses into straight-line graphs [16].

Example 9.1 demonstrates that the heat stored in the air is typically much less than in the fabric, and to all intents and purposes it can be neglected for most buildings. However, as mentioned in Chapter 4, the BEMS room temperature sensor will primarily monitor the air temperature, and the fabric temperature will take longer to rise than the air temperature. So the dry resultant temperature will still be uncomfortable at the start of the week if there has not been sufficient preheating to warm the fabric.

Example 9.1 developed equation (9.18), a straightforward first-order equation describing the preheating:

$$pQ_{des} = \frac{H_{des}}{F_2}(t_{ai} - t_{ao}) + \Sigma m^* C_p \frac{dt_{ai}}{dT} \qquad (9.18)$$

It also describes the cooling down process, when the heating is off and $pQ_{des} = 0$:

$$0 = \frac{H_{des}}{F_2}(t_{ai} - t_{ao}) + \Sigma m^* C_p \frac{dt_{ai}}{dT} \qquad (9.19)$$

This can be rearranged to give

$$\frac{1}{\tau}\int_{T_1}^{T_2} dT = \int_{t_d}^{t_p} \frac{dt_{ai}}{(t_{ai} - t_{ao})}$$

where $\tau = F_2 \, \Sigma m^* C_p / H_{des}$ is the time constant (s). The time and temperature limits of integration are as shown in Fig. 9.1. Integration of the equation yields

$$\frac{T_2 - T_1}{\tau} = \ln\left[\frac{t_d - t_{ao}}{t_p - t_{ao}}\right] \qquad (9.20)$$

Similarly, equation (9.18) may be rearranged for integration:

$$\frac{1}{\tau}\int_{T_2}^{T_3} dT = \int_{t_p}^{t_d} \frac{dt_{ai}}{p(t_{ai} - t_{ao})_{des} - (t_{ai} - t_{ao})}$$

yielding

$$\frac{T_3 - T_2}{\tau} = -\ln\left[\frac{p(t_{ai} - t_{ao})_{des} - (t_d - t_{ao})}{p(t_{ai} - t_{ao})_{des} - (t_p - t_{ao})}\right] \qquad (9.21)$$

This is the preheat curve with the heating on full output for the preheat duration. If the optimum start controller did not reduce the heating at the occupancy time of T_3 (when hopefully the temperature is t_d), then the inside temperature would increase to the steady-state equilibrium temperature with the full heat output. This can be determined from equation (9.18). Steady-state conditions prevail when the fabric has soaked up all its heat and the air temperature is constant, so

$$\Sigma m^* C_p \frac{dt_{ai}}{dT} = 0$$

and

$$pQ_{des} = \frac{H_{des}}{F_2}(t_{ai\,ss} - t_{ao})$$

where $t_{ai\,ss}$ is the steady-state temperature (°C) giving

$$t_{ai\,ss} = \frac{F_2 \, pQ_{des}}{H_{des}} + t_{ao}$$

$$= pF_2(t_c - t_{ao})_{des} + t_{ao}$$

$$= p(t_{ai} - t_{ao})_{des} + t_{ao} \qquad (9.22)$$

To avoid achieving this excess temperature the heating must be controlled at reduced output from T_3. A compensator loop would do this adequately.

And note that in the integrations above, t_{ao} has been assumed constant during the unoccupied period, $T_3 - T_1$. Temperature t_{ao} will most likely drop significantly during this period, but the average steady-state t_{ao}, $t_{ao\,ss}$, to which the cooling curve is tending can be ascertained in a similar fashion to

$t_{ai\,ss}$. Temperature $t_{ao\,ss}$ will lag behind t_{ao} and will differ from it as $t_{ao\,ss}$ is the outside temperature perceived through the fabric which influences t_{ai}.

9.5 Preheat time

The preheat time can be determined from the intersection of the cooling curve and the preheat curve. Equations (9.21) and (9.22) can be used to derive the intersection, but first the equations are made less cumbersome by using the following terms:

$$\Delta t = t_d - t_{ao} \quad \Delta t_{des} = (t_{ai} - t_{ao})_{des} \quad \Delta t_p = t_p - t_{ao}$$

$$\Phi = \text{load factor} = \frac{\Delta t}{\Delta t_{des}}$$

$$x = \frac{\text{unoccupied time}}{\tau} = \frac{T_3 - T_1}{\tau}$$

These terms make the cooling curve

$$\frac{T_2 - T_1}{\tau} = \ln\left[\frac{\delta t}{\delta t_p}\right] \tag{9.23}$$

and the preheat curve

$$\frac{T_3 - T_2}{\tau} = -\ln\left[\frac{p\Delta t_{des} - \Delta t}{p\Delta t_{des} - \Delta t_p}\right] \tag{9.24}$$

Equation (9.24) added to equation (9.23) gives

$$x = \ln\left[\frac{\Delta t}{\Delta t_p}\right] - \ln\left[\frac{p\Delta t_{des} - \Delta t}{p\Delta t_{des} - \Delta t_p}\right] \tag{9.25}$$

or

$$e^x = \frac{\Delta t(p\Delta t_{des} - \Delta t_p)}{\Delta t_p(p\Delta t_{des} - \Delta t)}$$

from which Δt_p can be derived:

$$\Delta t_p = \frac{p\Phi\Delta t_{des}}{\Phi + e^x(p - \Phi)}$$

or

$$t_p = t_d - \Phi\Delta t_{des}\left[\frac{(p - \Phi)(e^x - 1)}{\Phi + e^x(p - \Phi)}\right] \tag{9.26}$$

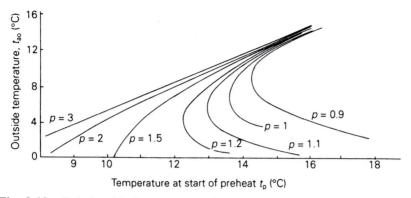

Fig. 9.12 Relationship between t_p and t_{ao}.

This equation can be used as the basis of a BEMS optimizer algorithm. Alternatively, the preheat time can be used for an algorithm. This is determined by substituting equation (9.26) into equation (9.24), producing

$$\frac{T_3 - T_2}{\tau} = \ln\left[\frac{pe^x}{\Phi + e^x(p - \Phi)}\right] \qquad (9.27)$$

As the two equations show, the preheat time varies with the load factor, Φ. This variation is graphed in Fig. 9.12, which is based on equation (9.26). A value of $\tau = 40\,000$ s (11.1 h) was chosen for the graphs in Fig. 9.12 with $T_3 - T_1 = 15$ h.

9.6 Plant size

As the plant size ratio, p, reduces, so a distinct bend develops. To understand this characteristic better, it is useful to relate the preheat temperature to the preheat time by examining equation (9.24) and the corresponding graph (Fig. 9.13).

Equation (9.24) also exhibits the bending, becoming more marked at lower values of plant size ratio, p. The turning points of the curves are the lowest temperatures that the building can drop to for the heating system to be able to achieve the required temperature by occupancy. Any lower than these temperatures and the building will be cold at the start of occupancy.

For a given plant size ratio, p, unoccupied time, τx, the turning-point temperature, or the minimum t_{ai} or t_p for successful preheating, can be determined from differentiating equation (9.26) and equating to zero to find the minimum value. The variable in equation (9.26) is Φ, the load factor, which in turn is a function of t_{ao}. So it is necessary to differentiate equation (9.26) with respect to t_{ao}:

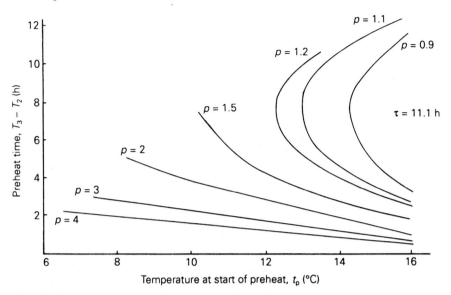

Fig. 9.13   Preheat curves for various plant sizes.

$$t_p - t_{ao} = \frac{p\Phi \Delta t_{des}}{\Phi + e^x(p - \Phi)} \tag{9.26}$$

noting that

$$\frac{d\Phi}{dt_{ao}} = \frac{-1}{(t_d - t_{ao})_{des}} = \frac{-1}{\Delta t_{des}}$$

and for the minimum turning point $dt_p/dt_{ao} = 0$, so we get

$$\Phi = \frac{pe^{x/2}(e^{x/2} - 1)}{e^x - 1} \tag{9.28}$$

Substituting this into equation (9.26) for the preheat temperature at the turning point is a little complex. Substitution into equation (9.27) is easier:

$$\frac{T_3 - T_2}{\tau} = \ln\left[\frac{pe^x}{\Phi + e^x(p - \Phi)}\right] \tag{9.27}$$

$$= \ln p + x - \ln[\Phi(1 - e^x) + pe^x]$$

$$= \ln p + x - \ln\left[\frac{pe^{x/2}(e^{x/2} - 1)}{e^x - 1}(1 - e^x) + pe^x\right]$$

$$= \ln p + x - \ln[-pe^x - pe^{x/2} + pe^x]$$

so

$$\frac{T_3 - T_2}{\tau} = \frac{x}{2}$$

or

$$T_3 - T_2 = \frac{T_3 - T_1}{2} \tag{9.29}$$

The turning point will always be in the middle of the unoccupied period. The preheat temperature at this point is found from equations (9.26) and (9.28):

$$\Delta t_p = p\Delta t_{des}\frac{e^{x/2} - 1}{e^x - 1}$$

$$t_p = p\Delta t_{des}\frac{e^{x/2} - 1}{e^x - 1} + t_{ao} \tag{9.30}$$

Once the inside temperature has fallen below this value of t_p, then it is not possible to achieve the required temperature, t_d, by the start of occupancy, T_3.

There will be no turning point outside the design temperature range, that is as long as t_{ao} is not lower than $t_{ao\ des}$, provided that the plant size ratio, p, is such that $\Phi \geqslant 1.0$, or

$$\frac{pe^{x/2}(e^{x/2} - 1)}{e^x - 1} \geqslant 1.0$$

yielding

$$p \geqslant \frac{e^x - 1}{e^{x/2}(e^{x/2} - 1)}$$

For $\tau = 11.1$ h and $T_3 - T_1 = 15$ h, the plant size would have to be

$$p \geqslant \frac{2.863}{1.97(1.97 - 1)}$$

$$\geqslant 1.5$$

Equation (9.24), the preheat temperature and time equation of Fig. 9.13, is the locus of the preheat temperature and preheat time. Turning the graphs in Fig. 9.13 anticlockwise through 90° relates the loci to the conventional time and design temperature of Fig. 9.1. Figure 9.14 shows some of the loci for $\tau = 11.1$ h and $\tau x = 15$ h.

As t_{ao} decreases, the preheat time increases. But the preheat temperature, t_p, reduces until the turning point then it increases. The locus eventually goes back to time T_1, at a preheat temperature of t_d, i.e. the heating should not be switched off. The value of Φ when the locus reaches T_1 is when

$$T_3 - T_2 = T_3 - T_1 = \tau x$$

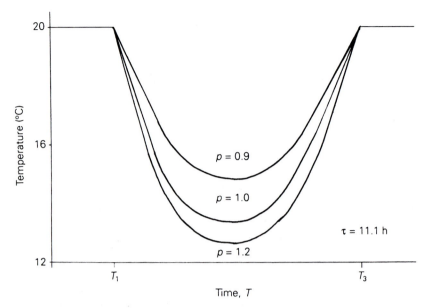

Fig. 9.14 Preheat loci.

Putting this into equation (9.27):

$$x = \ln\left[\frac{pe^x}{\Phi + e^x(p - \Phi)}\right]$$

$$e^x = \frac{pe^x}{\Phi + e^x(p - \Phi)}$$

$$\Phi = p$$

When this load is achieved, then

$$\Phi = p = \frac{\Delta t}{\Delta t_{des}}$$

$$= \frac{t_d - t_{ao}}{(t_{ai} - t_{ao})_{des}}$$

$$= \frac{(t_d - t_{ao})H_{des}}{F_2 Q_{des}}$$

Putting this value of p into the first-order heating equation (9.18) yields

$$\frac{(t_d - t_{ao})H_{des}}{F_2 Q_{des}} = \frac{H_{des}}{F_2}(t_{ai} - t_{ao}) + \Sigma mC_p\frac{dt_{ai}}{dT} \qquad (9.18)$$

As $t_d = t_{ai} = t_p$ at T_1, the heating system is only capable of dealing with the steady-state heat loss and cannot cope with any stored heat loss, so

$$\Sigma m C_p \frac{dt_{ai}}{dT} = 0$$

If the locus has no bending point, the plant size is large and it can cope with all intermittent heating loads within the unoccupied time $T_3 - T_1$.

9.7   Weekend shutdown

In the first-order model we have used to derive the preheat curve, the preheat time can be of any duration within the unoccupied period, $T_3 - T_1$. Some commercial BEMS optimizer algorithms use search periods, during which the BEMS is determining when to start the heating. Typical search periods are about 8 h. Unfortunately, during cold weather and/or with a small plant size ratio, p, this may not be sufficient to achieve comfort conditions by occupancy. The unoccupied period, τx, influences the preheat period and this changes over weekends and holidays. This is best seen in the following example.

Example 9.2
Calculate the preheat time and preheat temperature for a weekday and a weekend with the following conditions and data:

$$p = 1.2, \quad \tau = 11.1 \text{ h}, \quad \Phi = 1.0, \quad \Delta t_{des} = 20 \text{ K}$$

Occupancy ends at 6 pm and commences at 9 am.

Solution
For the weekday:

$$x = \frac{15.0}{11.1} = 1.351, \quad e^{1.351} = 3.861$$

Equation (9.27) gives

$$\frac{T_3 - T_2}{11.1} = \ln\left[\frac{1.2 \times 3.861}{1.0 + 3.861(1.2 - 1.0)}\right]$$

$$= \ln(4.63/1.77)$$

$$T_3 - T_2 = 11.1 \times 0.96$$

$$= 10.67 \text{ h}$$

From equation (9.26) we have $\Delta t_p = 13.6$. As $\Phi = 1.0$ then $t_{ao} = 0°C$, giving $t_p = 13.6°C$.
    For the weekend:

$$x = \frac{15.0 + 48}{11.1} = 5.676, \quad e^{5.676} = 291.8$$

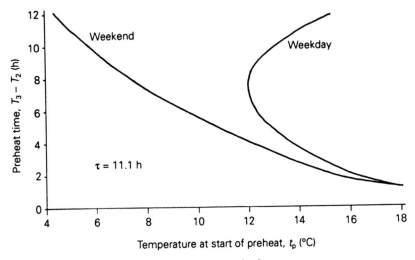

Fig. 9.15   Weekend and weekday preheat loci.

Equation (9.27) gives

$$\frac{T_3 - T_2}{11.1} = \ln\left[\frac{1.2 \times 291.8}{1.0 + 291.8(1.2 - 1.0)}\right]$$

$$= \ln(350/59.4)$$

$$T_3 - T_2 = 11.1 \times 1.77$$

$$= 19.65 \text{ h}$$

From equation (9.26) we have $\Delta t_p = 0.4$. As $\Phi = 1.0$ then $t_{ao} = 0°C$, giving $t_p = 0.4°C$.

For Example 9.2 the weekend preheat period in cold weather lasts almost a day. In practice the heating would be brought on, at a reduced level, by a frost protection configuration in the BEMS before t_p reached 0.4°C.

The preheat loci for the Example 9.2 is shown in Fig. 9.15. There is no turning point in the weekend locus, but there is a turning point in the week-day locus. From equation (9.28) the weekday turning point is

$$\Phi = \frac{pe^{x/2}(e^{x/2} - 1)}{e^x - 1} \tag{9.28}$$

$$= \frac{1.2\ e^{0.676}(e^{0.676} - 1)}{e^{1.351} - 1}$$

$$= 0.8$$

So $t_{ao} = 4°C$. There is a turning point for the weekend locus but it is outside the design temperature range Δt_{des}:

$$\Phi = 1.13 \quad \text{and} \quad t_p = -2.7°C$$

9.8 Influence of thermal mass

A building's thermal mass, or thermal capacity, greatly affect its response. The greater the thermal capacity, the greater the time constant, τ. Time constant τ appears in equation (9.26) for t_p in the term x, which equals $(T_3 - T_2)/\tau$:

$$t_p = t_d - \Phi\Delta t_{des}\left[\frac{(p - \Phi)(e^x - 1)}{\Phi + e^x(p - \Phi)}\right] \tag{9.26}$$

and in the preheat time equation (9.27):

$$\frac{T_3 - T_2}{\tau} = \ln\left[\frac{pe^x}{\Phi + e^x(p - \Phi)}\right] \tag{9.27}$$

As τ increases, so x decreases. This raises the turning-point temperature and makes the preheat locus shallower. This is demonstrated in Fig. 9.16 and it can also be seen by examining equation (9.30):

$$t_p = p\Delta t_{des}\frac{e^{x/2} - 1}{e^x - 1} + t_{ao} \tag{9.30}$$

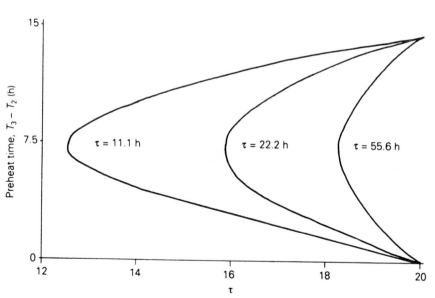

Fig. 9.16 Variation in preheat locus with time constant, τ.

Example 9.3

Calculate the turning-point temperature and the corresponding value of t_{ao} for building A and plant with the following details:

$$p = 1.2,\ \tau = 11.1\ \text{h},\ t_d = 20°\text{C},\ \Delta t_{des} = 20\ \text{K}$$

Occupancy ends at 6 pm and commences at 9 am.

Compare these with building B which has a time constant of 22.2 h.

Solution

For building A, $x = 1.351$, $e^x = 3.863$ and $e^{x/2} = 1.97$
For building B, $x = 0.676$, $e^x = 1.97$ and $e^{x/2} = 1.4$

To determine t_{ao} we can use equation (9.28):

$$\Phi_A = \frac{1.2 \times 1.97 \times (1.97 - 1)}{3.863 - 1} \qquad t_{ao} = 20 - 20 \times 0.8$$

$$= 0.8 \qquad\qquad\qquad = 4°\text{C}$$

$$\Phi_B = \frac{1.2 \times 1.4 \times (1.4 - 1)}{1.97 - 1} \qquad t_{ao} = 20 - 20 \times 0.69$$

$$= 0.69 \qquad\qquad\qquad = 6.2°\text{C}$$

Remember that the turning point is at the same preheat time $(T_3 - T_1)/2$. The heavier building B reaches this point at a higher temperature, t_{ao}, than the lighter building A.

To determine t_p we can use equation (9.26):

$$t_{pA} = t_d - 0.8 \times 20 \left[\frac{(1.2 - 0.8)(3.863 - 1)}{0.8 + 3.863(1.2 - 0.8)} \right]$$

$$t_{pA} = 12.2°\text{C}$$

$$t_{pB} = 20 - 0.69 \times 20 \left[\frac{(1.2 - 0.69)(1.97 - 1)}{0.69 + 1.97(1.2 - 0.69)} \right]$$

$$t_{pB} = 16.0°\text{C}$$

Building B, the building with the greater thermal capacity, has a higher turning point (16°C) and it occurs at a higher outside temperature (6.2°C). The turning-point time is independent of τ and is the same for both buildings. So it can be concluded that for a given outside temperature, t_{ao}, the heavier building will have to have its heating switched on earlier. As the plant sizes are the same, the energy consumed by the heavier building will be greater.

9.9 Emitter output

In equation (9.12), used to develop the optimizer model, the heat input from
the boilers and the output from the emitters were

$$p_1 \dot{m} C_p(t_f - t_r) = p_2 B(t_m - t_{ai})^n$$

Then to simplify the solution of the equation it was assumed that the boilers
were at maximum output during the preheat period and therefore they would
dictate the heating system output. This was held to be a reasonable assumption, except when p_1 was much larger than p_2 and the emitters would limit
the heat output. This case is considered in Example 9.4.

Example 9.4
Determine how the output of a radiator heating system varies with load if
$p_1 = 1.5$ and $p_2 = 1.2$. The steady-state heat loss is

$$Q_{des} = \frac{H_{des}}{F_2}(t_{ai} - t_{ao})_{des}$$

$$= 100 \text{ kW}$$

where $(t_{ai} - t_{ao})_{des}$ = design temperature difference (K)
$$= 20 - 0$$

Solution
Assuming design conditions of $t_m = 75°C$ and $t_{ai} = 20°C$, then B = 0.546 as

$$0.546 \times (75 - 20)^{1.3} = 100$$

The greatest design load would be when the inside of the building got as low
as the outside temperature of 0°C. When heating commenced, the output
from the emitters would be

$$1.2 \times 0.546 \times (75 - 0)^{1.3} = 180 \text{ kW}$$

The maximum output from the boilers is

$$1.5 \times 100 = 150 \text{ kW}$$

So the boilers restrict the emitter output to a constant 150 kW until

$$1.2 \times 0.546 \times (75 - t_{ai})^{1.3} = 150$$

$$t_{ai} = 9.7°C$$

From $t_{ai} = 9.7°C$ to 20°C the emitter output will reduce from 150 kW to
120 kW (Fig. 9.17).

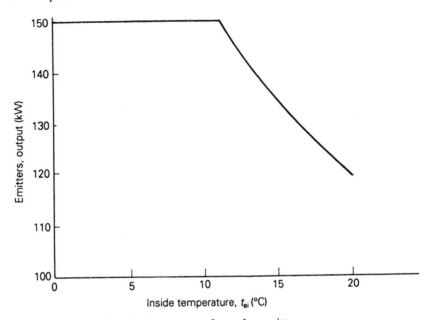

Fig. 9.17   Variation in heat output from the emitters.

When the emitter output is varying, the preheat differential equation (9.18) is altered to

$$p_2 B(t_m - t_{ai})^{1.3} = \frac{H_{des}}{F_2}(t_{ai} - t_{ao}) + \Sigma m^* C_p \frac{dt_{ai}}{dT}$$

The left-hand side can be simplified using a binomial expansion:

$$p_2 B t_m^{1.3}\left[1 - \frac{1.3 t_{ai}}{t_m}\right] = \frac{H_{des}}{F_2}(t_{ai} - t_{ao}) + \Sigma m^* C_p \frac{dt_{ai}}{dT}$$

and from the design conditions $B = \dfrac{H_{des}}{F_2}\dfrac{\Delta t_{des}}{(t_m - t_d)^{1.3}}$

Putting in the time constant, τ, gives

$$\frac{p_2 \Delta t_{des}\, t_m^{1.3}}{(t_m - t_d)^{1.3}}\left[1 - \frac{1.3 t_{ai}}{t_m}\right] = (t_{ai} - t_{ao}) + \tau\frac{dt_{ai}}{dT}$$

$$p^* \Delta t_{des}\left[1 - (1.3 t_{ai}/t_m)\right] = t_{ai} - t_{ao} + \tau\frac{dt_{ai}}{dT}$$

$$p^* \Delta t_{des} = t_{ai}(1 + 1.3 p^*/t_m) - t_{ao} + \tau\frac{dt_{ai}}{dT} \tag{9.31}$$

where $p^* = \dfrac{p_2 t_m^{1.3}}{(t_m - t_d)^{1.3}}$

Integrating equation (9.31) yields

$$(T_3 - T_2)\,\Gamma = -\ln\left[\frac{p^*\Delta t_{des} - (\Gamma t_d - t_{ao})}{p^*\Delta t_{des} - (\Gamma t_p - t_{ao})}\right] \tag{9.32}$$

where $\Gamma = \{1 + 1.3p^*/t_m\}$

≈ 1

as $1.3p^* \ll t_m$. This may be seen from Example 9.4, where

$$p^* = \frac{1.2 \times 75^{1.3}}{(75 - 55)^{1.3}}$$

$$= 1.8$$

so

$$1.3 \times 1.8 = 2.34 \ll 75$$

and

$$\Gamma = 1 + (2.34/75)$$

$$= 1.03$$

So equation (9.32) becomes

$$\frac{(T_3 - T_2)}{\tau} = -\ln\left[\frac{p^*\Delta t_{des} - (t_d - t_{ao})}{p^*\Delta t_{des} - (t_p - t_{ao})}\right]$$

$$= -\ln\left[\frac{p^*\Delta t_{des} - \Delta t}{p^*\Delta t_{des} - \Delta t_p}\right] \tag{9.33}$$

This is the same as equation (9.24), except that p has been replaced by p^*. Figure 9.18 shows the preheat loci for various values of p^*, and it can be seen that the turning point has been greatly reduced by the effectively larger plant size.

9.10 Birtles and John's equation

One of the most successful empirical algorithms for optimum start in BEMSs is based on Birtles and John's equation:

$$\ln(T_3 - T_2) = A_{BJ}(t_p - t_d) + B_{BJ} \tag{9.8}$$

where A_{BJ} is a constant associated with the thermal weight of the building, and B_{BJ} is a constant associated with the time between switching on the heating and the interior starting to heat up.

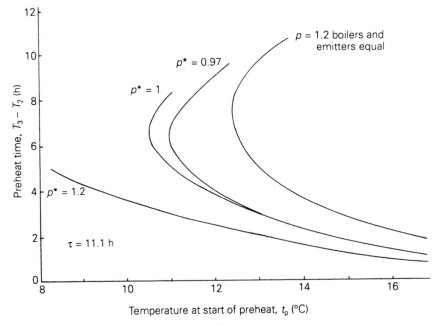

Fig. 9.18 Effects of varying the emitter output.

This performs well, except at low outside temperatures. So an outside temperature term was later added:

$$\ln(T_3 - T_2) = A_{BJ}(t_p - t_d) + B_{BJ} + C_{BJ}t_{ao} \tag{9.9}$$

where C_{BJ} is a constant.

Although this equation does not look like the first-order equations we have just been considering, the graph of equation (9.8), as shown in Fig. 9.8, is not so dissimilar. In fact, this graph can be quite closely fitted by a first-order equation where $\tau = 12.2$ h and $p = 1.5$. Indeed, it has been shown [13] that in a comparison with the first-order model, the constant A_{BJ} is approximately given by

$$A_{BJ} \approx \frac{\tau H_{des}}{p Q_{des}} \approx \frac{\tau}{p(t_{ai} - t_{ao})_{des}} \tag{9.34}$$

As Birtles and John's equation is empirical, one cannot determine the three constants. Algorithms based on the equation use default values which are adapted as the algorithm 'learns' from its past performance. A recursive least-squares method is often used for adaptation [4, 14].

9.11 Supervisory jacket

Similar to the commercial PID program (Chapter 6), the optimizer loop equations discussed above cannot be written straight into a commercial

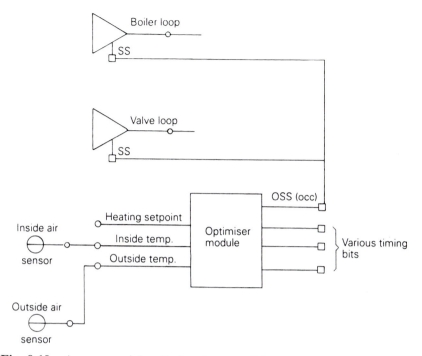

Fig. 9.19 A commercial optimizer loop module.

optimizer loop program. Abnormal data that may be erroneous or atypical (e.g. a boiler going to lockout or a sensor failure) could upset the self-adaptation process. This filtering process is carried out by a **supervisory jacket**, which is extra software around the fundamental optimizer algorithm.

An important feature of this supervisory jacket should be to disconnect the operation of the compensator loop from its connection to the heating system and to open the three-port valve fully to the boiler system. This will allow full heating during preheating. If the compensator were left connected, the heating system output would vary according to the outside temperature and the self-learning process of the optimizer would be 'confused'. On a mild day the compensator would restrict the heat output and the building would take longer than expected to achieve the required temperature. Likewise, on a cold day the compensator would allow full output of the heating, so the building would get up to temperature quicker than expected. The self-adaptation would be making large and erroneous changes.

Figure 9.19 shows the characteristics of a commercial optimizer module for heating. Here the optimizer module activates a boiler loop and a three-port valve loop at the optimum time by a digital output, OSS(OCC), optimum start/stop (occupancy). The connections to the loops are via the setpoint select bits (SS).

On the left of the module are analogue addresses relating to sensors, e.g. the inside and outside temperature sensors, and the heating setpoint. The optimizer algorithm is based on a first-order model, and in the configuration of the module the heating up and cooling down temperature slopes are required. These slopes can be obtained from graphs showing logged data of the building inside temperature. As they are difficult to estimate, it is possible to use the default values preset in the supervisory jacket. The default values will be large, but the self-adaptation will gradually set them to more accurate values.

Optimum stop is also available in the module, which will switch off the heating at the optimum time, so that the building will cool down to a comfortable target temperature by the end of occupancy.

9.12   Savings with optimizer loops

A prime claim in the argument that BEMSs save energy and money is that they have optimizer loops which are good savers. However, there is little analytical evidence as to the savings that can be expected, apart from the findings discussed in the first chapter.

An analysis of the expected savings from an optimizer, compared to continuous heating, has been carried out by Fisk [9, 15] using a first-order model of the building, similar to the model discussed in this chapter. The saving, S, is given by

$$S = \frac{T^*}{T + T^*}\left[1 + \frac{1}{\beta\alpha}\ln\{\beta\exp(-\alpha) - \beta + 1\}\right] \qquad (9.35)$$

where T = unoccupied period (h)
   T^* = occupancy period (h)
      = $24 - T$
   $\beta = \dfrac{Q_{occ}}{Q_{max}}$
   Q_{max} = maximum heat output from heating system
      = pQ_{des} (W)
   Q_{occ} = heat output during occupancy to maintain steady-state design inside temperature (W)
      $= \dfrac{Q_{des}(t_{ai\ des} - t_{ao})}{(t_{ai\ des} - t_{ao\ des})}$
      $= \Phi Q_{des}$
   $\alpha = \dfrac{T^*}{\tau}$
      τ = first-order time constant (h)

In the derivation of equation (9.35) the continuous heating used for comparison was with a heat input of Q_{occ}. The factor β is related to the plant size ratio, p, as can be seen by the following manipulation of the above definitions:

$$\beta = \frac{Q_{occ}}{Q_{max}} = \frac{\Phi Q_{des}}{p Q_{des}} = \frac{\Phi}{p}$$

where Φ is the load factor due to the weather, as used earlier.

As p increases so β reduces, which increases S, the savings.

Example 9.5

What are the savings for the single-storey, partitioned office block with a time constant of 5.9 h, considered in Example 9.4? The occupancy is from 9 am to 6 pm and the steady-state heat loss is 64 kW; the plant size ratio, p, is 1.25. Take the outside temperature as 7°C, with $\Delta t_{des} = 20$ K and $t_{ai\,des} = 20°C$.

Solution

From the data:

$$\Phi = \frac{20 - 7}{20} = 0.65$$

$$\beta = \frac{0.65}{1.25} = 0.52$$

$$\alpha = \frac{6\ pm - 9\ am}{5.9} = 1.53$$

Substituting these values into equation (9.35):

$$S = \frac{T*}{T + T*}\left[1 + \frac{1}{\beta\alpha}\ln\{\beta\exp(-\alpha) - \beta + 1\}\right] \qquad (9.35)$$

we get

$$S = \frac{9}{15 + 9}\left[1 + \frac{1}{0.52 \times 1.53}\ln\{0.52\exp(-1.53) - 0.52 + 1\}\right]$$

$$= 0.375\,[1 + 1.26 \times \ln\{0.52\exp(-1.53) - 0.52 + 1\}] = 12.7\%$$

The maximum saving that could be made would be by having an extremely large heating plant, so that β tends to zero:

$$S = \frac{T*}{T + T*}\left[1 + \frac{\exp(-\alpha) - 1}{\alpha}\right] = 18.3\%$$

Even this is not very close to the ratio of hours occupancy to the hours run with continuous heating, i.e. 9/24 = 0.375. This is due to the storage of heat in the fabric and its loss during unoccupied periods, as well as during occupancy.

References

[1]   Technical Report No. 54, Property Services Agency, 1973. Croydon, UK.
[2]   Birtles, A. and John, R. (1985) A new optimum start algorithm. *Building Services Engineering Research and Technology*, **6**(3).
[3]   Sharma, V., Hibbert, P. and Archer, P. (1982) A value for money guide to optimiser selection. *Building Services Environmental Engineering*, **5**(1), September.
[4]   Ogata, K. (1987) *Discrete-time Control Systems*, Prentice-Hall, Englewood Cliffs NJ.
[5]   Dufton, A. F. (1934) The warming of walls. *Journal of the Institute of Heating and Ventilating Engineers*, **2**(21).
[6]   McLaughlin, R. K., McLean, R. C. and Bonthron, W. J. (1981) *Heating Services Design*, Butterworths, Sevenoaks.
[7]   Bassett, C. R. and Pritchard, M. D. W. (1968) *Heating*, Longman, Harlow.
[8]   Clarke, J. A. *Energy Simulation in Building Design*, Adam Hilger, Bristol.
[9]   Fisk, D. J. (1981) *Thermal Control of Buildings*, Applied Science Publishers, London.
[10]   Crabb, J. A., Murdoch, N. and Penman, J. M. (1987) A simplified thermal response model. *Building Services Engineering Research and Technology*, **8**(1).
[11]   O'Callaghan, P. W. (1980) *Building for Energy Conservation*, Pergamon, Oxford.
[12]   CIBSE (1986) CIBSE Guide, Volume A, Section A9, Estimation of plant capacity, Chartered Institution of Building Services Engineers, London.
[13]   Levermore, G. J. (1988) Simple model for an optimiser. *Building Services Engineering Research and Technology*, **9**(3).
[14]   Iserman, R. (1981) *Digital Control Systems*. Springer-Verlag, Berlin.
[15]   Fisk, D. J. and Bloomfield, D. P. (1977) Optimisation of intermittent heating. *Building and Environment*, **12**; also BRE Current Paper CP 14/77.
[16]   Letherman, K. M. (1981) *Automatic Controls for Heating and Air Conditioning Principles and Applications*, Pergamon, Oxford.
[17]   CIBSE (1999) CIBSE Guide A, Environmental design, Chartered Institution of Building Services Engineers, London.

10
Control of air conditioning

Full air conditioning with windows fixed closed and tempered air from ceiling diffusers is now not necessarily the only design choice for new commercial, deep-plan buildings in Europe, although it is still required for high heat gain buildings. It is also still very common in warmer and more humid climates in America, the Near East and the Far East. In Europe the alternative design choices are **mixed-mode** air conditioning or **natural ventilation**. Mixed-mode air conditioning is where window opening, **thermal fabric storage** and **night ventilation** are involved, in addition to air conditioning, to save energy and to reduce the use of refrigerants in the cooling system. There are several mixed-mode systems and several variations, but in a number of them the cooled air is delivered from floor or low-level diffusers and extracted at higher level due to **displacement ventilation**. The cool air warms up around heat sources, e.g. occupants and computers, then it rises. **Chilled ceilings**, with cool water flowing through pipes bonded to ceiling panels, provide radiant cooling and convective cooling. **Chilled beams** are water-to-air heat exchangers placed near or in the ceiling; they can be used near to, or above, windows for coping with solar gains in summer. These low-energy air-conditioning systems are discussed in Chapter 12; this chapter considers full air-conditioning systems and the basic control of traditional, conventional air-conditioning plant.

10.1 Control from the supply air temperature

Consider the heater battery and ductwork shown in Fig. 10.1. An air-conditioning system would have a cooling coil and a humidifier as well, but we will start by examining the control of the heater battery alone. To

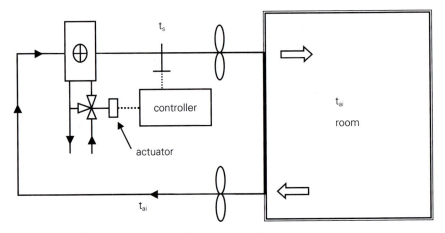

Fig. 10.1 Heater battery supplying warm air to a room.

simplify matters, there is full recirculation (as for preheating prior to occupation) hence there is no fresh air.

The heater battery is controlled by a modulating diverting three-port valve. A diverting valve is chosen instead of a mixing valve, used with the compensator control (Chapter 8), as a diverting valve does not require a pump. Confusingly perhaps, a mixing valve inserted in the return pipe from a heater battery can be used in a diverting application. On a large non-domestic building there may be a number of heater batteries and there would then be the extra cost of a pump for each mixing valve. Hence there is a tradition in the United Kingdom of using diverting valves for heating and cooling coils.

However, mixing valves provide better control at lower loads with a greater water flow, so they give a more even temperature distribution over the coil. The diverter valve reduces the flow of water considerably at low loads and stratification could occur. In parts of Europe and other countries with colder winters than the United Kingdom, mixing valves are used to reduce the stratification; this prevents the very cold air from freezing the colder part of the coil.

The stem of the valve is moved by the actuator in response to a signal from the BEMS outstation or stand-alone controller, which in turn receives a signal from the duct temperature sensor. The valve is effectively controlled by the temperature sensor, hence the heat output from the heater battery is effectively controlled by the temperature sensor. An equation relating the duct air temperature to the heat output is the **control law** or control equation.

In Example 10.1, a simple example with the sensor not in the best position (see later), the control equation is determined from the design conditions.

Example 10.1

A 10 kW heater battery as in Figure 10.1 is sized to provide a supply of hot air at 30°C to a room at 20°C. The control is proportional-only control with a proportional band of 2 K. Determine the control equation.

Solution

As the load on the heater battery reduces, the supply air temperature rises. If the heater battery's output is a maximum when the supply temperature is 30°C then it should be zero at 32°C; this is the setpoint (SP). The control equation is therefore

$$\dot{Q}_{in} = \dot{Q}_{in\ max}\left[\frac{SP - t_s}{PB}\right]$$

$$= 10\left[\frac{32 - t_s}{2}\right] \tag{10.1}$$

where t_s = the supply air temperature (°C)
          PB = the proportional band (K)
          $\dot{Q}_{in}$ = the heat input to room from heater battery (kW)
          $\dot{Q}_{in\ max}$ = the maximum heat input from the heater battery (kW)

Note that the design supply air temperature of 30°C is 10 K above the design room temperature. This is a typical value that is often taken as a rule of thumb for comfortable air supply. In cooling, the rule of thumb is 8 K below the room temperature.

To find how the room temperature varies with this control, it requires the room equation or the room's **heat balance equation**. This is the balance of the heat input from the hot air and the heat loss from the fabric and infiltration. It is from this steady-state balance equation that the heater battery is sized. The room air temperature, t_{ai}, can be determined from the control equation and the balance equation. This is done in Example 10.2.

Example 10.2

The heater battery and its control in Example 10.1 are designed to heat the room air temperature to 20°C when it is 0°C outside (the outside design temperature). If the heater battery is designed for a steady-state heat loss from the room of 10 kW, what is the room air temperature when it is 10°C outside? It is assumed there is full recirculation.

Solution

The heat balance for the room at design conditions is

$$[\Sigma(UA) + NV/3](t_{ai} - t_{ao}) = \dot{m}C_p(t_s - t_{ai}) = 10\ kW$$

Putting in the design values gives

$$[\Sigma(UA) + NV/3](20 - 0) = \dot{m}C_p(30 - 20) = 10 \text{ kW}$$

so

$$[\Sigma(UA) + NV/3] = 0.5 \text{ kW K}^{-1}$$

and

$$\dot{m}C_p = 1 \text{ kW K}^{-1}$$

Hence the heat balance equation at 10°C outside temperature becomes

$$0.5(t_{ai} - 10) = 1(t_s - t_{ai}) \tag{10.2}$$

and the control equation is

$$\dot{Q}_{in} = 10[(32 - t_s)/2]$$

In the steady state, the heat supplied by the heater battery is supplied to the room by the warm air, and this balances the heat loss from the room. So equation (10.1) equals equation (10.2):

$$0.5(t_{ai} - 10) = 1(t_s - t_{ai}) = 10[(32 - t_s)/2] \tag{10.3}$$

which can be solved for either t_s or t_{ai}. From the right-hand side of the equation we get

$$t_s = t_{ai} + 5(32 - t_s)$$

$$t_s = (t_{ai} + 160)/6$$

Substituting this into the left-hand side of equation (10.3) gives

$$0.5(t_{ai} - 10) = (t_{ai} + 160)/6 - t_{ai}$$

$$3(t_{ai} - 10) = t_{ai} + 160 - 6t_{ai}$$

$$8t_{ai} = 190$$

$$t_{ai} = 23.75°C$$

One can similarly work out the other inside air temperatures for the corresponding outside temperatures. The highest inside temperature that could occur would be 32°C, which occurs when the supply air temperature and the outside temperature are also 32°C, i.e. there is no error signal here as t_s is at the setpoint temperature. At this very high temperature there is no requirement for any heating, so the room air is simply recirculated without any heating. But it does show the poor control from the supply duct sensor with the large variation in possible room temperature (Fig. 10.2). This relationship between inside and outside temperature is linear, as the control equation

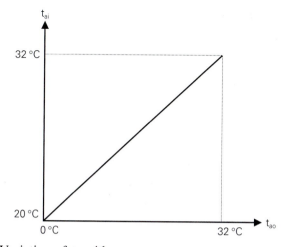

Fig. 10.2 Variation of t_{ai} with t_{ao}.

and the heat balance equation are linear. One can determine the line from the two extreme points, full heat output (t_{ai} = 20°C, t_{ao} = 0°C) and zero heat output ($t_{ai} = t_s = t_{ao}$ = 32°C). The equation that can be derived is

$$t_{ai} = 0.375t_{ao} + 20 \qquad (10.4)$$

Often, however, the heat output from the heater battery does not vary linearly with the movement of the diverter valve and consequently the supply temperature (see later). The supply duct temperature sensor control gives a large variation in room temperature, as the sensor is close to the heater battery, but it does give quick-responding control. To reduce the offset, PI control is sometimes used instead of proportional-only control.

10.2 Control from the room air temperature (return duct)

The control of the heater battery from the supply temperature is unsatisfactory; this is because the room temperature ranges from 20°C to 32°C as the outside temperature varies from 0°C to 32°C. However, the sensor is close to the heater battery which gives a fast response to the control. Nevertheless, it would seem more sensible to control directly from the room temperature. Often this is achieved by putting the sensor in the return air duct as the return air is at room temperature and the return duct often goes back to the plant room, so the sensor cable run is shorter than if it went to the room. If the sensor is placed in the room, there is the problem of choosing a representative position. Example 10.3 considers control from a sensor in the return air duct.

Example 10.3
Consider the proportional-only control of the heater battery in Example 10.2 but this time with the sensor in the return air duct. The proportional band is 2 K and the setpoint at the top of the proportional band is 21°C. What is the room temperature when the outside temperature is 10°C and what is the relationship between t_{ao} and t_{ai}? Use the heater, room and design data from Example 10.2.

Solution
From the control conditions we can see that this form of control is better than the previous supply temperature control, in that the room temperature is constrained between 21°C and 19°C, the extent of the proportional band.

As in Example 10.2, we need to establish the control equation and the balance equation. For the control equation, the heater battery will be at full output when the room temperature is at 19°C and at zero output when the room temperature is at 21°C. The control equation is therefore

$$\dot{Q}_{in} = \dot{Q}_{in\,max}\left[\frac{SP - t_s}{PB}\right]$$

$$= 10[(21 - t_{ai})/2] \tag{10.5}$$

The heat balance equation remains the same as in Example 10.2:

$$[\Sigma(UA) + NV/3](t_{ai} - t_{ao}) = \dot{m}C_p(t_s - t_{ai}) = 10 \text{ kW} \quad \text{(at design conditions)}$$

When t_{ao} is 10°C this becomes

$$0.5(t_{ai} - 10) = 1(t_s - t_{ai}) = \dot{Q} \tag{10.6}$$

As the control equation is in terms of t_{ai}, the supply temperature part of equation (10.6) is not required. Combining equations (10.5) and (10.6) gives

$$0.5(t_{ai} - 10) = 10[(21 - t_{ai})/2]$$

$$t_{ai} - 10 = 210 - 10t_{ai}$$

$$t_{ai} = 20°C$$

The systems are linear, so we can determine the linear relationship between t_{ai} and t_{ao} from the extreme points at the ends of the proportional band ($t_{ai} = 21°C$, $t_{ao} = 21°C$ and $t_{ai} = 19°C$, $t_{ao} = -1°C$). The last point has $t_{ao} = -1°C$ as the heating system at full output can raise the room temperature through 20 K. The equation of this line can be determined from the two endpoints of the control equation:

$$21 = 21m + c \tag{10.7}$$

$$19 = -1m + c \tag{10.8}$$

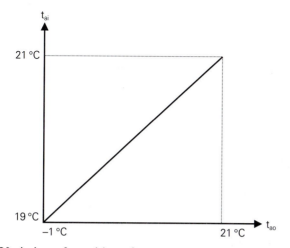

Fig. 10.3   Variation of t_{ai} with t_{ao} for room temperature control.

where m = the slope (dimensionless)
     c = the intercept (°C)
From equations (10.8) and (10.7) come $m = 0.091$ and $c = 19.1°C$, giving

$$t_{ai} = 0.091t_{ao} + 19.1 \qquad (10.9)$$

This is shown in Fig. 10.3.

10.3   Control from the reset room (return air duct) sensor

The return duct sensor control worked very well for the steady-state conditions in Example 10.3. But there is a time lag of the air travelling from the heater battery, through the room and out to the extract duct sensor. For a room with four air changes per hour of supply air, the transit time for the air would be 15 min. But it has a much better control of the room temperature than the sensor in the supply duct, as can be seen by comparing Figs 10.3 and 10.4. A good compromise is to use both supply duct and return duct sensors in a **reset** or **cascade** control (similar to the compensator in Chapter 8). Here the return duct sensor resets the setpoint of the supply control. Hence responsive heater battery control is achieved with close control of the room temperature.

Example 10.4
Consider a heater battery as in Example 10.3 but with reset control to maintain the room temperature between 19°C and 21°C. What is a suitable reset schedule for the supply air temperature? What is the room temperature when the outside temperature is 10°C and what is the relationship between

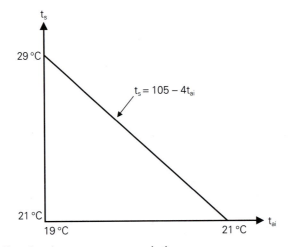

Fig. 10.4   Supply air temperature variation.

t_{ao} and t_{ai}? Use the heater battery and room heat loss design data from Example 10.3.

Solution
As the heater battery's local control is PI, it is assumed there is little offset in the steady state and it is assumed the setpoint temperature is achieved. The heater battery should be on its full output of 10 kW when the room temperature is 19°C, which means the supply air temperature, t_s, should be controlled to 29°C. Likewise when the room temperature is 21°C, the upper limit of control, there should be no heating requirement, i.e. $t_s = t_{ai} = 21°C$. We will use this point here, but a strong case could be made for having $t_s = 22°C$ or 23°C to avoid the occupants experiencing uncomfortable draughts at low loads. From these limiting points, the control equation can be determined as

$$t_s = 105 - 4t_{ai} \qquad (10.10)$$

which is shown in Fig. 10.4.

Note that unlike the previous supply control, where t_s increased as the load reduced and t_{ai} increased, here t_s reduces with decreasing load and this is intuitively better.

The heat balance equation is

$$[\Sigma(UA) + NV/3](t_{ai} - t_{ao}) = \dot{m}C_p(t_s - t_{ai}) = 10 \text{ kW} \quad \text{(at design conditions)}$$

or

$$0.5(t_{ai} - t_{ao}) = 1(t_s - t_{ai})$$

When $t_{ao} = 10°C$ this becomes

$$0.5(t_{ai} - 10) = 1(t_s - t_{ai}) \qquad (10.11)$$

Combining equations (10.10) and (10.11) gives

$$5.5t_{ai} = 105 + 5$$

$$t_{ai} = 20°C$$

The relationship between t_{ai} and t_{ao} can be ascertained from these equations as

$$t_{ai} = 19.1 + 0.09t_{ai}$$

which is the same as equation (10.9) for the return duct control. However, this reset control has a better dynamic control and its integral term on the supply duct PI control can cope better with the non-linear heat exchanger. It has been assumed that the heater battery output is linear, but this is a simplification (see later).

10.4   Recirculation

The examples so far have considered the system to be on full recirculation, with all the air from the room being ducted to the intake of the heater battery and no fresh air coming into the system. Real systems will be on full recirculation during preheat prior to occupation. However, during occupancy there will be a need for fresh air to maintain good air quality, so the outside supply damper will partly open, as will the extract damper to balance the airflows. The recirculation damper will partly close to allow the fresh air in. The damper arrangement is shown in Fig. 10.5. Many full air-conditioning systems use recirculation, although it is possible to use full fresh air with

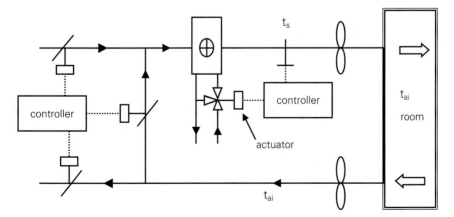

Fig. 10.5   Recirculation system.

heat recovery to maintain good indoor air quality and to reduce sick building syndrome. Full fresh air is often used with displacement ventilation and fan coil systems with possible heat recovery.

Example 10.5
Consider the reset supply air temperature control in Example 10.4 but now with 20% fresh air. What is the inside temperature, t_{ai}, when $t_{ao} = 10°C$? And what is the relationship between t_{ai} and t_{ao}?

Solution
With a fraction x of fresh air coming in and mixing with a fraction $(1 - x)$ of recirculated air, the heat input from the heater battery is

$$\dot{Q}_{in} = x\dot{m}C_p(t_s - t_{ao}) + (1 - x)\dot{m}C_p(t_s - t_{ai}) \qquad (10.12)$$

This can be simplified to

$$\dot{Q}_{in} = \dot{m}C_p\{(t_s - t_{ai}) + x(t_{ai} - t_{ao})\}$$

which corresponds to

$$\dot{Q}_{out} = \text{heat lost to room by} + \text{heat loss through}$$
$$\text{supply warm air} \qquad \text{air extracted}$$

Equating the heat loss from the room and the heat loss through extraction to the hot air supply:

$$\dot{m}C_p\{(t_s - t_{ai}) + x(t_{ai} - t_{ao})\} = [\Sigma(UA) + NV/3](t_{ai} - t_{ao}) + x\dot{m}C_p(t_{ai} - t_{ao})$$

The x terms cancel to leave

$$\dot{m}C_p(t_s - t_{ai}) = [\Sigma(UA) + NV/3](t_{ai} - t_{ao})$$

This is the heat balance for the room; the left-hand side should not be confused with the output of the heater battery. Also it is unfortunate that it is the same as the full recirculation balance equation, and implies that the same temperatures prevail as in the full recirculation case. In fact, this is partially true as they do, but only when the heater battery has the capacity to satisfy both the room heat requirement and the heating of the fresh air. When the heater battery does not have the capacity, the room temperature reduces.

Consider the following design conditions: $t_{ai} = 20°C$ when $t_{ao} = 0°C$ but the heater battery full output is still 10 kW based on full recirculation. Under these conditions the net heat input to the fresh air (when warmed it does contribute to heating the room but this is not included) is equal to the extract air heat loss:

$$x\dot{m}C_p(t_{ai} - t_{ao}) = 0.2 \times 1 \times (20 - 0)$$

$$= 4 \text{ kW}$$

The new room temperature under these design conditions (with the 20% fresh air) can be found from the balance equation:

$$10 \text{ kW} = [\Sigma(UA) + NV/3](t_{ai} - t_{ao}) + x\dot{m}C_p(t_{ai} - t_{ao})$$

$$= 0.5(t_{ai} - t_{ao}) + 0.2(t_{ai} - t_{ao})$$

$$t_{ai} - t_{ao} = 14.3 \text{ K}$$

The heater battery can raise the room temperature 14.3°C above the outside temperature when it is at full output. This gives $t_{ai} = 14.3°C$ when $t_{ao} = 0°C$ and $t_s = 24.3°C$.

However, when the outside temperature is 10°C the heater battery is capable of raising t_{ai} to 24.3°C, so we can use the reset control equation with the modified balance equation. The heat input from the heater battery is

$$\dot{Q}_{in} = x\dot{m}C_p(t_s - t_{ao}) + (1 - x)\dot{m}C_p(t_s - t_{ai}) \tag{10.12}$$

When $t_{ao} = 10°C$ with $\dot{m}C_p = 1 \text{ kW K}^{-1}$ and $x = 0.2$ then

$$\dot{Q}_{in} = 0.2(t_s - 10) + 0.8(t_s - t_{ai}) \tag{10.13}$$

The control equation is equation (10.10) as before:

$$t_s = 105 - 4t_{ai} \tag{10.10}$$

In the steady-state condition, $\dot{Q}_{in}$ equals the heat loss through the room and the hot air extracted:

$$[\Sigma(UA) + NV/3](t_{ai} - t_{ao}) + x\dot{m}C_p(t_{ai} - t_{ao}) = 0.5(t_{ai} - 10) + 0.2(t_{ai} - 10)$$

$$= 0.7(t_{ai} - 10) \tag{10.14}$$

The steady-state equations (10.13) and (10.14) are equal, and substituting for t_s from equation (10.10) gives

$$0.7(t_{ai} - 10) = 0.2(105 - 4t_{ai} - 10) + 0.8(105 - 4t_{ai} - t_{ai})$$

$$= 19 - 0.8t_{ai} + 84 - 4t_{ai}$$

$$5.5t_{ai} = 110$$

$$t_{ai} = 20°C$$

This is the same as for the full recirculation system in Example 10.4, but to account for the fresh air, the power output of the heater battery is 7 kW, which is higher than the 5 kW for full recirculation.

In this example the heater battery size has been left at 10 kW, as in the full recirculation system. But if a fresh air input is required as part of the design, the battery should be sized larger to account for the design fresh air.

Example 10.5 shows that fresh air input adds to the heater battery's load, but recirculation can be seen as a very good method of saving heat and energy. However, systems such as swimming pools need fresh air, in the UK, with its low water vapour content, to replace the humid air extracted from the pool hall. With the potential high humidities in pool halls there is the risk of serious condensation and in severe cases structural damage. Recirculation here would exacerbate the humidity problem although it would save heat. Hence a solution is to use full fresh air but with heat recovery via a heat exchanger such as a **run-around coil** or a **plate heat exchanger**. However, these do offer airflow resistance to fans, so there is an electrical penalty for saving heat in winter and in summer.

10.5   Basic cooling coil control

So far only heater batteries have been considered, but one of the prime purposes of air conditioning is to cool the air for summer comfort. Humidity control is also part of full air conditioning but this will be considered later.

The design airflow rates for an air-conditioning system are often based on the summer cooling condition rather than the winter heating condition, as the summer cooling load for a building is often greater than the winter heating load. This can be appreciated by considering a window, the key element in heat loss and solar gain. For a double-glazed window with a U-value of 3.6 W m^{-2} K^{-1} and design $t_{ai} = 21°C$ for design t_{ao} of $-3°C$ then the design heat loss is 86.4 W m^{-2}. In the summer under design UK conditions the solar radiation on an east-facing window is over 600 W m^{-2} at 8 am [1, 25]. It is even larger for a south-facing window in spring and autumn. Although this is a simple example, cooling loads tend to be larger than heating loads [2], so the air supply is based on the summer cooling load.

Control of the cooling coil is very similar to control of the heating coil. The arguments for reset control being the best of the three methods also apply to the cooling coil. Here, however, the design conditions are usually taken as a supply temperature of 13°C to maintain the room at 21°C. This latter temperature would be the dry resultant temperature, but most air-conditioning duct and room sensors are air temperature sensors. So the control of air temperature is considered here. Another difference from the heater battery control is that the solar load is not easily related to outside temperature, much as some authors and lecturers would like it to be. The new CIBSE Weather and Solar Data Guide [3] gives the data on solar and other climatic parameters of interest to the services engineer, including hourly actual values for a number of sites. The solar load in the following examples is regarded as a steady-state constant load, as are the occupancy, lighting and equipment loads considered for the cooling load assessment.

10.5.1 Cooling system

The cooling is provided by a refrigeration plant (evaporator, condenser and compressor) with either direct refrigerant evaporating in the cooling coil or by an indirect system providing chilled water from the evaporator. Figure 10.13 shows the production of chilled water for a cooling coil from a chiller's evaporator and the cooling of the condenser with water from a cooling tower. No control valves are shown, only temperatures to indicate typical values. The supply air comes off the cooling coil at between 11°C and 13°C, with fan and duct gains [4] adding typically 1–2 K to the supply temperature by the time the air reaches the room or zone. This cooling plant has its own set of controls and a reasonably controlled water supply temperature can be expected.

Note that because of the different temperature differentials between chilled and hot water, chilled water flow rates are up to five times greater for the same thermal output as heating systems. For a heater or cooler coil, the heat exchange is not linear so the **log mean temperature difference** (LMTD) must be used in the heat balance equation [5]:

$$\dot{m}C_p(t_f - t_r) = CUA \left[\frac{\Delta t_{max} - \Delta t_{min}}{\ln\left(\dfrac{\Delta t_{max}}{\Delta t_{min}}\right)} \right] \tag{10.15}$$

where C = the heat exchanger correction factor for the coil
   U = the heat transmission coefficient, or U-value, of the coil (W m^{-2} K^{-1})
   A = the heat transfer area of the coil (m^2)
   Δt_{max} = the maximum temperature difference between the two fluids at one end of the heat exchanger (K)
   Δt_{min} = the minimum temperature difference between the two fluids at the other end of the heat exchanger (K)

For a typical cooling system design (t_s = 13°C when t_{ao} = 28°C; chilled water flow t_f = 6°C and return t_r = 11°C) and regarding the coil as a counterflow heat exchanger, we have Δt_{max} = 17 K and Δt_{min} = 7 K. The LMTD on the right then becomes 11.2. For a typical heating system design (t_s = 30°C when t_{ao} = 0°C; hot water flow t_f = 80°C and return t_r = 70°C) we have Δt_{max} = 70 K and Δt_{min} = 50 K. The LMTD on the right then becomes 59.4.

So for the same thermal power, the left-hand side of equation (10.15) is the same for the cooling and heating, with the temperature differential ($t_f - t_r$) for the hot water twice that of the chilled water. But the cooler coil is often larger than the heating coil (because of the lower LMTD value), consequently A is larger for the cooler coil. Without going into the minutiae of the values, equation (10.15) does indicate that the chilled water flow rate could be up to five times that of hot water, with its consequent implications for pump energy consumptions.

Example 10.6
A cooling coil is to be controlled to supply cool air to an air-conditioned zone of a building. The zone has a design sensible cooling load of 10 kW, due to solar gain, occupancy, equipment and conduction and infiltration. And the room is designed to be maintained at 21°C when the design outside air is at 28°C. The design fresh air proportion is 20%. Select a suitable control for this coil and determine the inside temperature when the outside temperature is 20°C. The cooling coil is now sized to equal the design steady-state heat gain and the fresh air.

Solution
Note that the design supply air temperature should be 13°C, 8 K below the room temperature of 21°C, 8 K being the design rule of thumb for comfort when the cool air is delivered from a conventional ceiling diffuser.

A suitable control for this coil is by a room, or return duct, sensor signal resetting the coil's PI control (which has its own supply air temperature sensor). As with the heating coil, the design temperatures are for plant sizing and the control schedule will have a band of control around the design inside air temperature. As the room air temperature rises above 21°C, the cooling coil will increase its output. Corresponding to the heater battery examples, it is proposed that the cooling coil has its maximum output when t_{ai} is 23°C. This means that the supply air should be 15°C to be 8 K below the room temperature. At 15°C the supply air is above the design value of 13°C but the design conditions assume that the room temperature will be maintained constant, which is not practical with proportional control. One could maintain the supply temperature at 13°C at full load but risk the chance of discomfort. A temperature of 15°C will be used in this example. As with the heater battery, the supply air temperature will change as the room temperature drops. When the room temperature is 21°C there is no cooling requirement and the supply temperature is also 21°C. So the control equation is given by solving the linear equation for these two points

$$21 = 21m + c$$
$$15 = 23m + c$$

where m is the slope and c is the intercept (°C). This gives

$$t_s = 84 - 3t_{ai}$$

The heat balance equation for the cooling coil air is like equation (10.12) except for the sign of $(t_s - t_{ao})$ and $(t_s - t_{ai})$ as the process is now cooling instead of heating:

$$\dot{Q}_{in} = x\dot{m}C_p(t_{ao} - t_s) + (1 - x)\dot{m}C_p(t_{ai} - t_s) \qquad (10.16)$$

Under design conditions the cooling coil must absorb the heat gains of 10 kW and it must also cope with the warm air coming in and later extracted, which is effectively cooling the outside environment. A proportion of air, x, is taken in through the supply damper at temperature, t_{ao}, and it is later extracted back to the atmosphere at temperature, t_{ai}, giving cooling power to the atmosphere of $x\dot{m}C_p(t_{ao} - t_{ai})$. So the loads on the chiller via the cooling coil are the heat gain to the room, 10 kW, and the cooling of the outside environment. Hence

$$\dot{Q}_{in} = 10 + x\dot{m}C_p(t_{ao} - t_{ai})$$

Combining this with equation (10.16) at design conditions gives

$$\dot{Q}_{in} = 10 + 0.2\dot{m}C_p(28 - 21) = 0.2\dot{m}C_p(28 - 13) + 0.8\dot{m}C_p(21 - 13)$$

so

$$\dot{m}C_p = 1.25 \text{ kW K}^{-1}$$

giving

$$\dot{Q}_{in} = 0.25(t_{ao} - t_s) + 1.0(t_{ai} - t_s)$$

In the steady-state condition, the heat gains due to infiltration and conduction – solar, $\dot{Q}_s$, occupancy, $\dot{Q}_o$, and equipment, $\dot{Q}_e$ – are equal to 10 kW. Depending on the outside temperature, the supply fresh air, infiltration and conduction loads may be negative, i.e. they help the cooling process. At the design conditions,

$$10 \text{ kW} = [\Sigma(UA) + NV/3](t_{ao} - t_{ai}) + \dot{Q}_s + \dot{Q}_o + \dot{Q}_e$$

With the same U-values and infiltration rates as the previous examples, we have

$$10 = 0.5(28 - 21) + \dot{Q}_s + \dot{Q}_o + \dot{Q}_e$$
$$10 = 3.5 + \dot{Q}_s + \dot{Q}_o + \dot{Q}_e$$

giving

$$\dot{Q}_s + \dot{Q}_o + \dot{Q}_e = 6.5 \text{ kW}$$

So the balance equation for the coil and room is

$$\dot{Q}_{in} = 0.25(t_{ao} - t_s) + 1.0(t_{ai} - t_s) = 0.5(t_{ao} - t_{ai}) + 6.5 + 0.25(t_{ao} - t_{ai})$$
$$\qquad (10.17)$$

From the control and balance equations, the problem can be solved for $t_{ao} = 20°C$:

$$t_s = 84 - 3t_{ai}$$

$$\dot{Q}_{in} = 0.25(20 - t_s) + 1.0(t_{ai} - t_s) = 0.75(20 - t_{ai}) + 6.5 \quad (10.17)$$

Combining them gives

$$0.25(20 - 84 + 3t_{ai}) + 1.0(t_{ai} - 84 + 3t_{ai}) = 0.75(20 - t_{ai}) + 6.5$$

$$0.25(-64 + 3t_{ai}) + 1.0(4t_{ai} - 84) = 0.75(20 - t_{ai}) + 6.5$$

$$0.75t_{ai} - 16 + 4t_{ai} - 84 = 15 - 0.75t_{ai} + 6.5$$

$$5.5t_{ai} = 121.5$$

$$t_{ai} = 22.1°C$$

From Example 10.6 the cooling coil input, $\dot{Q}_{in}$, the supply air temperature, t_s, and the inside air temperature, t_{ai}, can also be determined for various outside temperatures. It is found that these vary linearly with outside temperature. The equations for the inside air temperature and $\dot{Q}_{in}$ are

$$t_{ai} = 0.0909t_{ao} + 20.28 \quad (10.18)$$

$$\dot{Q}_{in} = 0.68t_{ao} - 8.66 \quad (10.19)$$

Hence the equation for the supply air temperature is

$$t_s = 84 - 3(0.0909t_{ao} + 20.28)$$

$$t_s = 23.16 - 0.273t_{ao}$$

Although the temperatures and cooling inputs are linear, it is difficult to simply ascertain the endpoints of the lines as with simple proportional control. This is because the system has now become more complex.

10.5.2   Free cooling and recirculation

Note that with the use of fresh air, the cooling coil is assisted at low outside temperatures. In fact, cooling is not required until the outside temperature, t_{ao}, rises above a certain value. As the cooling system (compressor, evaporator and condenser) is not required to operate when the outside air is sufficient for cooling, the cooling is **free**. The supply, extract and recirculation dampers should be adjusted to increase the amount of fresh air as the inside temperature rises before the cooling system is switched on to save energy by using **free cooling**. (Note that the term 'free cooling' is also applied to using the chiller and its refrigerant with the compressor switched off in a thermosyphon cycle [6]. This also uses the cool outside air via the air-cooled condenser or water-cooled condenser and cooling tower.)

But how much free cooling can one achieve? This can be determined from the balance equations for the room, or zone, and the supply air system. From these the control equation can be determined for the air dampers.

For Example 10.6 the input from the cooling coil is zero and equation (10.16) gives

$$0.25(t_{ao} - t_s) + 1.0(t_{ai} - t_s) = 0 \qquad (10.20)$$

The heat gains to the room are absorbed by the supply fresh air, the fabric and infiltration losses, so

$$[\Sigma(UA) + NV/3](t_{ao} - t_{ai}) + x\dot{m}C_p(t_{ao} - t_s) = 6.5 \text{ kW}$$

$$0.5(t_{ao} - t_{ai}) + 0.25(t_{ao} - t_{ai}) = 6.5$$

giving

$$(t_{ao} - t_{ai}) = 8.67 \qquad (10.21)$$

The control equation is required to relate t_s and t_{ai}. Although the cooling coil is not cooling, the control equation does give the point at which the cooling is off but about to come on if the temperature rises slightly. So

$$t_s = 84 - 3t_{ai} \qquad (10.22)$$

From equations (10.20), (10.21) and (10.22) we get

$$0.25(t_{ai} - 8.67 - 84 + 3t_{ai}) + t_{ai} - 84 + 3t_{ai} = 0$$

$$5t_{ai} = 107.17$$

$$t_{ai} = 21.4°C$$

From equation (10.21) we have

$$t_{ao} = t_{ai} - 8.67$$

$$t_{ao} = 12.7°C$$

A check on this is that the cooling coil should have no cooling effect, i.e. $Q_{in} = 0$, so from equation (10.19):

$$0 = 0.68t_{ao} - 8.66$$

$$t_{ao} = 12.7°C$$

This confirms that the fresh air can maintain a satisfactory temperature in the room up to an outside temperature of 12.7°C. Above this the cooler would have to be switched on or the fresh air supply proportion increased. The full amount of fresh air is 100% ($x = 1.0$). In this situation the fresh air is the full supply air quantity, so the supply air temperature, t_s, is the same as the outside air temperature, t_{ao}. Hence the heat balance for the room is due to the outside cool air and fabric and infiltration heat losses equalling the solar, occupant and equipment heat gains:

$$mC_p(t_{ai} - t_{ao}) + [\Sigma(UA) + NV/3](t_{ai} - t_{ao}) = 6.5 \text{ kW}$$

$$1.75(t_{ai} - t_{ao}) = 6.5$$

$$t_{ai} - t_{ao} = 3.7 \text{ K}$$

The full fresh air could satisfactorily remove the heat gains to the room until the time that the outside air temperature is 3.7 K below the room temperature, t_{ai}. However, the control schedule for the cooling coil from the past examples implies that action should be taken when the room temperature reaches 21°C and t_{ao} has therefore reached the limit of free cooling, which is 17.3°C. So, using the previous example schedules, the chiller should be switched on when t_{ai} reaches 21°C.

The fresh air helps the cooling process all the time t_{ao} is less than t_{ai}, so it seems logical to control the air dampers with a signal from an outside air temperature sensor. But we cannot just use a temperature difference signal – the room might require heating because t_{ai} is low, and with t_{ao} lower than t_{ai} we would not want the outside air proportion to be increased. So the room air temperature is also used to control the fresh air dampers. Example 10.7 provides an illustration.

Example 10.7
For Example 10.6 consider the cooling by modulation of the fresh air supply. Determine what the outside air temperature is at the minimum and full fresh air positions. The control of the fresh air supply is by a signal of room air temperature from the return duct sensor. Suggest a control schedule for the fresh air modulation.

Solution
A suggested schedule is that the fresh air is a minimum (the 20% mentioned in the Example 10.6) when $t_{ai} \leqslant 21$°C and the fresh air is at a maximum when t_{ai} is 22°C. Above this room temperature the cooler coil would be switched on. With x as the fresh air proportion, the control equation is obtained from solving for these two points (where m is the slope and c the intercept):

$$0.2 = 21m + c$$

$$1 = 22m + c$$

giving $m = 0.8$ and $c = -16.6$. Hence the fresh air control equation is

$$x = 0.8t_{ai} - 16.6 \tag{10.23}$$

The heat balance equation for the room is

$$x \dot{m} C_p(t_{ai} - t_{ao}) + [\Sigma(UA) + NV/3](t_{ai} - t_{ao}) = 6.5 \text{ kW}$$

where the heat gain is 6.5 kW, as in the Example 10.6. Substituting the values in the heat balance equation gives

$$(1.25x + 0.5)(t_{ai} - t_{ao}) = 6.5 \tag{10.24}$$

Combining equations (10.23) and (10.24):

$$\{1.25(0.8t_{ai} - 16.6) + 0.5\}(t_{ai} - t_{ao}) = 6.5$$

$$(t_{ai} - 20.25)(t_{ai} - t_{ao}) = 6.5 \tag{10.25}$$

When $t_{ai} = 21°C$ then $x = 0.2$ (the bottom limit of the control equation); equation (10.25) gives $t_{ao} = 12.3°C$. At the top limit, where t_{ai} is 22°C and $x = 1.0$, equation (10.25) gives $t_{ao} = 18.3°C$.

Example 10.7 shows the importance of the free cooling. The chiller can be kept off until the outside temperature has reached 18.3°C, assuming the internal solar gains are at the design maximum (the design conditions were that the occupant, equipment and solar gains were 6.5 kW). With a reduced solar gain the free cooling could be extended even further. It is therefore not surprising that in moderate summer climates such as in the United Kingdom, where mean summer temperatures are about 17°C, the chiller is rarely required to run at full output. And with careful building design, mechanical ventilation or even natural ventilation can suffice.

In Examples 10.1 to 10.6 the cooler coil schedule required the chiller to be switched on when t_{ai} reached 21°C. But this is when the fresh air damper is beginning to open, and when $t_{ai} = 21°C$ it is fully open and the free cooling is at its maximum. Therefore it would be sensible to bring the chiller on when the free cooling cannot cope with the full heat gains to the room; this would be when the room temperature, t_{ai}, is 22°C. With the same control band, the cooling coil would be at its maximum when $t_{ai} = 24°C$.

As a protective override, so that the dampers are not open if t_{ao} is greater than t_{ai}, an outside temperature sensor signal can also be used as a logic signal. This is a necessary precaution because, when the chiller is switched on at say 22°C (Example 10.7), the fresh air will be helping the chiller. But when $t_{ao} > t_{ai}$ it will be adding to the chiller's load. Such a logic signal could be written for a BEMS outstation with a logic function and comparator function, or as a computer statement:

```
IF t_ao ≥ t_ai THEN outside supply fresh air damper at
minimum
           AND extract damper at minimum
           AND recirculation damper at maximum
```

The other dampers would also have to be adjusted in parallel with the fresh air damper, as shown by the AND statements. If the outside sensor were not installed then there might be occasions when the chiller would be working to cool both the room and the full fresh air supply, as shown by Example 10.8.

Example 10.8
If the outside temperature in Example 10.7 had been 24°C and the room was being cooled by the cooling coil with a full fresh air supply, what would have been the room temperature? Assume there was no outside air temperature control configured in the system. Also assume that the occupant, equipment and solar gains remained at 6.5 kW.

Solution
We can use the control equation for the fresh air damper from Example 10.7:

$$x = 0.8t_{ai} - 16.6 \tag{10.23}$$

But we cannot use heat balance equation (10.24) for the room with free cooling:

$$x\dot{m}C_p(t_{ai} - t_{ao}) + [\Sigma(UA) + NV/3](t_{ai} - t_{ao}) = 6.5 \text{ kW}$$

or

$$(1.25x + 0.5)(t_{ai} - t_{ao}) = 6.5 \tag{10.24}$$

It does not hold now as $(t_{ai} - t_{ao})$ is negative, so there is no cooling effect from the fresh air or the fabric and infiltration loss. The chiller has to be on, so we have to use balance equation (10.16) for the room with the cooling air from the chiller:

$$\dot{Q}_{in} = x\dot{m}C_p(t_{ao} - t_s) + (1 - x)\dot{m}C_p(t_{ai} - t_s)$$
$$= [\Sigma(UA) + NV/3)](t_{ao} - t_{ai}) + 6.5 + x\dot{m}C_p(t_{ao} - t_{ai}) \tag{10.16}$$

The control equation for the chilled supply air, t_s, with its now revised schedule (full output at 24°C and zero output at 22°C) is

$$t_s = 88 - 3t_{ai} \qquad (22°C \leqslant t_{ai} \leqslant 24°C) \tag{10.25}$$

and the fresh air damper control equation is

$$x = 0.8t_{ai} - 16.6 \qquad (21°C \leqslant t_{ai} \leqslant 22°C) \tag{10.23}$$

The chiller will come on when $t_{ai} = 22°C$, when $x = 1.0$. From Example 10.7 this is when $t_{ao} = 18.3°C$. So the fresh air proportion is 1.0, which makes equation (10.16) become

$$\dot{Q}_{in} = 1.25(t_{ao} - t_s) = 0.5(t_{ao} - t_{ai}) + 6.5 + 1.25(t_{ao} - t_{ai})$$

Substituting the control equation, equation (10.25) into this equation:

$$1.25(t_{ao} - 88 + 3t_{ai}) = 1.75(t_{ao} - t_{ai}) + 6.5$$

$$1.25t_{ao} - 110 + 3.75t_{ai} = 1.75t_{ao} - 1.75t_{ai} + 6.5$$

$$5.5t_{ai} = 116.5 + 0.5t_{ao}$$

$$\therefore \qquad t_{ai} = 21.2 + 0.09t_{ao} \qquad (10.26)$$

The outside fresh air should be reduced to a minimum when t_{ai} and t_{ao} are equal. Beyond this point the outside air becomes a heat gain and does not add free cooling. From equation (10.26) this temperature is when $t_{ai} = t_{ao}$, so

$$0.91t_{ai} = 21.2$$

$$t_{ai} = 23.3°C$$

When $t_{ao} = 24°C$ equation (10.26) gives

$$t_{ai} = 21.2 + 0.09(24)$$

$$t_{ai} = 23.4°C$$

Example 10.8 indicates that the fresh air supply should be reduced to a minimum when the outside air is no help in cooling the building and therefore of no help to the chiller. But this will not be a fixed inside or outside temperature as the inside gains will vary, especially Q_s. So it would be wise to include an outside temperature sensor for determining when to close the fresh air supply damper to minimum.

10.6 Deadband

Having considered the control of the heater battery and the cooling coil, it is necessary to ensure that the heater battery does not overheat the zone so that the chiller and cooling coil have to operate. Likewise, the heater battery should not come into operation if the cooling coil overcools. In the examples up to now, the heating was controlled for

$$19°C \leqslant t_{ai} \leqslant 21°C$$

the free cooling for

$$21°C \leqslant t_{ai} \leqslant 22°C$$

and the cooling coil for

$$22°C \leqslant t_{ai} \leqslant 24°C$$

It would be better to introduce a **deadband** of say 1 K between the heating going off and the free cooling coming on to stop any possible waste of

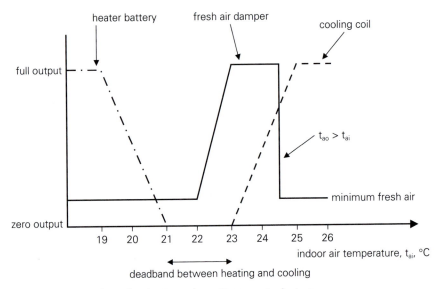

Fig. 10.6　Heating, fresh air and cooling control strategy.

energy. A deadband of 1 K or 0.5 K should be sufficient, e.g. starting the fresh air damper control at 22°C and the cooling coil at 23°C (Fig. 10.6). This type of air conditioning and its control is often called **sequence control** as the cooling and heating coils are operated in sequence instead of together, which can happen in **dew point** or reheat control (see later).

10.7　Humidity control

Only sensible heating and cooling have been considered so far, but in many climates humidity control is the most important aspect of air conditioning. As with sensible heating and cooling, humidification has its own control equations and **moisture content** balance equations. **Relative humidity** rather than the moisture content of the air is the important factor in air-conditioning control, so the **psychrometric chart** is a useful way to relate the dry bulb air temperature and the moisture content to the relative humidity.

Molecules of water exist as vapour in the air and they exert their own pressure in addition to the pressure of the dry air. Hence the **relative humidity**, φ (%), is defined as the following ratio [7]:

$$\varphi = 100p_v/p_s \qquad (10.27)$$

where p_v = water vapour pressure in air (kPa)

p_s = saturated vapour pressure at same temperature (kPa)

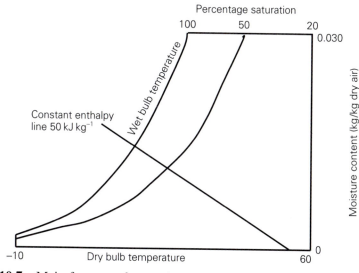

Fig. 10.7 Main features of a psychrometric chart.

The **saturated vapour pressure** (SVP), symbol p_s, is the maximum pressure the vapour can exert at that temperature and occurs when the air cannot contain any more vapour – it is saturated. SVP is a function of the temperature in the vapour and the surrounding air. As the temperature rises, so the air can absorb more vapour and the SVP rises. At 20°C the SVP is 2.337 kPa, which compares with the atmospheric pressure of 101.325 kPa. Further values of SVP can be found in the CIBSE Guide [7]. A psychrometric chart is often used for air-conditioning calculations (Fig. 10.7).

The vertical axis is the moisture content of the humid air, g, the mass of water vapour present in 1 kg of **dry air**. This is sometimes called the **humidity ratio**, the **specific humidity** or the **absolute humidity**. The horizontal axis is the **dry bulb temperature** and the curved line is the saturation moisture content line, or simply the **saturation line**. This yields another term, the **percentage saturation**, μ, which is the ratio of the moisture content in the air, g, to the moisture of the air when saturated, g_s:

$$\mu = 100g/g_s \tag{10.28}$$

The moisture content of the air can be approximately calculated from the vapour pressure, p_v, and the atmospheric pressure, p_{atmos}, via Dalton's law of partial pressures and the general gas law, assuming that the vapour and dry air behave as perfect gases [2]:

$$g = 0.622p_v/(p_{atmos} - p_v) \tag{10.29}$$

For saturated air at 20°C, from the figures given earlier, the moisture content is therefore

$$g = 0.622 \times 2.337/(101.325 - 2.337)$$

$$= 0.0147 \text{ kg/kg dry air}$$

In order to account for both the sensible heat in the humid air and the latent heat of evaporation of the water vapour, **enthalpy** is used. The specific enthalpy, h, of humid air (specific as it is related to 1 kg of dry air) is given as follows [7]:

$$h = h_a + gh_g$$

where h_a and h_g are the specific enthalpies of dry air and water vapour respectively. In numerical terms it is

$$h = 1.005t + g(1.89t + 2501) \qquad \text{kJ/kg dry air} \qquad (10.30)$$

where t is the dry bulb temperature above 0°C, 1.005 kJ kg^{-1} K^{-1} is the specific heat capacity of dry air, 1.89 kJ kg^{-1} K^{-1} is the specific heat capacity of the water vapour, and 2501 kJ kg^{-1} K^{-1} is the latent heat of evaporation of the water vapour at 0°C. So the specific enthalpy is the total heat energy, related to 0°C, of 1 kg of humid air.

Equation (10.30) can be rearranged to give

$$h = (1.005 + 1.89g)t + 2501g \qquad (10.31)$$

or

$$h = C_{pav}t + 2501g \qquad (10.32)$$

where C_{pav} = the humid specific heat capacity
　　　　　= $1.005t + 1.89g$
　　　　　= the sensible heat of the moisture and dry air
A practical average value for air conditioning is $C_{pav} = 1.025$ kJ kg^{-1} K^{-1}.

From equation (10.32) it can be seen that the latent heat part of the enthalpy is considerable, and this has important energy implications for the control of humidity (Example 10.9). Dehumidification and humidification are both part of the control of humidity, like heating and cooling to control the dry bulb temperature. Typically, dehumidification is required in the summer and humidification in the winter. The usual method of dehumidification is when the air is cooled, by a cooling coil, to its dew point so that water condenses out. The air can then be heated by a heater battery, the **after-heater**, to obtain a suitable supply air temperature and humidity. As with sensible heating and cooling, dehumidification and humidification need to be separated by a deadband.

Example 10.9

Calculate the cooling power required to dehumidify $1 \text{ m}^3 \text{ s}^{-1}$ of air at 70% RH and 21°C dry bulb temperature to 50% RH and 21°C dry bulb temperature. How much would this reduce the dry bulb temperature if this same cooling power were applied to sensible cooling only, with the air initially at a dry bulb temperature of 21°C?

Solution

The calculations can be done from the psychrometric chart but it is more accurate to use psychrometric tables [7]. Although relative humidity is regularly quoted, the psychrometric chart has percentage saturation, the more correct measure of humidity. From the CIBSE tables the air at 21°C, 70% saturation has a relative humidity that is actually 70.52%. The enthalpy is 49.06 kJ kg^{-1}. At 50% RH the enthalpy is 41.08 kJ kg^{-1}. Although the flow rate of the air is in $\text{m}^3 \text{ s}^{-1}$, the mass flow rate is needed because the enthalpy and moisture content are in terms of the mass of dry air. The conversion uses the **specific volume**, v ($\text{m}^3 \text{ kg}^{-1}$), the volume of air containing 1 kg of dry air and its associated moisture content. From the CIBSE Guide [7] this is $0.8476 \text{ m}^3 \text{ kg}^{-1}$ at 20°C and 70% saturation. The relationship between density, ρ, and specific volume is

$$\rho = (1 + g)/v$$

so

$$\rho = (1 + 0.011)/0.8476$$

$$= 1.193 \text{ kg m}^{-3}$$

not much different from the standard value of 1.2 kg m^{-3}. Traditionally, density is used for airflow measurements in pressure drop devices, such as orifice plates, but specific volume is commonly used for psychrometric calculations in air conditioning.

So the mass flow of air is 1.18 kg s^{-1} and the cooling power to dehumidify the air from 70% to 50% is

$$1.18 \times (49.06 - 41.08) = 9.4 \text{ kW}$$

We can check the humidity from equation (10.31):

$$h = (1.005 + 1.89g)t + 2501g \tag{10.31}$$

where $g = 0.011 \text{ kg/kg}_{da}$ (kg_{da} = mass of dry air) and $t = 21°C$, giving $h = 49.05 \text{ kJ kg}^{-1}$. For sensible cooling only the final air temperature, t_{final}, would be given by

$$\dot{m} C_p (21 - t_{final}) = 9.4$$

$$1.193 \times 1.02 \times (21 - t_{final}) = 9.4$$

$$t_{final} = 13.3°C$$

Table 10.1   The heat output of a person (sensible/latent)

Dry bulb temperature (°C)	15	20	22	25
Heat output seated at rest (W)	100/15	90/25	80/35	65/55
Heat output during light work (W)	110/30	100/40	90/50	70/70

Example 10.9 shows that dehumidification and humidification can require significant energy inputs, but in most cases the relative humidity does not have a large effect on the overall thermal comfort [4]. A 40% RH change can be approximately compensated by a 1 K change in dry bulb temperature in a temperate climate.

Although the outside air has to be humidified or dehumidified for comfort, a significant source of moisture, and the consequent latent load, comes from the occupants. A seated person at rest when the dry bulb air temperature is 20°C produces 90 W of sensible heat and 25 W of latent heat [9]. When the person does light work, these values rise to 100 W and 40 W respectively (Table 10.1).

10.7.1   Dew point control

To maintain a room or zone at the required dry bulb and relative humidity, in summer the supply air must absorb the water vapour and cool the zone, i.e. it must absorb the latent and sensible loads. To do this the supply air condition must lie on the **room ratio line**, also known as the room process line or the **room sensible heat ratio** (RSHR); see Fig. 10.8. Here R is the required room condition, the RSHR line can then be drawn and S, the supply condition, must lie on the line. The heat balance for the supply air is

$$\dot{Q}_{\text{sensible}} = \dot{m}C_{\text{pas}}(t_R - t_S) = \dot{m}(h_X - h_S)$$

$$\dot{Q}_{\text{latent}} = \dot{m}h_{\text{fg}}(g_R - g_S) = \dot{m}(h_R - h_X)$$

The right-hand side is in terms of enthalpy. C_{pas} is the specific heat of humid air (typically $1.02 \text{ kJ kg}^{-1} \text{ K}^{-1}$) and h_{fg} is the latent heat of evaporation (2454 kJ kg^{-1} at 20°C).

To control the room air to the required condition, the cooler coil must be controlled to supply the air at the correct condition. In dew point control, the intake air to the cooler coil is cooled below its dew point to near that of the cooling coil, the ADP (Fig. 10.9). The line joining O, C and ADP is the coil process line. The control is via a temperature sensor after the cooling coil. An ordinary dry bulb sensor is used because it is assumed that the air is

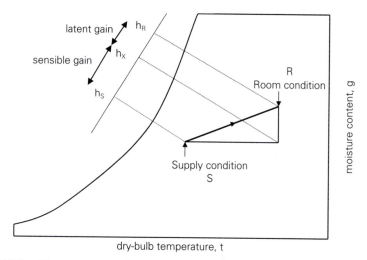

Fig. 10.8 The room ratio line.

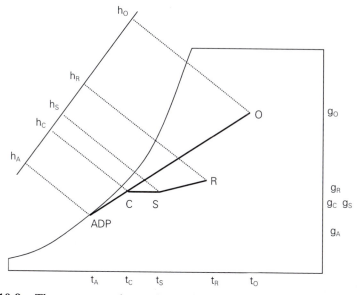

Fig. 10.9 The apparatus dew point and supply air condition.

almost saturated just after the cooling coil, and the sensor can therefore measure the dew point effectively.

Figure 10.9 shows the summer outside air, O, without any recirculation for simplicity. The heat exchange process is not 100% efficient, as expressed by the **coil contact factor**:

$$\beta = \frac{g_O - g_C}{g_O - g_A} = \frac{h_O - h_C}{h_O - h_A} = \frac{t_O - t_C}{t_O - t_A} \qquad (10.33)$$

The ADP depends on the design of the cooling coil and the airflow rate but a typical value is 0.7 to 0.8. In fact, the coil is like a counterflow heat exchanger and its ADP is between the chilled water supply and return temperatures. Typically it would have a value between 6°C and 11°C.

To control the room condition adequately, the supply air from the cooler coil needs to be reheated by the **after-heater** to ensure the supply air is on the room line ratio. The after-heater is controlled by a signal from a temperature sensor after the heater battery; this signal can be reset from a sensor in the return air duct measuring the room temperature. If the supply air is not on the room line ratio, it will absorb the sensible and latent heat gains and move on a parallel room line to the design room line. Consequently, the ultimate room condition will not be the design room condition. This is why the reheater is required, to get the supply air onto the room line.

10.7.2   Sequence control and dehumidification

It is wasteful of energy to have both heating and cooling. So often the cooling and heating coils are controlled in sequence, i.e. either one or the other is operating for heating and cooling, with a deadband between them. Both the cooler and heater battery are controlled by a signal from the same temperature sensor after the heater battery. This may well be reset from a signal from a temperature sensor in the return duct. As the cooling coil is controlled on the sensible heat to be absorbed, this does mean that the RH of the room can rise as the ADP rises to ADP' (Fig. 10.10).

There is some dehumidification depending on the ADP of the cooling coil but it is not controlled on its latent cooling. When the humidity does rise too high, e.g. above 60%, then a high-limit humidity sensor can be installed in the return duct to override the sequence control and open the diverting valve on the cooling coil so that it operates at maximum output. The after-heater also operates so that the system becomes dew point controlled to dehumidify the air, until the RH sensor measures a lower RH, e.g. 50%. Then the system returns to sequence control.

Face and bypass dampers (Fig. 10.11) can also help to control the humidity. Here the cooler coil is smaller than the supply duct so that some of the air can bypass the cooler coil. However, the resistance of the damper with the coil behind it is much larger than the resistance of the bypass damper, and control of the airflow can be difficult. To make control easier, a parallel blade damper can be used for the bypass, and an opposed blade damper for the coil [10].

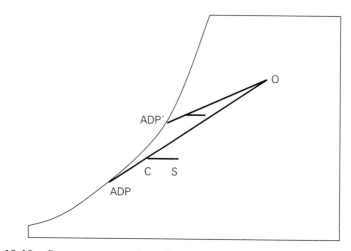

Fig. 10.10 Sequence control cooling.

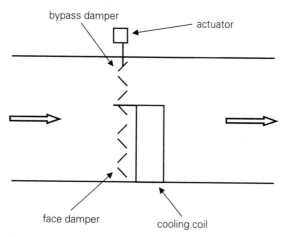

Fig. 10.11 Face and bypass dampers.

Evaporative cooling is another aid to humidity control, especially in warm, dry climates. Water, at the ambient temperature, is sprayed into the supply air (Fig. 10.12) and sensible heat in the air is used to evaporate the water, making the supply air moister but cooler. No heat is supplied or extracted to the air; it is an adiabatic process and the enthalpy of the air does not change, hence the evaporative cooling process on the psychrometric chart follows a

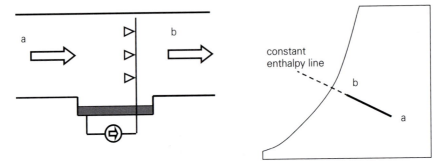

Fig. 10.12   Evaporative cooling.

constant enthalpy line. This is useful in hot, dry climates but it does not work when the outdoor humidity rises to higher levels.

Humidification may well be required in winter as the cold outside air does not contain much water vapour and when heated it can be very dry and uncomfortable. A steam humidifier in the duct or the room can raise the humidity. Other forms of humidifier [11] use atomizing jets, but it is important to take care over water treatment and to avoid unhealthy conditions. Older alternatives to the steam humidifier are sprayed cooling coils and water spray devices [2]. However, with the danger of legionella, they are rarely used in modern installations. The control is via a humidity sensor in the return duct or the room itself. Humidity sensors need a lot of maintenance and calibration as they can be subject to drift and contamination. Capacitive sensors are available using thin polymer films. The more expensive dewpoint sensors use a chilled mirror [10].

10.8   Chilled water control

Often with their own packaged controls, chillers can be interfaced to the BEMS but the packaged controls are rarely altered by the BEMS as there are pressure and temperature controls for safe operation. Constant switching of the compressor by the BEMS will be detrimental to the compressor and its motor. However, the BEMS often controls the chilled water temperature to the cooler coil, and the flow water temperature does have an influence on the efficiency, or coefficient of performance, of the chiller. A typical chiller and its component temperatures are shown in Fig. 10.13.

The **coefficient of performance** (COP) of the chiller can be approximately worked out from the Carnot cycle [12, 2] on the entropy–temperature diagram (Fig. 10.14). Entropy (J K^{-1}) measures the 'quality' of the heat; it is the quantity of heat divided by the absolute temperature. Entropy is useful in thermodynamics and measures the disorder of the molecules of a substance.

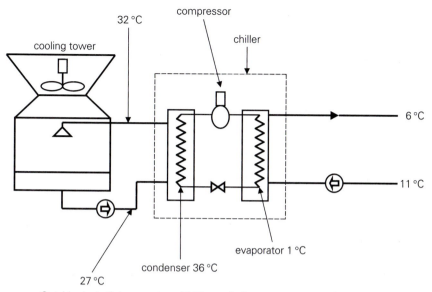

Fig. 10.13   Typical chilled water and chiller temperatures.

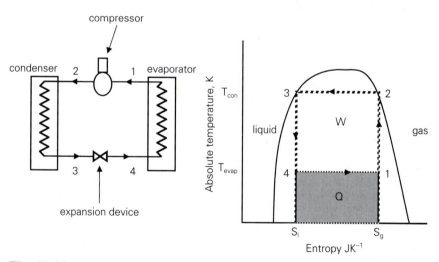

Fig. 10.14   A reversed Carnot cycle for refrigeration.

In Fig. 10.14 the phase change of the refrigerant is from gas (right) to liquid (left). The gas is more disordered with its molecules moving fast in random directions, hence its entropy, S_g, is greater. The perfect Carnot cycle is shown with heat $Q + W$ being ejected at the condenser temperature, T_{con} (K), and heat Q being absorbed at the evaporator temperature, T_{evap} (K). The COP is the ratio of the heat absorbed to the work done; (W):

$$\text{COP} = \frac{Q}{W}$$

$$= \frac{T_{evap}(S_g - S_1)}{(T_{con} - T_{evap})(S_g - S_1)}$$

$$= \frac{T_{evap}}{(T_{con} - T_{evap})} \qquad (10.34)$$

For the typical temperatures shown in Fig. 10.13 (a conventional air-conditioning system supplying air at 13°C off the cooling coil), the COP is

$$\text{COP} = (273 + 1)/(36 + 273 - 1 - 273)$$

$$= 7.83$$

But this is a perfect cycle and in reality the practical COP is approximately half this:

$$\text{COP}_{pract} = 3.92$$

This means that 3.9 kW of heat can be absorbed by the evaporator for every 1 kW of electricity supplied to the compressor. To accurately work out the practical COP of a chiller, it requires details of the evaporator, condenser, compressor and refrigerant plus a pressure–enthalpy chart. However, this Carnot COP is a useful rule of thumb. For instance, it shows that if the condenser temperature rises by 1 K then the COP reduces to 3.81, a decrease of 0.11. Likewise if the evaporator temperature decreases by 1 K then the COP reduces to 3.79, a decrease of 0.13.

This has several control implications. If the supply air temperature is allowed to rise, the evaporator temperature can be raised and the COP will rise. This is an advantage for displacement ventilation, where cool air is introduced at floor level just below the occupied temperature and hence above the normal air-conditioned temperature of 13°C.

10.9   Non-linearity

The control of the cooling coil and heater battery is interesting as there is non-linear heat exchange between the air and the chilled or hot water. The heater battery can be considered as a **counterflow heat** exchanger. (If the battery were connected with its flow water at the air inlet side, a parallel flow heat exchanger, it would not be so efficient and the air could not be heated to as high a temperature.) Using equation (10.15) the heat balance at the battery is

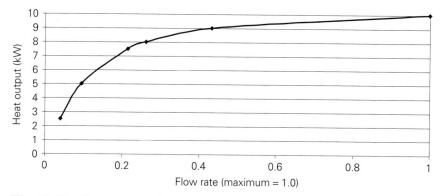

Fig. 10.15   Heat output of a heater battery.

$$\dot{m}_{\text{water}} C_{p\,\text{water}}(t_{\text{f}} - t_{\text{r}}) = CUA\left[\frac{\Delta t_{\max} - \Delta t_{\min}}{\ln\left(\dfrac{\Delta t_{\max}}{\Delta t_{\min}}\right)}\right] = \dot{m}_{\text{air}} C_{\text{pair}}(t_{\text{s}} - t_{\text{inlet}})\quad(10.35)$$

The left-hand side refers to the hot water and the right-hand side to the warmed air, where t_{inlet} is the temperature of the air into the heater battery and t_{s} is the supply air temperature out of the battery. Figure 10.15 shows this non-linear output for varying water flows through the battery. Here the design conditions are that the heater battery provides 10 kW of heating to supply air to the room at 28°C when it is 0°C outside, and 80% of the room air is recirculated. The design room air temperature is 20°C and the heat input to the air is 10 kW. The hot water design flow and return temperatures are 80°C and 70°C respectively. Putting these values into equation (10.35) and solving it for various loads using Microsoft Excel Solver produces Fig. 10.15.

Equation (10.35) is difficult to use for matching the valve to the battery output, to linearize it, as the equation cannot give a zero output. A practical valve can shut off the flow, although there may be some slight leakage with age and quality of design. Petitjean [13] suggests that instead of an equation based on the LMTD equation, producing an equal percentage characteristic, a more practical valve equation is

$$V = V_{\max}\left[\frac{\Phi}{(1/s) - (1 - \Phi)}\right]$$

When the valve stroke position, s, is zero then the flow, V, is also zero.

For good control, a linear relationship between the control signal and the heat output is required, so a small change in the control valve stem position does not produce a very large or very small change in the heat. The valve

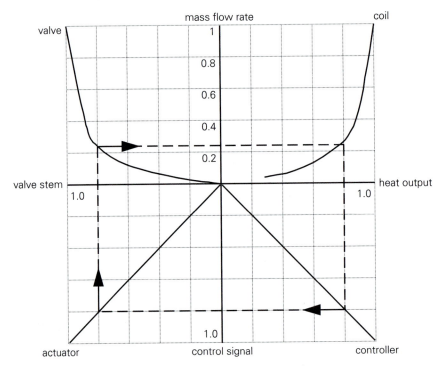

Fig. 10.16   Valve matching to linearize the heater battery output.

can be matched in non-linearity to make the overall response of the coil linear (Fig. 10.16). This is often called the spider's web diagram and can be used to design a valve for a coil. Each quadrant shows a characteristic of each control element. Those for the controller and actuator are linear but the valve is the mirror image of the battery output. For 80% heat output, the controller signal to the actuator is 80% and the valve stem is 80% of its full travel. However, the valve only lets through just over 20% mass flow rate, corresponding to the required flow rate through the coil to produce 80% heat output.

10.10   Variable air volume (VAV) systems

Varying the speed of a fan as its load changes will alter its power consumption, so compared to a conventional a.c. system with fixed fan speed, it can save energy [14]. An alternative is to raise the supply air temperature, for a reducing cooling load, but this would cause more chilled water to be diverted back to the chiller, not the most efficient energy-saving method. Directly raising the chilled water temperature would save more energy and it would have to be considered carefully as a strategy in conjunction with fan

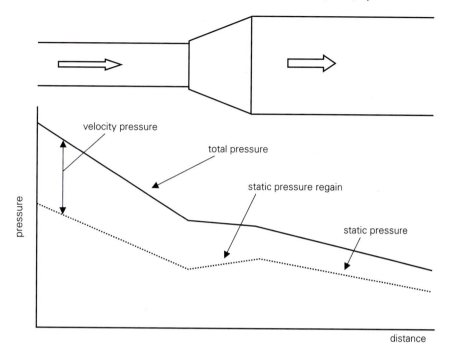

Fig. 10.17   Static pressure regain.

speed control. However, the a.c. fan runs during the full year of occupancy, whereas in temperate climates the chiller may not run when fresh air is providing the free cooling.

Before analysing VAV systems in detail, consider the different types of pressure present in a ducted air system. In air systems the velocity pressure, $\frac{1}{2}\rho v^2$, is significant, whereas in water systems it is comparatively small and seldom considered. So for air systems the total pressure consists of the dynamic pressure and the static pressures:

$$\text{total pressure} = \text{static pressure} + \text{velocity pressure} \qquad (10.36)$$

The static pressure generated by the fan overcomes the friction losses of the air moving down the ducts. If there is no flow then the static pressure is the pressure which is pushing against the duct walls. The losses due to fittings in the duct are proportional to the velocity pressure.

Equation (10.36) is a form of the energy conservation equation. For a length of duct, the total pressure at the entry equals the total pressure at the end of the duct minus the friction losses. If the duct increases in diameter (Fig. 10.17) then the velocity pressure decreases but there is no loss in energy, so the static pressure increases; this is the **static pressure regain**. In

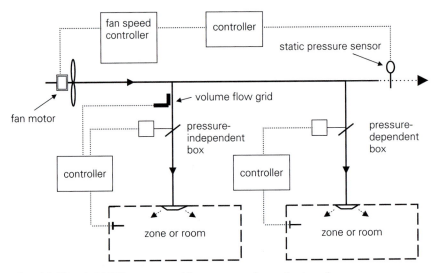

Fig. 10.18   A VAV system with pressure-dependent and pressure-independent boxes.

practice there is some pressure loss at an expansion, called the **recovery factor**, typically between 0.7 and 0.9.

The static regain method is the standard method of sizing high-velocity duct systems, often found in VAV systems as the main ducts. The other main duct-sizing design method is **equal friction**. This is used on secondary, medium- and low-velocity ducts in VAV systems and also on the return duct system. Although the static regain method is recommended for high-velocity systems [15], it does lead to larger duct sizes than the constant friction method. As static regain can occur in VAV systems, it means that the total pressure (velocity pressure plus static pressure) must be considered in VAV calculations even though the system is controlled by a static pressure signal.

Figure 10.18 shows a VAV system with pressure-dependent and pressure-independent VAV terminal units or boxes. Both types of box are shown to demonstrate the differences, although it is unlikely there would be both types in one system. The pressure-independent box has a flow grid upstream of the damper with holes facing into the airstream to measure the total pressure, and holes facing downstream to measure the static pressure (Fig. 10.19). The difference is the velocity pressure, and a thermistor arrangement (acting as a hot wire anemometer) converts the velocity pressure into an electric signal proportional to the flow rate [16]. The box controller then produces an error signal by comparing the measured flow rate with the required flow rate, as determined by a proportional-only or PI signal from the room

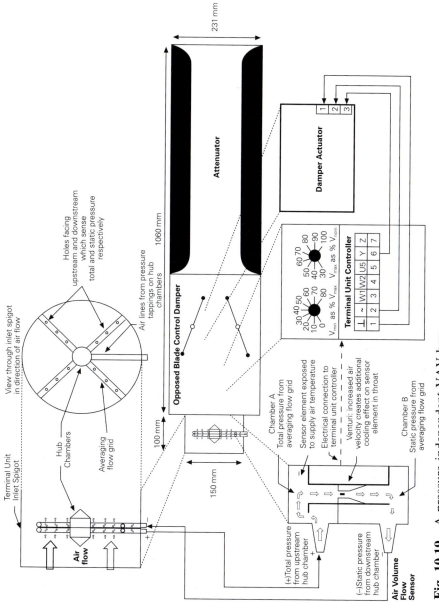

Fig. 10.19 A pressure-independent VAV box.

temperature sensor. This is a form of reset or cascade control so common in HVAC systems. However, the damper actuator has its speed controlled by a proportional-only controller, so that it is effectively an integral controller of damper angle:

$$\frac{d\vartheta}{dT} = K_p e$$

$$\vartheta = K_p \int e \, dT$$

where ϑ is the damper angle, K_p is the proportional gain and e is the error in flow rate. There is often a deadband around the setpoint angle of the damper to reduce hunting and instability. Typically for a $0 \, V_{dc}$ to $10 \, V_{dc}$ control signal to the actuator, 6 V represents the setpoint and the deadband is ± 0.23 V, which for a box with a full flow of $210 \, l \, s^{-1}$ would represent $\pm 8 \, l \, s^{-1}$. Between 6.23 V and 10 V the damper would open, and from 5.77 V to 2 V the damper would close. A pressure-dependent box has no flow grid and generally there is a pressure-regulating damper upstream to maintain a reasonably constant pressure and achieve the required flow rate into the room.

The speed of the fan is controlled via a fan speed controller, often a static frequency inverter (an efficient way of controlling an induction motor). A signal is sent from the static pressure sensor to a PI controller (Fig. 10.18).

A static pressure sensor is used as it is cheap to implement. The total pressure or the velocity pressure is measured using a **pitot static tube** or a flow grid. The pitot tube's positioning is crucial as the velocity, and therefore the velocity pressure, varies across the duct. This means it is easier to use a static pressure sensor to assess the pressure in the system. As the boxes close down with reducing cooling loads, the pressure is built up in a constant volume, constant speed system. The system will ride up the fan curve (Chapter 13). However, the static pressure sensor picks up the increase in pressure and the fan speed is reduced.

10.10.1   Synchronous and asynchronous systems and savings

VAV systems vary the fan speed according to the load on the individual boxes. Most loads, such as occupancy, equipment and lighting gains, are fairly constant but solar loads do vary during the day. VAV systems can capitalize on such load diversity by adjusting the fan output according to the overall load. To understand how VAV systems perform, consider two extreme situations [17].

In a totally synchronous system the loads vary together (Fig. 10.20). Here the fan has to vary by a large amount because the load varies so considerably. The diversities of the boxes vary synchronously. Diversity, x,

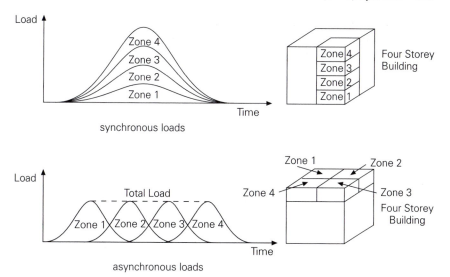

Fig. 10.20 Loads: (a) synchronous and (b) asynchronous.

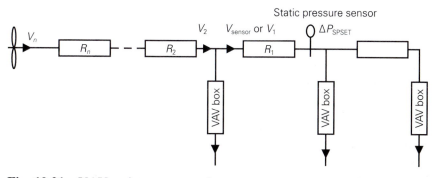

Fig. 10.21 VAV resistance network.

is the ratio of the actual volume flow $V(T)$ at time T to the maximum design value V_{des}. For a synchronous system with n boxes we have

$$x_1(T) = x_2(T) = x_3(T) = \ldots = x_i(T) = \ldots = x_n(T)$$

However, there is little variation in the fan speed for a totally asynchronous system. For an n-box asynchronous system we have

$$\sum_{i=1}^{i=n} x_i = \text{constant}$$

The slight variation in the fan speed will arise as the index duct resistances to the different zone boxes will be different (Fig. 10.21). However, the fan size in the asynchronous system may not be as large as for a constant volume system, where the peak loads may have been added together at the design stage without any consideration of diversity. However, the asynchronous system is a constant volume system, although the variable-speed fan system is flexible for possible changes in the building zones or refurbishment.

Most installed VAV systems lie somewhere between synchronous and asynchronous. However, the greatest savings are made with the synchronous systems. The savings calculations for VAV systems are very similar to the calculations for variable-volume water systems (Chapter 13). The difference between water systems and air systems is that velocity pressure and static pressure regain are not considered for water systems but they can be important for air systems.

VAV fans are often of the backward-curved centrifugal type and the fan total pressure can be approximated to a simple parabolic equation:

$$\Delta P = \Delta P_0 \left(n^2 - (V/V_0)^2 \right) \tag{10.37}$$

where n is the fan speed ratio $(0 \leqslant n \leqslant 1)$; $n = 0$ indicates zero speed and $n = 1$ indicates full speed. V is the volume flow rate, ΔP is the total pressure drop, ΔP_0 and V_0 are the curve intercepts on the total pressure and volume axes. The velocity, v, is related to V through the duct diameter. Resistances and network analysis are considered in Chapter 13, but note that the operating point of the VAV system is where the system total pressure resistance $(R_{static} + \frac{1}{2}\rho A^{-2})$ intercepts the fan total pressure curve.

The maximum possible savings occur when the VAV system goes to zero volume flow, an unlikely possibility but a useful indication of the minimum speed of the fan. As with water systems (Chapter 13), the position of the control sensor – for VAV systems this will be the static pressure sensor – is crucial to the magnitude of the savings. The total pressure of the fan at part load is

$$\Delta P_{SPSET} + \frac{1}{2}\rho \frac{V^2_{sensor}}{A^2_{sensor}} + \sum_{i=1}^{n} R_i V_i^2$$

where Fig. 10.21 shows the values. As the static pressure sensor is moved further away from the fan along the index run, ΔP_{SPSET} reduces and the pressure drop along the index run, $\sum R_i V_i^2$, increases. At zero volume flow, an extreme condition used to give an indication of the likely savings in a VAV system, the total pressure reduces to ΔP_{SPSET}, the control setpoint. The fan speed can be found by equating this to equation (10.37). With the full-load fan pressure as ΔP_0 (derived in Chapter 13), the potential power saving of a VAV system is given by

$$\frac{2}{3}\left(\frac{\Delta P_0}{\eta_{\text{DES}}}\right)V_{\text{des}} - \left[\frac{\Delta P_{\text{SPSET}} + \frac{1}{2}\rho\frac{V_{\text{sensor}}^2}{A_{\text{sensor}}^2} + \sum_{i=1}^{n}R_iV_i^2}{\eta_{\text{PL}}}\right]V_{\text{n}} - \chi \qquad (10.38)$$

where ΔP_{SPSET} = the static pressure setpoint

V_{des} = the full-load design flow rate

V_{n} = the part-load total volume flow at the fan

η_{DES} = the total efficiency (fan and motor) at the design full load

η_{PL} = the total part-load efficiency

χ = the power into the fan when there is zero flow rate and hence zero power into the air

The last two terms in equation (10.38) give the part-load power, but when there is zero flow ($V_{\text{n}} = 0$) then the middle term becomes zero and χ is the power into the fan system. As the static pressure sensor moves away from the fan, ΔP_{SPSET} reduces and the potential savings increase. However, there is more likelihood of starving some boxes as ΔP_{SPSET} reduces and the fan can reduce to lower speeds, giving lower total pressures.

Starvation can also occur on start-up of a VAV system as most of the boxes will be open and those nearest to the fan will take a greater proportion of the air. As VAV systems often serve open-plan areas with few walls for positioning sensors, it can be a problem to choose the locations for the room temperature sensors. It is possible to use aspirated sensors, which draw room air through them, but they are primarily sensing warm air near the ceiling. Appropriate offsets will be required to relate them to the room air temperature. Open-plan areas may also allow air from one zone to diffuse into an adjacent zone. If there are different setpoints to the boxes serving the two zones, or different loads in the two zones, then the stray air will affect the control of the boxes. An extreme example to illustrate this is where one zone is being heated (by a reheat coil in the box) and the other box is cooling its zone. Cool air straying into the warmed zone will increase the heat output from the heating box and any warm air straying into the cooled zone will increase its cooling output – a recipe for instability.

Maintenance of the minimum fresh air to a zone can also be a problem with VAV systems when the main air-handling plant's fresh air intake is at a minimum (the fresh air damper is at its minimum opening), and the zone box is at a minimum flow rate as the cooling load is low. A compromise between energy efficiency and fresh air supply has to be determined and this may well depend on having box controllers linked to the BEMS to determine the situations in the zone boxes.

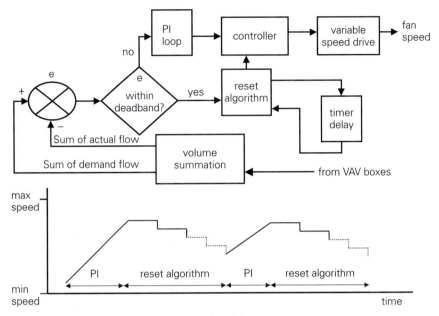

Fig. 10.22   Tung and Deng reset algorithm.

10.10.2   Box polling

With the increasing use of small BEMS outstations or serially interfaceable controllers for VAV boxes, it is possible to assess, or **poll**, the boxes for their cooling demands [18]. A simple control strategy is to maintain the fan speed so there is only one box fully open. If more than one box is fully open then the fan speed is increased until only one box is fully open, or preferably just below fully open, so that this box is also satisfied. If all the boxes are less than fully open, the fan speed can be reduced until one approaches fully open. Limits are required to ensure that malfunctioning boxes or poorly designed zones do not dictate the control of the supply fan. One system that was implemented in fact used three boxes on their low-limit alarms as the signal to increase fan speed by raising the static pressure upwards by 5% every 60 s [19, 20].

Figure 10.22 shows the control algorithm of Tung and Deng [23] where standard PI control is used until the summated flow is within a 'deadband' of the required total flow. Here the static pressure is reset down until the summated flow error moves out of the deadband.

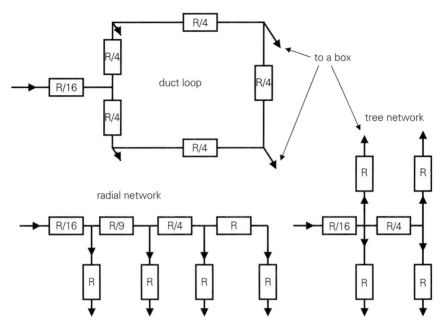

Fig. 10.23   Networks: duct loop, radial and tree.

10.10.3   Duct loops

Duct loops are often used in VAV and other air systems to save energy and ensure a balanced air delivery. Like electric ring mains, the supply duct forms a loop as shown in Fig. 10.23. The size of the duct in the loop, for a symmetrical loop, is based on half the volume flow going round each side of the loop. In some cases the duct loop is given a constant diameter as this makes the fabrication and assembly that much easier and may have cost benefits [21]. Compared to a radial distribution, with one branch at each node (Fig. 10.23) there is a lower resistance and pressure drop for the duct loop.

By network analysis, using the resistance relationship between pressure drop and volume flow, $\Delta P = RV^2$ (Chapter 13), the pressure drop across the duct loop can be calculated assuming each box has a volume flow, V, through it. The loop is considered at full output, i.e. all the boxes are of similar performance and are fully open. The pressure drop is calculated by going round the index circuit to the halfway point of the loop (not including the final resistance of the box and ductwork after it). It is given by

$$\left(\frac{R}{16}\right)(4V)^2 + \left(\frac{R}{4}\right)(2V)^2 + \left(\frac{R}{4}\right)V^2 = 2.25RV^2$$

In comparison the pressure drop across the radial system is

$$\left(\frac{R}{16}\right)(4V)^2 + \left(\frac{R}{9}\right)(3V)^2 + \left(\frac{R}{4}\right)(4V)^2 + RV^2 + RV^2 = 5RV^2$$

The tree network, with two branches from each node, also has a low resistance of

$$\left(\frac{R}{16}\right)(4V)^2 + \left(\frac{R}{4}\right)(2V)^2 + RV^2 = 3RV^2$$

Besides their low resistances, duct loops have several other advantages [22]:

- Provide flexibility for future design.
- Take advantage of diversity.
- Operate at lower and more uniform static pressure than a radial system for any given flow.
- Produce less noise at the first terminal.

A comparison has been made between duct loop and tree distribution systems [17, 23]. Two building layouts are considered, one large and one small (Fig. 10.24).

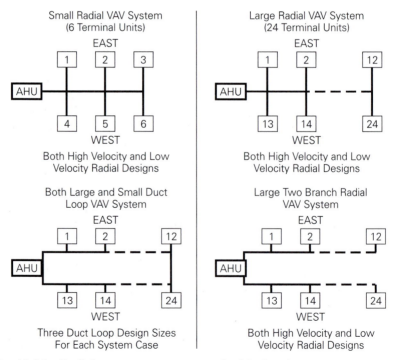

Fig. 10.24   Radial tree systems compared with duct loop systems.

Table 10.2   Radial tree and duct loop designs at full load

System	Volume flow rate $(m^{-3} s^{-1})$	Fan power (W)	Fan power saving % (c.f. high-velocity system)	Fan power saving % (c.f. low-velocity system)	Maximum duct diameter (mm)
HVS radial	4.8	2 732	0	−27	710
LVS radial	4.8	2 156	21	0	800
500 mm DLS	4.8	2 171	21	−1	500
560 mm DLS	4.8	2 074	24	4	560
630 mm DLS	4.8	2 016	26	6	630
HVL radial	19.2	13 084	0	−25	1250
LVL radial	19.2	10 433	20	0	1400
2BHV radial	19.2	14 411	−10	−38	1000
2BLV radial	19.2	9 913	24	5	1120
900 mm DLL	19.2	9 730	25	7	900
1000 mm DLL	19.2	8 939	32	14	1000
1120 mm DLL	19.2	8 442	35	19	1120

HVS = high-velocity small system, LVS = low-velocity small system, DLS = duct loop small system, HVL = high-velocity large system, LVL = low-velocity large system, DLL = duct loop large system, 2BLV = two-branch low-velocity large system, 2 BHV = two-branch high-velocity large system.

Each has a cooling load of 80 W m^{-2} and is divided into 10 m × 10 m zones each supplied by a large box supplying 0.8 m^3 s^{-1} to a number of outlets via an octopus connector and flexible ducting. The static pressure at the inlet to each VAV box is 67 Pa at full flow, which covers the box and its downstream connections. A maximum pressure drop of 250 Pa was taken across the main air-handling unit (cooler coil, heater battery and filters) which, as with many air systems, is where most of the pressure drop occurs. High- and low-velocity radial tree systems were considered. The high velocities were between 10 and 15 m s^{-1}. The low-velocity systems were designed to a pressure drop of 1 Pa m^{-1}. This pressure drop was also used for the start of the duct loop systems but the duct loop had a constant diameter duct. Each half took half the maximum flow with a neutral point halfway round the loop. Part-load calculations for a duct loop with different flows in each half require a numerical method of solution such as the spreadsheet Excel Solver. Table 10.2 shows the comparison of a number of radial tree and duct loop designs at full load. Only air power requirements are considered, i.e. fan and motor efficiencies are not included. The duct loop can save up to 35% of fan power compared to a high-velocity radial tree. At part-load conditions (considered as the west-facing zones having only 50% (0.4 m^{-3} s^{-1}), similar savings were maintained but reduced by up to 2% for each case.

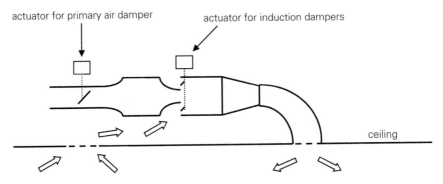

actuator for primary air damper   actuator for induction dampers

ceiling

Fig. 10.25   An induction VAV box.

The primary control strategy for VAV systems is to control the fan speed. However, if the fan speed is maximum and a number of boxes are fully open then the system is not coping with the cooling load. The supply air temperature can be reduced to deal with the increased load. Similarly there is a low limit to the fan speed below which air distribution is poor and fresh air amounts may be limited. So here the supply air temperature can be increased [10, 24].

10.10.4   Induction and fan-assisted VAV boxes

A number of VAV terminal units, or boxes, combine primary cool air with warmer recirculated air induced into the box through a ceiling grille by the primary air. The primary air has to be at a higher pressure than an ordinary VAV system to ensure induction. However, the primary air can be cooler than the normal 13°C, and is suitable for use with off-peak ice and chilled water storage systems. The air into the room is almost constant, independent of the load, and so ensures good room air distribution even at low loads. The induction box has a volume flow grid for the primary air, to give pressure-independent operation, and the primary air damper is controlled by a signal from the room temperature sensor. The primary air damper is also regulated in conjunction with the recirculation damper. As the primary damper closes, the recirculation damper opens (Fig. 10.25).

Instead of relying on induction, a fan can be incorporated into the fan-assisted terminal (FAT) to mix and move the primary and recirculated air through the supply diffuser into the room (Fig. 10.26). The series fan runs continuously and ensures there is good air circulation in the room, especially at low cooling loads and during heating from the reheat coil inside the box. The advantages and control are similar to the induction unit above.

Rather than using a continuously running series fan, a small parallel fan can be incorporated into the recirculation entry of the FAT (Fig. 10.27) to

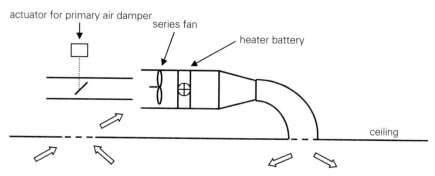

Fig. 10.26 A FAT with series fan.

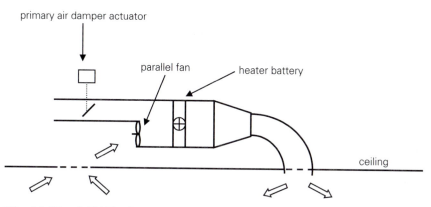

Fig. 10.27 A FAT with parallel fan.

operate at low cooling loads or heating loads when the primary airflow is low. When the cooling load is reasonably large and airflows are high, the parallel fan does not run.

10.11 Fan coil units

Fan coil units (FCUs) come in a variety of types but all have heating and cooling coils with their own fans circulating room air through the coils. The units are situated in the zone either as an underwindow unit or in the ceiling void. A great advantage over an all-air system, like VAV, is that the thermal power is primarily transmitted through water, and this uses much less energy than air transmission (Chapter 11). However, humidity cannot easily be controlled as the heating and cooling coils are controlled in sequence with a deadband between their operation. The chilled water to the FCUs' cooling coils is often supplied from a central chiller but the heating can similarly come from a central low-temperature hot water system or an electric heater can be used. Balancing and control of the water systems and control of the central plant are little different to normal systems.

Either two-pipe or four-pipe water supplies are provided to FCUs. A two-pipe system is used for changeover systems when either chilled or hot water is passed through the coil. For temperate climates, like the UK climate, this is an unsatisfactory system as there is no clear-cut seasonal change requiring a change to all cooling or all heating. Often in spring and autumn one day may well require both heating and cooling. Four-pipe systems are therefore more satisfactory.

An interesting feature of fan coil units is the coil, which is a conventional finned-tube heat exchanger for a two-pipe system, but is a shared fin block for a four-pipe system. Typically the coil has four rows of copper tubing, 9–16 mm outside diameter [10], with aluminium plate fins, three rows being for cooling and the fourth for heating. This requires the use of special three-port valves on both the supply and return pipes to shut off the cooling when heating is required, and vice versa.

The fresh air supply either comes locally through a vent in the wall behind an underwindow FCU or via a ducted system from an air-handling unit (AHU). For the ACU the supply may be to the FCU itself or to a separate supply diffuser in the room or zone, the FCU just using recirculated room air. With the central AHU, humidity can be dealt with there and the FCU used for sensible cooling. Condensate drainage is necessary where the cooling coil runs wet when central dehumidification is not provided. In warm climates, residual condensate from intermittent operation of the system can give rise to odours.

10.11.1   FCU control

There are three ways to control FCUs:

- Fan control
- Waterside control
- Airside control

For typical FCUs there are three fan speeds: medium with an airflow of say 100%, low with 83% airflow, and high with 118% airflow. Invariably these are manual settings, and for ceiling systems the setting would be fixed at the commissioning stage. For free-standing FCUs, occupants may well have access to the units and so be able to change the speed setting. This is especially applicable in hotels and cellular offices. As noise is related to the fan speed, most FCUs are set to medium or low speed. Connection to a BEMS allows the FCU fan speeds to be adjusted automatically, e.g. for preheating or precooling before occupancy starts.

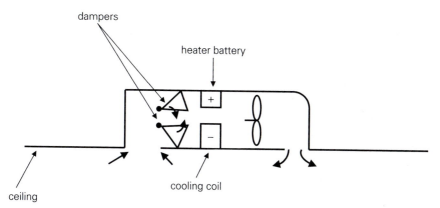

Fig. 10.28 An airside-controlled fan coil unit.

Modulated control of fan speed from a room temperature sensor is rare because of noise disturbance to the occupants, although on/off control is more frequent, especially in warmer and tropical climates.

Waterside control is more popular. Here the fan speed is kept constant (hence airflow and noise are uniform and not disturbing to the occupants), but the heating or cooling is regulated by two- or three-port modulating valves or on/off valves and the control signals come from return duct or room temperature sensors.

However, the valves are of small diameter relating to the small-diameter tubes in the heat exchanger. Consequently, dirt and initial debris from installation work needs to be carefully flushed out so there is no metal residue to cause 'wire drawing' of the valves.

Airside control, using dampers to modulate the airflow through separate heating and cooling coils (Fig. 10.28) is gaining popularity, as it avoids the valve problem above. However, two-port valves with on/off control are still needed for the coils.

References

[1]   CIBSE (1983) CIBSE Guide, Section A9, Estimation of plant capacity, Chartered Institution of Building Services Engineers, London.
[2]   Legg, R. (1991) *Air conditioning systems design, commissioning and maintenance*, Batsford, London.
[3]   CIBSE (2000) CIBSE Guide, Weather and Solar Data, Chartered Institution of Building Services Engineers, London.
[4]   Jones, W. P. (1997) *Air Conditioning Applications and Design*, 2nd edn, Edward Arnold, London.
[5]   Holman, J. P. (1997) *Heat Transfer*, 8th edn, McGraw-Hill, New York.

[6]   Pearson, S. F. (1989) Thermosyphon cooling. *Proc. Inst. Refrigeration*, **6**(1).

[7]   CIBSE (1975) CIBSE Guide, Section C1, Properties of humid air, Chartered Institution of Building Services Engineers, London.

[8]   Awbi, H. B. (1991) *Ventilation of Buildings*, Spon, London.

[9]   CIBSE (1986) CIBSE Guide, Section A7, Internal heat gains, Chartered Institution of Building Services Engineers, London.

[10] CIBSE (2000) CIBSE Guide, Automatic control systems, Chartered Institution of Building Services Engineers, London.

[11] ASHRAE (1996) *ASHRAE Handbook*, Chapter 20, HVAC systems and equipment, American Society of Heating, Refrigerating and Air-Conditioning Engineers, Atlanta GA.

[12] Eastop, T. D. and Watson, W. E. (1992) *Mechanical Services for Buildings*, Longman, Harlow.

[13] Petitjean, R. (1994) *Total hydronic balancing: a handbook for design and troubleshooting of hydronic HVAC systems*, Tour & Andersson, Boras, Sweden.

[14] BRECSU (1996) Variable flow control. BRECSU General Information Report 41, Building Research Establishment, Watford, UK.

[15] Pita, E. G. (1981) *Air conditioning principles and systems: an energy approach*, John Wiley, New York.

[16] Khoo, I., Levermore, G. J. and Letherman, K. M. (1998) Modelling of variable air volume terminal units: part 1, steady-state models. *BSERT*, **19**(3).

[17] Khoo, I., Levermore, G. J. and Letherman, K. M. (1997) Duct loops and VAV modelling and control. Paper presented at CLIMA 2000, Brussels, September 1997.

[18] Englander, S. L. and Norford, L. K. (1992) Saving fan energy in VAV systems: part 2, supply fan control for static pressure minimisation using DDC zone feedback. *ASHRAE Trans.*, **98**(1).

[19] Warren, M. and Norford, L. K. (1993) Integrating VAV zone requirements with supply fan operation. *ASHRAE Journal*, **35**(4).

[20] Tung, D. S. L. and Deng, S. (1997) Variable-air-volume air-conditioning system under reduced static pressure control. *BSERT*, **18**(2).

[21] P. Day, private communication.

[22] Chen, S. and Demster, S. (1996) *Variable Air Volume Systems for Environmental Quality*, McGraw-Hill, New York.

[23] Khoo, I., Levermore, G. J. and Letherman, K. M. (1996) Duct looping in VAV systems. *BSERT*, **18**(10).

[24] BSRIA (1999) *Control Strategies*, BSRIA, Bracknell, UK.

[25] CIBSE (1999) CIBSE Guide A, Environmental design, Chartered Institution of Building Services Engineers, London.

11
Natural ventilation and its control

11.1 Introduction

Displacement ventilation, chilled ceilings and mixed-mode air conditioning can save a considerable amount of energy and CO_2, reducing the atmospheric pollution associated with fossil fuels and greenhouse gases. In these buildings there is usually the option for the occupants to use natural ventilation by opening windows. However, natural ventilation is the key element for ventilation in some buildings, and summer cooling and detailed strategies, including summer night ventilation, are employed. Where natural ventilation is the sole means of summer cooling, it is important to have good control and careful design to reduce the heat gains, especially from solar radiation. This is because natural ventilation is variable, due to the wind and inside–outside temperature difference, and it can only provide about 20 W m^{-2} of cooling. With a very high ventilation rate and an inside–outside temperature difference of 6 K, this can at best rise to 50 W m^{-2}, although this is an optimistic value [1].

Some studies show that occupants prefer natural ventilation. A study of 480 office staff, covering all business sectors, found that 89% preferred buildings that were not air conditioned [1]. It also reported that the most important factors in the building design were good daylight and ventilation by opening windows. However, a similar survey reported that 75% of respondents stated a desire for air conditioning but nearly 50% expressed concern that air conditioning and poor lighting adversely affected their health [1]. It has also been suggested that naturally ventilated buildings have lower **sick building syndrome** scores than air-conditioned buildings [2, 3]. More recent work suggests there is little difference between air-conditioned and naturally ventilated buildings [4]. However, natural ventilation is increasingly

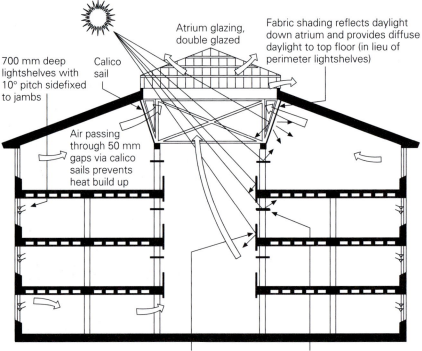

700 mm deep lightshelves with 10° pitch sidefixed to jambs

Calico sail

Atrium glazing, double glazed

Fabric shading reflects daylight down atrium and provides diffuse daylight to top floor (in lieu of perimeter lightshelves)

Air passing through 50 mm gaps via calico sails prevents heat build up

Light coloured perforated metal balustrades and smoke reservoirs reflect daylight down atrium

Fabric lightshelves provide element of reflected daylight

Fig. 11.1   Anglia Polytechnic Learning Resource Centre. (Reprinted, with permission, from [1, 5])

being used in buildings where a few years ago standard air conditioning would have been employed.

Most naturally ventilated buildings are shallow plan as natural ventilation from an open window is generally considered to be effective in ventilating up to 6 m into a room. For a shallow-plan building (shallow plan is effectively defined by the limits of natural ventilation) this effectively means the width is up to 15 m (6 m from windows on each side of the building and a 3 m access corridor in between). The heuristic rules for natural ventilation are as follows [1]:

- Single-sided single-opening effective to a depth of 2 times the floor-to-ceiling height.
- Single-sided double-opening effective to a depth of 2.5 times the floor-to-ceiling height.
- Cross-ventilation (window openings on different, generally opposite, walls) effective up to 5 times the floor-to-ceiling height.

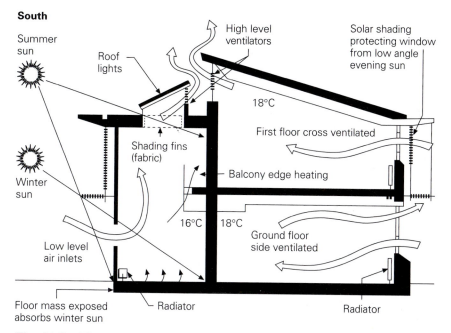

Fig. 11.2 The street of John Cabot City Technology College. (Reprinted, with permission, from [1, 6])

To use natural ventilation in deeper-plan buildings, a designer can use

- Atria
- Streets
- Towers
- Funnels or Chimneys
- Windcatchers

Figure 11.1 shows an atrium at the Anglia Polytechnic Learning Resource Centre which enables **stack-induced** cross-ventilation in this 30 m wide building [1, 5]. (Stack effect ventilation is discussed in detail later on.)

A street is generally longer and narrower than an atrium and like an old city street. Figure 11.2 shows a street and its contribution to ventilation, although there could be cross-ventilation through the bottom room, which is not shown [1].

Towers can also provide good routes for naturally ventilated air and they can use the stack effect. Often these towers contain staircases so that they are functional for access and ventilation (Fig. 11.3) [1, 7]. The ventilation relies on the doors to the tower being left open, although they close in the event of a fire (see later). Reducing the width of the stack ventilation route still further produces funnels or chimneys (Fig. 11.4) [1, 8].

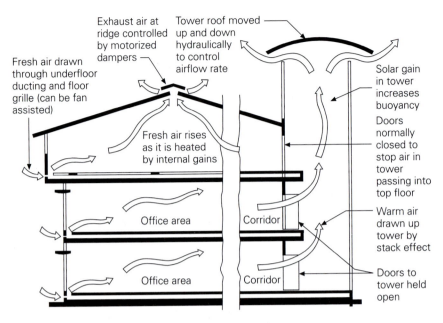

Fig. 11.3 Ventilation tower at the Inland Revenue building, Nottingham. (Reprinted, with permission, from [1, 7])

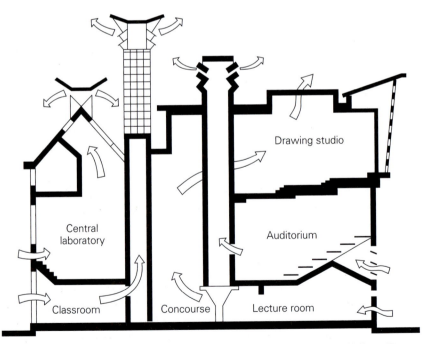

Fig. 11.4 Ventilation funnels or chimneys at the Queen's Building, De Montfort University. (Reprinted, with permission, from [1, 8])

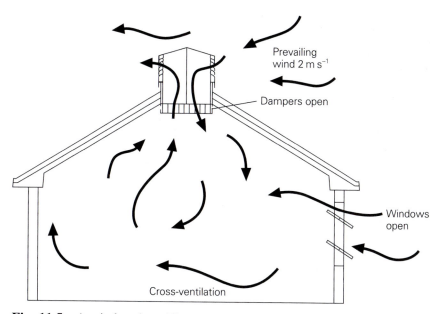

Fig. 11.5   A windcatcher. (Courtesy Monodraught)

All these devices are part of the building design and primarily they rely on stack ventilation. Precautions are often taken in the design to try to ensure that wind does not cause reverse flow down the ventilators. There are often dampers and actuators on the outlets and sometimes the inlets (as some of the figures show) to regulate the flow and counteract any excess ventilation due to the wind. However, a wind ventilator such as the 'windcatcher' (Fig. 11.5), uses both the wind effect and the stack effect to ventilate, taking wind from any direction. The damper is controlled by a signal from an inside room temperature sensor.

There is often careful limitation of solar gains by overhangs and solar shading (Figs 11.2 and 11.3). Exposed concrete ceilings and floors can be useful in absorbing heat gains and damping out fluctuations in temperature (see later).

11.2   Airflow through openings

As with pipes and ducts in mechanical air and water systems, window openings and ventilators, etc., have a resistance to the flow of air through them, and an equation like Darcy's equation. But with natural ventilation the flow rate, often given the symbol Q (m^3 s^{-1}), replaces the pressure drop on the left-hand side:

$$Q = k\Delta P^n \tag{11.1}$$

k = the flow coefficient, like the inverse of the resistance (m^3 s^{-1} Pa^{-n})
ΔP = the pressure difference across the opening (Pa)
n = the flow exponent
For purpose-built openings with a cross-sectional dimension greater than 10 mm [9, 25], such as ventilators, trickle vents in windows and open windows themselves, the flow is turbulent and $n = 0.5$. In the CIBSE Guide this equation is written as

$$Q = AF(2\Delta P/\rho)^{0.5} \tag{11.2}$$

which implies the standard relationship between the velocity pressure and the friction loss at a component, as can be seen by rearranging equation (11.2):

$$\Delta P = (1/F^2)0.5\rho v^2 \tag{11.3}$$

where v = the air velocity (m s^{-1})
        = Q/A
    A = the area of the opening (m^2)
    F = a flow factor
The flow factor F is conventionally termed the discharge coefficient, C_d, and equation (11.2) becomes

$$Q = AC_d(2\Delta P/\rho)^{0.5} \tag{11.4}$$

C_d is akin to a pressure loss coefficient in ducted air systems and indicates that, due to turbulence and the airflow through the opening, the opening is effectively contracted. The theoretical value of C_d for a sharp-edged opening is 0.61, and it is common practice when measuring practical openings to calculate an equivalent area assuming a discharge coefficient of 0.61.

In very narrow openings the flow tends to be nearer to laminar flow and the index n approaches 1.0. For cracks and background leakage a practical value is between 0.6 and 0.7 [9, 25] and C_d is replaced by LC_i where L is the crack length (m) and C_i is the infiltration coefficient. The motive pressure for ventilation and infiltration is due to the wind, **wind-driven**, and/or warm air in the building rising and exiting at a higher level as cold, denser outside air enters at a lower level in the building, **stack effect**.

11.3 Wind-driven ventilation

If a building has relatively sharp corners, the pattern of airflow around it due to the wind is independent of the wind speed. Hence the pressure on the windward side of the building is related to the kinetic energy of the air released as the air is slowed down and stopped by the building's wall. The kinetic energy of the air is related to its dynamic pressure, $0.5\rho v^2$. But in

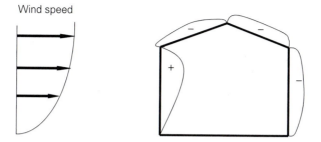

Fig. 11.6   Wind pressures around a building.

practice the air is not completely stopped and a dimensionless pressure coefficient, C_p, is required to calculate the actual pressure of the wind, P_w:

$$P_w = C_p(0.5\rho v_{ref}^2) \tag{11.5}$$

where v_{ref} is the mean wind speed at the reference height, usually taken as the height of the building (m) and P_w is relative to the static pressure of the undisturbed wind. Typical values of C_p are between +0.9 and −0.9 [10] but the flow of air through a building (Fig. 11.6) goes from high pressure (the windward side) to low pressure (suction, leeward side and also surfaces parallel to the wind flow).

So the pressure difference between the windward and leeward sides of a building is

$$\Delta P_w = \Delta C_p(0.5\rho v_{ref}^2) \tag{11.6}$$

where ΔC_p is the difference in pressure coefficients between the windward and leeward sides. Values of C_p, to derive ΔC_p, are obtained empirically, largely based on the results of wind tunnel studies [11] and values are given for simple, common building geometries [9, 10, 25]. Typically ΔC_p is about 1.0 but it can be as low as 0.1 for buildings in a sheltered location.

11.3.1   Wind speed data

The wind speed can be obtained from meteorological data as given in the CIBSE Weather and Solar Data Guide [12]. This also gives the probability of various wind speeds via the Weibull distribution (Fig. 11.7). The equations for this distribution are

$$f(v) = \frac{k}{A}\left(\frac{v}{A}\right)^{k-1}\exp-\left(\frac{v}{A}\right)^k \tag{11.7}$$

$$F(v) = \exp-\left(\frac{v}{A}\right)^k \tag{11.8}$$

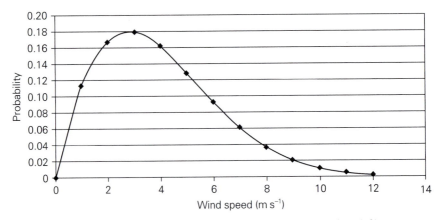

Fig. 11.7   Weibull distribution of wind speed ($A = 4.47$, $k = 1.8$).

where v = the speed to the nearest integral value (m s^{-1})
  $f(v)$ = frequency of occurrence of any speed v between $v - 0.5$ (m s^{-1})
   and $v + 0.5$ (m s^{-1})
  $F(v)$ = probability of the wind speed exceeding v
  A = a location parameter related to the mean speed (m s^{-1})
  k = a 'spread' parameter; a smaller k causes a wider spread about
   the mean

For Heathrow, near London, $A = 4.47$ m s^{-1} and $k = 1.80$. With v at 5 m s^{-1} equation (11.7) shows that a wind of this speed (wind speeds rounded to the nearest 1 m s^{-1}) will blow for 13% of the time, and equation (11.8) shows that a wind speed of 5 m s^{-1} will be exceeded for 29% of the time. The data given in the CIBSE Guide is over the 20 years from 1976 to 1995.

Before this data can be used in equation (11.6) the wind speed must be corrected for the height of the building to give v_{ref}. The meteorological data relates to the speed at a standard 10 m height in an open Met Office site. The following equation is used to correct for height and location [12, 25]:

$$v = v_{m} K_{s} z^{a} \qquad (11.9)$$

where v = mean wind speed at height z
  v_{m} = mean wind speed at 10 m height in open country
  K_{s} = parameter relating the wind speed to the nature of the terrain
  z = height above ground
  a = exponent relating wind speed to height above ground

Values of K_{s} and a are given in Table 11.1; note that wind speed is averaged over the hour.

Table 11.1 Wind speed terrain and height parameters

Terrain	K_s	a
Open, flat country	0.68	0.17
Country with scattered wind breaks	0.52	0.2
Urban	0.35	0.25
City	0.21	0.33

Example 11.1
What is the wind pressure due to the average wind speed in an urban area near Heathrow? The average wind speed at Heathrow is 3.5 m s^{-1}. Assume a pressure coefficient of 0.5 and a building height of 5 m above the ground, and assume summer conditions prevail with an outside air density of 1.2 kg m^{-3}. Calculate the volume flow of air through open windows on the windward and leeward sides, each of area 0.5 m^2 and with discharge coefficient 0.61. Assume $\Delta C_p = 1.0$ and assume the wind-driven pressure drop across each opening is the same.

Solution
From equation (11.9) the wind speed for the urban area around Heathrow at 5 m height is

$$v = v_m K_s z^a$$

$$= 3.5 \times 0.35 \times 5^{0.25}$$

$$= 1.83 \text{ m s}^{-1}$$

The wind pressure from equation (11.5) is

$$P_w = C_p(0.5\rho v_{ref}^2)$$

$$= 0.5 \times 0.5 \times 1.2 \times (1.83)^2$$

$$= 1 \text{ Pa}$$

Here ρ is the density of the outside air, and here summertime conditions are assumed with the standard air density of 1.2 kg m^{-3}. The pressure is very small compared to mechanical ventilation, where 1 Pa would overcome the friction in a typical duct 1 m long. So the window openings have to be large for a decent ventilation rate to cool the building.

As to the airflow due to the wind pressure, the pressure across the building's two windows is given by equation (11.6) as

$$\Delta P_w = \Delta C_p(0.5\rho v_{ref}^2) \tag{11.6}$$

Assuming the pressure difference is divided equally between the two windows, put half the pressure difference into the flow equation (11.4) and obtain

$$Q = AC_d(\Delta P_w/\rho)^{0.5} \qquad (11.4)$$

Combining this with equation (11.6) gives

$$Q = AC_d(0.5\Delta C_p)^{0.5}v_{ref} \qquad (11.10)$$
$$= 0.5 \times 0.61(0.5 \times 1.0)^{0.5} \times 1.83$$
$$= 0.39 \text{ m}^3 \text{ s}^{-1}$$

If this served an office 3 m high, 15 m deep between the windows, and with a 5 m wall length then it would require an air change rate of 6.24 h^{-1}.

Note that in Example 11.1 the volume flow rate, and consequently the air change rate, is directly proportional to the wind velocity:

$$Q = AC_d(0.5\Delta C_p)^{0.5}v_{ref} \qquad (11.10)$$

From inspection of tables for C_p [9, 25], a typical average for ΔC_p for low-rise buildings is 0.8. With C_d typically 0.61 then equation (11.10) approximates to

$$Q \approx 0.4Av_{ref}$$

11.3.2 Natural ventilation networks

In Example 11.1 the window openings were of equal area, but this is un-likely in practice. To examine the effect of different opening sizes, we can consider the openings as part of a fluid flow network, like pipe or duct flow. Here the pipe or duct joining the openings is the building itself, a large and effectively frictionless pipe or duct (Fig. 11.8).

As with pipe and duct networks, the inlet and outlet resistances (R_i and R_o) can be replaced by an equivalent resistance, R_{eq}. In a natural ventilation network a resistance, R, is given by

$$R = \frac{1}{2}\rho\frac{1}{A^2C_d^2}$$

11.3.2.1 *Openings in series*

Hence, for an inlet and outlet resistance **in series** (Fig. 11.4), the equivalent area and resistance are given by

$$R_{eq} = R_i + R_o$$
$$\frac{1}{2}\rho\frac{Q^2}{A_{eq}^2C_d^2} = \frac{1}{2}\rho\frac{Q^2}{A_i^2C_d^2} + \frac{1}{2}\rho\frac{Q^2}{A_o^2C_d^2} \qquad (11.11)$$

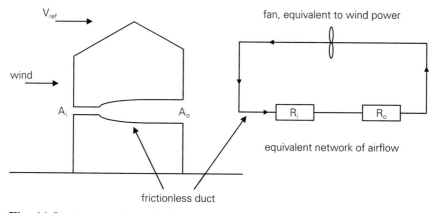

Fig. 11.8   A natural ventilation network.

For large openings $C_d = 0.61$, and assuming the density of the air is the same at the inlet and the outlet, we have

$$\frac{1}{A_{eq}^2} = \frac{1}{A_i^2} + \frac{1}{A_o^2} \tag{11.12}$$

So the smaller of the two openings dominates the flow. However, for a constant wind pressure on the building, equation (11.11) gives the volume flow, Q, in the following way:

$$\text{constant pressure} = \frac{1}{2}\rho\frac{Q^2}{A_i^2 C_d^2} + \frac{1}{2}\rho\frac{Q^2}{A_o^2 C_d^2}$$

so

$$Q \propto \frac{A_i A_o}{\sqrt{(A_i^2 + A_o^2)}} \tag{11.13}$$

or

$$Q \propto \frac{A_i x}{\sqrt{(1 + x^2)}} \tag{11.14}$$

where x = ratio of area of openings
$= A_o/A_i$

As Fig. 11.9 shows, Q varies almost linearly with x, which means that if one of the opening areas is controlled linearly then the volume flow rate will almost be controlled linearly. An actuator on a sash window could effect this by simply raising and lowering the sash. However, other windows may not be so linear in their opening areas.

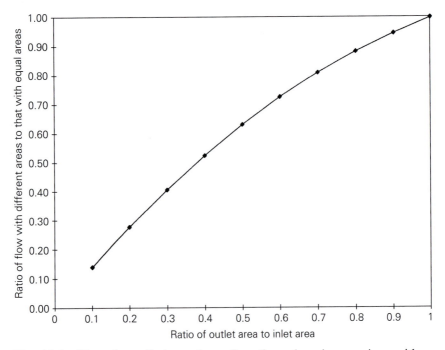

Fig. 11.9 Natural ventilation volume flow through series openings with different areas.

If one of the window openings closes, i.e. its area goes to zero, then there is single-sided ventilation [25, 9] given by

$$Q = 0.025 A v_{\text{ref}}$$

Other combinations of openings are given in the CIBSE Guide [25].

11.3.2.2 *Openings in parallel*

If there are two openings in the same wall, reasonably close together, then they may be regarded as being in parallel. As with pipe and duct resistances, the equivalent resistance is

$$\frac{1}{R_{\text{eq}}^2} = \frac{1}{R_1^2} + \frac{1}{R_2^2}$$

or

$$\frac{2A_{\text{eq}}^2 C_{\text{d}}^2}{\rho} = \frac{2C_{\text{d}}^2}{\rho}(A_1^2 + A_2^2)$$

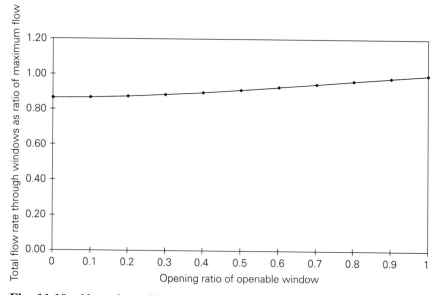

Fig. 11.10 Natural ventilation volume flow with two outlets of equal area in parallel but only one controlled.

giving

$$A_{eq}^2 = A_1^2 + A_2^2$$

Consider the two parallel openings A_1, A_2 (resistances R_1, R_2) of equal area, $A_1 = A_2 = A$. The equivalent area is then

$$A_{eq} = A \sqrt{2} = 1.414A$$

So although the openings are both of area A the equivalent area is less than $2A$ because of the square law. This has implications for the flow rate in a building. Consider the simple case of an inlet and outlet on opposite sides of a building. If the outlet consists of two window openings of the same area, equal to the inlet area, A_i, then from equation (11.13) we have

$$Q \propto \frac{A_i \sqrt{A_1^2 + A_2^2}}{\sqrt{A_i^2 + A_1^2 + A_2^2}}$$

If $A_1 = A_2 = A_i$ and only the A_2 window is controlled in its opening, i.e. $A_2 = xA_i$ where $0 \leqslant x \leqslant 1$, then

$$Q \propto \frac{A_i \sqrt{(1 + x^2)}}{\sqrt{(2 + x^2)}} \tag{11.14}$$

which is shown graphically in Fig. 11.10. This demonstrates that the openable window has very little control over the volume flow rate. It also has

Table 11.2   Wind pressure coefficients for a square building [16]

Location[a]	Wind angle (0° is directly onto the building)							
	0°	45°	90°	135°	180°	225°	270°	315°
Face 1	0.4	0.1	−0.3	−0.35	−0.2	−0.35	−0.3	−0.1
Face 2	−0.2	−0.35	−0.3	0.1	0.4	0.1	−0.3	−0.35

[a] Face 1 is the front face and face 2 is the back face.

implications for occupants trying to control their own ventilation; there is an interaction between window openings. Invariably if there is an attempt to control the natural ventilation automatically by window opening, it is usually applied to the top part of the window which may be tilted. The CIBSE Natural Ventilation Applications Manual [1] gives a number of case studies with different methods of controlling ventilation. This manual also gives guidance on the opening characteristics of various window types.

Although the natural ventilation network (Fig. 11.4) has a 'duct' system with a 'fan' driving the air, the wind-driven ventilation varies with the wind speed. The driving pressure of the wind is the velocity pressure on the building's facades. With a combined pressure coefficient ΔC_p from the windward side to the leeward side of the building (i.e. across the frictionless duct or network), this driving pressure is $\frac{1}{2}\Delta C_p \rho v_{ref}^2$ and in terms of the resistance to the airflow is $\frac{1}{2}\rho(Q^2/A^2 C_d^2)$. Combining these equations gives

$$Q = AC_d \sqrt{(\Delta C_p)}\, v_{ref} \tag{11.15}$$

where A is the equivalent area of all the series and parallel openings in the building. As A, C_d and ΔC_p are constants, the volume flow varies linearly with the wind speed, although A can be varied to control the flow. This assumes that the openings are reasonably large. When they are small or just cracks, the index in the flow–pressure equation reduces as the flow tends to laminar.

However, this is for a simple geometry with the wind blowing normal to the windward facade. But as the wind direction changes, the pressure coefficients also change. However, ΔC_p does not become zero when the wind blows across the surface, as Table 11.2 shows. The data in this table is for a square building in a semi-sheltered location. Here ΔC_p varies between 0.6 and 0.45. Even when the wind is blowing at 90° to the surface there is still a non-zero pressure coefficient. This can be appreciated from Bernouilli's equation (Chapter 13), where the velocity pressure in the wind causes a reduction in the static pressure, sucking air out from the openings (note that both faces at 90° have negative pressure coefficients).

Buildings in practice are more complex than the simple models above. They have many ventilation and infiltration paths, hence they need to be modelled by multizone, multipath networks. Simplified models for multizone, multipath networks are the BRE model [13] and the Lawrence Berkeley Laboratory (LBL) model [14], and network models considering flow paths through buildings in series and in parallel. Awbi [15] and Liddament [16] review these models.

11.4 Stack ventilation

Apart from wind pressure creating natural ventilation, there is also stack ventilation. Air density decreases with increasing absolute temperature. At normal temperature (20°C) and standard atmospheric pressure the density is 1.2 kg m⁻³, but at other temperatures, t (°C), the general gas law gives the density as

$$\rho = 1.2\frac{(273 + 20)}{(273 + t)}$$

A building with warm air inside can be considered as a column of warm air with an equal volume of cold, heavier air outside, envisaged as columns contained in a frictionless duct, with the weight of the air represented as fans (Fig. 11.11). The hot inside air can be considered as being instantly cooled to the outside temperature once through the top resistance, and the heating of the cold air that enters can be treated in a similar way.

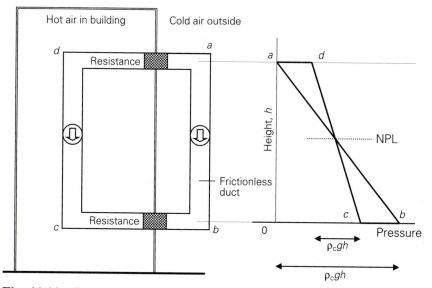

Fig. 11.11   Stack ventilation as a frictionless duct concept.

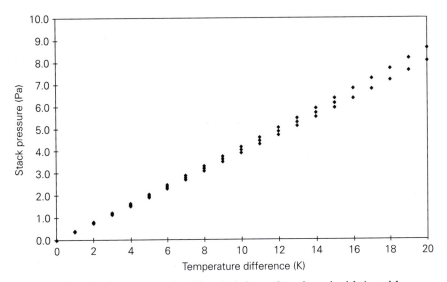

Fig. 11.12 Stack pressure for 10 m height and various inside/outside temperature differences.

The cold air will enter and drive out the warm air if there are openings provided for this to happen. From Fig. 11.11 the pressure difference driving this air movement is

$$\Delta P = gh(\rho_c - \rho_h) \quad \text{Pa} \tag{11.16}$$

which, from the general gas law becomes

$$\Delta P = 3462h \left[\frac{1}{(t_c + 273)} - \frac{1}{(t_h + 273)} \right] \tag{11.17}$$

where t_c and t_h are the cold and hot air temperatures and h is the height of the air columns. Figure 11.12 shows the stack pressures that can be generated, as calculated from equation (11.17) with $t_h = 20°C$, $25°C$ and $30°C$ and t_c going as low as $0°C$. As Fig. 11.12 shows, equation (11.17) is almost linear. This can be shown by approximating it to

$$\Delta P = 12.68h \left[\frac{t_h - t_c}{273 + t_h + t_c} \right] \tag{11.18}$$

where $t_c t_h \ll (273)^2$ and so has been omitted. For $t_c = 20°C$ and $t_h = 30°$ this reduces to give the approximate relationship

$$\Delta P \approx 0.039h(t_h - t_c) \tag{11.19}$$

This is the driving pressure pushing cool air into the building and also pushing out the warm interior air. It can be considered as a network with a frictionless duct, as in the wind-driven ventilation. The resistances to the driving pressure are the same as in the wind-driven network. So for an inlet of area A_i and an outlet of area A_o we have

$$gh(\rho_c - \rho_h) = \frac{1}{2}\rho_c \frac{Q^2}{A_i^2 C_d^2} + \frac{1}{2}\rho_h \frac{Q^2}{A_o^2 C_d^2}$$

If the mean air density, ρ, is used on the right-hand side of this equation, the resistances used in the wind-driven ventilation can apply, so the areas can be replaced by the equivalent series resistance area (11.12):

$$\frac{1}{A_{eq}^2} = \frac{1}{A_i^2} + \frac{1}{A_o^2}$$

and

$$0.039h(t_h - t_c) \approx \frac{1}{2}\rho \frac{Q^2}{A_{eq}^2 C_d^2} \qquad (11.20)$$

Example 11.2
Consider an office at 25°C, the same as the outside air temperature, with an opening of 0.58 m² in the outside facade, 1.85 m above ground level, and another opening to an atrium. The atrium has an opening 13.35 m above the ground. Assuming the atrium is at a uniform temperature of 28°C, what is the area of the atrium opening to produce an airflow of 0.5 m³ s⁻¹ (which corresponds to an air change rate of 5 h⁻¹)? Calculate the heat extraction by the natural ventilation and also the atrium temperature when the outside temperature drops to 24°C, assuming the heat gains extracted remain constant.

Solution
The average density of the air between 25°C and 28°C is 1.18 kg m⁻³. From

$$0.039 \times 11.5 \times (28 - 25) \approx \frac{1}{2}1.18\frac{0.5^2}{A_{eq}^2 0.61^2} \qquad (11.20)$$

we obtain $A_{eq}^2 \approx 0.295$, so

$$\frac{1}{0.295} = \frac{1}{0.58^2} + \frac{1}{A_o^2}$$

$$A_o \approx 1.54 \text{ m}^2$$

At steady state the heat removed by the natural ventilation $\dot{H}$ is

$$\dot{H}\dot{m}C_p(t_{ai} - t_{ao}) \qquad (11.21)$$

where $\dot{m}$ is the mass flow rate of the air (kg s⁻¹), $\dot{m} = 1.84$ kW.

To find the resultant inside temperature if the outside temperature changes, this requires the balance equations for the system. It entails using equation (11.21) for the heat transfer and equation (11.20) for the stack ventilation. Both are interdependent; as the inside–outside temperature difference increases, the ventilation rate increases (Fig. 11.8).

$$0.039 \times 11.5 \times (t_{ai} - 24) \approx \frac{1}{2} \times 1.18 \frac{\dot{m}^2}{\rho A_{eq}^2 0.61^2} \tag{11.20}$$

Putting in A_{eq}, etc., gives

$$\dot{m}^2 \approx 0.11(t_{ai} - 24) \tag{11.20}$$

and the heat balance is

$$\dot{m} = \frac{1.8}{(t_{ai} - 24)} \tag{11.21}$$

Combining equations (11.20) and (11.21) we get

$$1.8 \approx 0.098(t_{ai} - 24)^3 \tag{11.22}$$

Note that equation (11.22) would have been rather complex if equation (11.20) had not been simplified and left in terms of the reciprocals of the absolute temperatures. A numerical method, or Excel's Solver, would be needed to solve it. Equation (11.22) gives $t_{ai} \approx 27°C$. Using numerical analysis, as mentioned above, gives $t_{ai} = 27°C$ and also shows that if the heat extraction remains constant then t_{ai} is 3 K above t_{ao}. Although equation (11.20) shows that stack pressure is approximately linear with the temperature difference, the equivalent area of the openings, A_{eq}, has a significant influence (Fig. 11.13). Here the inlet and outlet areas are set equal (at 0.78 m²) for the design airflow rate of 0.5 m³ s⁻¹ and the outlet area is automatically controlled; the inlet area, under the control of the occupants, is left fully open. Figure 11.13 shows the non-linearity of the relationship between atrium, or stack, temperature and outlet area, with the outside air temperature remaining constant at 25°C. Once the outlet area gets below 30% open, the stack temperature rises dramatically. Figure 11.13 is based on the approximate equation, from the heat extraction and pressure difference:

$$(t_{atrium} - t_{ao})^3 \approx \frac{\dot{H}^2}{A_{eq}^2}$$

$$(t_{atrium} - t_{ao}) \approx \dot{H}^{2/3} \left[\frac{A_o^2 + A_i^2}{A_i^2 A_o^2} \right]^{1/3} \tag{11.23}$$

This steady-state treatment is rather simplistic in that it assumes the atrium has a uniform temperature, 28°C initially in Example 11.2, and it ignores heat transfer to the air in its passage through the office and atrium. More

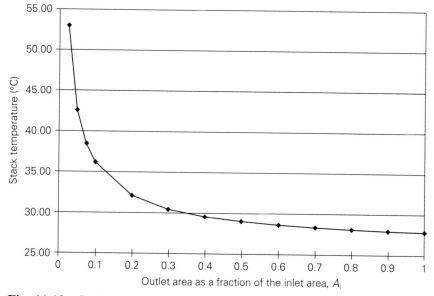

Fig. 11.13 Stack temperature variation with outlet area; the outside air temperature remains constant at 25°C.

exact models will be required for a number of openings from different floor levels into the atrium. For multi-storey buildings with atria, streets, funnels or towers to help the ventilation, the stack effect will be greater at the lower floor openings than the higher openings. Consequently, the ventilation openings will be larger for the upper floors. In the Anglia Polytechnic Learning Resource Centre the window opening area for the top floor is five times greater than for the bottom floor.

11.4.1 Control of stack ventilation

Control of a stack ventilation system can be examined by considering Example 11.2 or the natural ventilation examples in Figs 11.1 to 11.4, where louvres and even a tower roof are used to control the ventilation. Control can be achieved by opening and closing the openings depending on the room temperature, but because natural ventilation relies on driving forces which are not fully controllable, close control of the internal environment is impractical. Following the above discussion, the room temperature is taken as the mean of the outside temperature and the atrium, or exhaust, temperature.

Consider Example 11.2 where the area of the opening, A_o, is controlled by a proportional control algorithm on the room temperature, t_{ai}, which is taken as the average of the outside temperature and the atrium temperature:

$$t_{ai} = \frac{t_{at} + t_{ao}}{2}$$

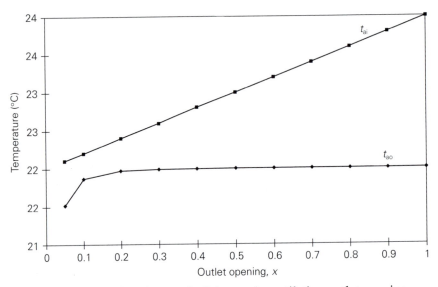

Fig. 11.14 Proportional control of the stack ventilation outlet opening, from room temperature.

where t_{at} is the the atrium temperature. The proportional control equation is

$$x = 1.0\left[\frac{t_{ai} - 22}{2}\right] \tag{11.24}$$

where x is the fraction that the opening is open; $x = 1$ for fully open. The setpoint is 22°C and the proportional band is 2 K. So it is fully open at 24°C and shut at 22°C. The ventilation equation, based on equation (11.23), is

$$t_{at} - t_{ao} = 2.02\left[\frac{A_o^2 + A_i^2}{A_o^2 A_i^2}\right]^{1/3} \tag{11.23}$$

Combining equations (11.23) and (11.24) then rearranging gives

$$t_{ao}^3 = 8x^3 + 10\,648 - \left(\frac{1.03}{A_i^2}\right)\left[\frac{x^2 + 1}{x^2}\right]$$

Figure 11.14 shows the results of this relationship. The control is surprisingly good but it is over a very shallow range of outside air temperature. It almost seems a constant value but it is varying very slowly. In order to enhance stack ventilation **atria**, it is possible to design **streets**, especially **funnels** and **towers**, to catch the sun. This increases the funnel temperature and therefore increases the stack temperature difference without increasing the office temperature.

This is an evolution from the nineteenth century, when fires and chimneys were used to enhance the stack ventilation. Extract could be through the kitchen area, where there was plenty of heat from fires, or through special chimneys with fires provided. However, in the modern stack or funnel, glass is often used for the solar trap. In Fig. 11.3 the ventilation tower is partly constructed of glass blocks.

11.4.2   Neutral pressure level

Some publications refer to a **neutral pressure level** (NPL) for stack ventilation. In Example 11.2, which is similar to the example in the CIBSE NV Applications Manual, the NPL was used to obtain the solution. The NPL is where the inside and outside pressure gradients cross (Fig. 11.11). The stack ventilation can be considered as a duct circuit, with fans representing the static pressure of the hot and cold columns of inside and outside air, and the openings as resistances. One could also consider that there are perfect heating and cooling systems at the top and bottom of the circuit, respectively, to allow the air to be recirculated in this imaginary duct system. Putting point *a* as the datum point of zero pressure, the circuit pressure drops are shown. The intersection of lines *ab* and *cd* is the neutral point or NPL. This depends on the resistances of the openings. Here the resistance of *bc* is larger than *ad*, so the neutral point is nearer to *ad*. Above the NPL the pressure is forcing the air out of the building, but below it the pressure is sucking the air into the building. However, if another opening were to be made above the existing NPL, but below the exit opening, air would still be drawn into the new opening and out of the upper exit. But the NPL would rise to between the new inlet and the exit. The NPL can be used to calculate the volume flows through multiple openings [1].

11.5   Combined stack and wind

Depending on the outside weather conditions, natural ventilation has a variable power to cool. Stack pressure can get up to 9 Pa (Fig. 11.8) and wind-driven ventilation can match or surpass this, depending on the wind speed and the site. In suburbia, for a 10 m high building and a wind speed of 2.5 m s^{-1}, the pressure difference is about 4 Pa [1]. Stack ventilation depends on

$$0.039h(t_h - t_c) \tag{11.19}$$

whereas wind-driven ventilation is proportional to $\frac{1}{2}\Delta C_p \rho v_{ref}^2$. With a conservative ΔC_p of 0.5 the wind-driven pressure becomes $0.3 v_{ref}^2$, which for a reference velocity of 4 m s^{-1} (a typical summer value for the London area) gives a pressure of 4.8 Pa. The stack pressure for a 10 m height and a 5 K temperature difference gives 2 Pa, which is equivalent to a velocity of 2.5 m s^{-1}.

Although wind-driven ventilation can surpass stack ventilation, the conservative method for designing natural ventilation is to consider a hot, cloudless day in the height of summer (typical anticyclonic weather conditions) with little if any wind. Hence the natural ventilation is designed for stack ventilation only. However, there are few times when the wind speed is actually zero. Commonly used cup anemometers, such as the Munroe anemometer used in the past by the UK Met Office, are not very accurate at low wind speeds, as there is some inertia to get the cups rotating. Studies with ultrasonic anemometers are showing that there are non-zero wind speeds in typical anticyclonic conditions [17]. Hence designers need to consider both wind-driven and stack-driven ventilation and its control.

Calculations for combined wind and stack ventilation simply add the wind pressure to the stack pressure to calculate the combined ventilation rate. This is of course if the wind direction is such as to assist the stack ventilation, i.e. the low-level inlet is facing windward and the high-level extract is leeward.

If the atrium, street, funnel or tower has openings in it at various levels, such as from different floor levels, then for basic manual calculations they can be treated as separate flow paths and the flow rates may be combined (Fig. 11.15) [1].

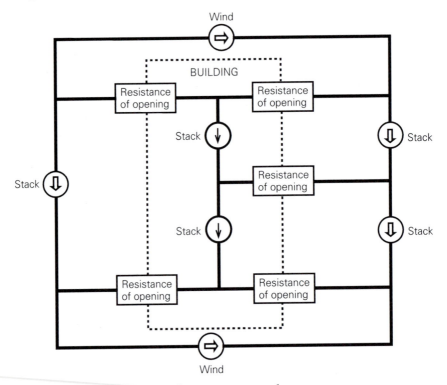

Fig. 11.15   A wind and stack pressure network.

When the wind effect opposes the stack ventilation then it is a good approximation to assume [18] that when $Q_w > Q_s$ (where Q_w is the volume flow due to the wind effect and Q_s is the volume flow due to the stack effect) then only the wind flow is considered, i.e. $Q = Q_w$. Likewise when $Q_s > Q_w$ then $Q = Q_s$. This applies to conventional openings such as windows; it does not apply to 'windcatchers' that allow air in and out of the same unit and utilize the wind and the stack effect together.

11.6 Control for comfort with natural ventilation

It is now generally considered that the close control of a fully air-conditioned zone is not necessary with a naturally ventilated building, not that close control of temperature can be achieved in a naturally ventilated building anyway. In a naturally ventilated building with openable windows, occupants feel they have more control and are tolerant of higher temperatures than would be acceptable for an air-conditioned building. Hence an inside dry resultant temperature of 25°C is considered acceptable in a naturally ventilated office whereas 21°C would apply to an air-conditioned office.

11.6.1 Adaptive comfort theory

Most comfort standards [19] are based primarily on Fanger's work in the 1970s, which relates to steady-state conditions in a closely controlled climate chamber [20]. Although these standards are suitable for most designs, some field studies have found discrepancies [21] and in the early 1970s the **adaptive comfort theory** (ADT) was developed [22]. ADT implies that the outside weather conditions have an influence on the inside comfort conditions, as can be seen from hotter climates having higher design inside temperatures. In other words, if it is hotter outside, the occupants adapt to a hotter temperature inside. Occupant actions can also influence comfort. If a change occurs such as to produce discomfort, people react in ways which tend to restore their comfort [23]. Other factors also have an influence; there are three main groups [24]:

- Acclimatization (which is physiological).
- Adjustment (e.g. adjusting clothing or opening windows to relieve discomfort).
- Habituation and expectation (people in hot countries tolerate hotter temperatures).

The ADT may be expressed in terms of the following equation [23]:

$$t_c = at_{out} + b$$

where t_c = the desired comfort temperature
         a, b = empirically derived constants

t_{out} = an outside temperature index based on the daily average outside temperature and the running average temperature of previous days

This is also reflected in the CIBSE design conditions [25] that naturally
ventilated buildings are considered comfortable provided the dry resultant
temperature does not rise above 25°C for more than 5% of the occupied time
and above 28°C for more than 1% of the time. (It has been found that the
28°C condition is often satisfied if the 25°C criterion is satisfied [26], and
the 28°C condition has consequently been left out of the CIBSE Guide [25].)
This is a summer design criterion, but designers must also consider winter
conditions and ensure that windows seal well, so the ingress of cold air can
be controlled. Stack ventilation is even more effective in cold weather!

11.7   Natural ventilation control

A critical factor in natural ventilation is the availability of windows that the
occupants can open and close at their discretion. However, naturally ventilated buildings often incorporate either high-level windows or ventilators,
which are under central BEMS control, as in the PowerGen building [23].
Note that occupant-controlled windows can be welcomed by the occupant
operating the window, yet perhaps not appreciated so much by other occupants affected by the window opening but not able to control it. This problem is minimized if the window unit has independent high- and low-level
openings. The low-level windows can provide ventilation for the local occupants, and the high-level windows for occupants further into the building.
This may require an actuator on the high-level openings, perhaps operated
by an infrared device. The upper-level windows can then be used in an
automatic night-cooling schedule. Manual control can mean that windows
are left open or closed during unoccupied hours, with possible overheating
or overcooling and discomfort consequences the next day.

Although occupant opening of windows is difficult to predict, the AIVC
reports a study [27] from which the following approximate equations have
been developed. For a cloudy day it quotes

$$O = 1.6 + 1.2t_{ao}$$

valid for 5°C $\leq t_{ao} \leq$ 15°C and 3% $\leq O \leq$ 17%, where O is the percentage
of open windows. And for a sunny day (sunny not being defined) it quotes

$$O = 1.6 + 1.98t_{ao}$$

valid for 2°C $\leq t_{ao} \leq$ 25°C and 3% $\leq O \leq$ 48%

Multiple trickle vents within the main window heads can provide draught-
free background ventilation in the winter as the incoming air velocities are
low. And ventilators can be provided with heater coils to give heated fresh
air when necessary.

Security concerns will probably influence the type of window opening available, and automatic control of high-level windows above the main window may be necessary for secure night-cooling ventilation.

High-level windows can be operated with actuators just like valves and dampers, with chain, helical cable, piston, and rack-and-pinion actuators. The actuator must be sized to cope with weight of window and wind forces.

When the windows are automatically controlled, **anemometers** may be needed to sense the wind speed so that a control signal can close the windows as the wind speed increases. A **wind vane** sensor may also be used so that unwanted reverse flows of exhaust air may be avoided by shutting the exhaust vents on the windward side and opening vents on the leeward side. As wind speed and direction can fluctuate quickly, frequent sampling is recommended, i.e. a 1 s sampling rate. Averaging techniques (Chapter 14) can then be used to provide a signal for stable control. These techniques also apply to solar sensors and, to a lesser extent, rain sensors.

Rain sensors may be necessary to enable windows to be closed when rain ingress may occur. Similarly, **solar sensors** may be used to aid the control of natural ventilation, but the solar control algorithm should not be too fast acting so that passing clouds do not cause hunting. A suitable integral term would suffice for this. **Occupancy sensors** (infrared and other movement sensors) and **air quality sensors** (CO_2, humidity and odour sensors) may also be used to vary the ventilation.

If there is mechanical cooling, as in a **mixed-mode** building, then nearby windows should be kept shut. This can be achieved either by staff training or microswitches on the windows linked to the cooling system. Microswitches are used in the 17-storey BSI headquarters [1]. A displacement ventilation system with perimeter fan coil units and personally openable fanlights (baffled with rainscreen louvres) has microswitches which turn off a fan coil unit when the adjacent window is opened.

To enable cross-ventilation the internal doors between offices and corridors may need to be kept open. If they are fire doors then **emergency-released electromagnets** can be installed. The Inland Revenue headquarters building in Nottingham uses electromagnets on all the fire doors between the offices and the ventilation towers [1].

11.7.1 Control of windcatchers, funnels and towers

Often ventilators are incorporated into naturally ventilated buildings. They are sited at roof level, direct the wind (catching it from any direction) into the building and allow out the internal air (Fig. 11.5). And they can operate independently of the occupant-controlled windows. Control is by an opposed-blade damper receiving a signal from an internal sensor. A typical control strategy is proportional control with the damper being fully closed at 16°C and then opening by 10% for each 1 K rise in room temperature.

The ventilator uses both stack and wind-driven effects, although it is considered that wind-driven ventilation predominates even at low wind speeds. Measurements taken at one site show an air change rate of 1.24 h^{-1} at a wind speed of 1.7 m s^{-1}, rising to 5.2 h^{-1} at a speed of 4.5 m s^{-1} during a very hot spell of summer weather (about 29°C dry bulb temperature) [28]. The wind speeds here were measured with an ultrasonic anemometer.

A glazed panel can be incorporated into the top surface of the ventilator to heat the air in the ventilator and so slightly enhance the stack effect locally. Note that maintenance of the control dampers needs to be considered, especially when they are at a high level.

The tower roofs for the Inland Revenue building in Nottingham (Fig. 11.3) were controlled to open in the occupied period if the following conditions [32] were satisfied:

- wind speed < 8 m s^{-1}
- internal temperature at top of tower > 27°C
- average corridor temperature > 25°C
- t_{ao} > 12°C
- rain intensity < intermediate

But 'intermediate' is not defined. There were similar rules for the ridge vents on the top floor.

The funnels, or chimneys, at De Montfort University's Queen's Building (Fig. 11.4) serve a number of deep-plan areas with conventional motorized dampers in the extract funnels, synchronized with motorized dampers on low-level air inlets [29]. Fans in the chimneys of the two auditoria are turned on to help the ventilation if

$$\text{wind speed} < 1.5 \text{ m s}^{-1} \quad \text{and} \quad t_{ao} > 20°C$$

The dampers had to be adapted from standard products so they would seal under a low pressure differential [1]. Access and maintenance of the dampers also has to be considered, especially in funnels and chimneys.

To control the airflow, the Ionica headquarters in Cambridge uses automatically controlled doors in the base of its chimneys on the atrium roof [1]. The control schedule is according to the natural ventilation requirements and the prevailing wind speed. The occupants were given airline-style instruction sheets and seminars on the control of their windows and blinds, so they could be constructive when helping to control their environment.

11.8 Night ventilation

For stack ventilation to be effective in warm and hot weather, equation (11.18) suggests that the air drawn in from outside must be colder than the air inside. This means that in hot weather, when the outside air is at

25–35°C, natural ventilation will not be very successful but wind-driven air will still be welcome. In countries with moderate climates, however, the night-time temperature is significantly below the daytime temperature, i.e. there is significant **diurnal variation** in temperature.

11.8.1 Diurnal temperature variation

When there is a large diurnal temperature variation, natural ventilation during the night is very useful in removing the heat of the day. But the building has to be designed with sufficient thermal mass to soak up the heat during the day to prevent high daytime temperatures and to release the heat at night. Careful thought should be given to shading, to reduce the direct entry of solar radiation through windows as the solar irradiance in the United Kingdom on a vertical surface can be up to 700 W m^{-2}. During clear summer days, although the daytime temperature will be high, the night-time temperature will be somewhat lower, due to longwave loss to the clear night sky. A typical difference could be as much as 12 K [12]. This gives the opportunity to use night ventilation to get rid of residual heat from the building with lower air temperatures. With careful design of the building, this residual heat will have been absorbed into its fabric, to be released at night.

11.8.2 Thermal mass and time constants for air and fabric

As natural ventilation has such a low cooling potential, about 20 W m^{-2}, the mass of the building fabric must be used to absorb as much of the heat gains as possible. It is considered that a well-designed naturally ventilated building can just about keep its peak internal temperature below the peak outside air temperature. However, in seven naturally ventilated buildings, actual measurements indicated that for external temperatures of about 20°C the internal temperatures would be 4–5 K higher [30]. In an office refurbishment where the window area was reduced and the internal ceiling exposed for increased thermal storage, the peak internal temperature was kept just below the peak outside temperature in a recent hot summer period [31].

Thermal mass has been considered in Chapter 9 on optimizers. It was shown that the heat stored in the air contained within the building was insignificant compared to the heat in the fabric (walls, ceilings and floors). This is simply due to the higher density of the fabric hence its greater ability to absorb heat. The simple model (9.18) was a first-order model:

$$pQ_{des} = \frac{H_{des}}{F_2}(t_{ai} - t_{ao}) + \sum(m^*C_p)\frac{dt_{ai}}{dT}$$

On the left-hand side of the equation is the heat input (W) with a plant size ratio of p. The first term on the right-hand side is the heat loss through the fabric and ventilation, and the summation term is due to the heat flow into the fabric. T is time. Although this is for heating, it can be applied to summer conditions where the heat input is due to the various gains: solar Q_s, internal Q_i, lighting Q_l, and equipment Q_e. Also if t_{ao} is higher than t_{ai} then the heat loss becomes a heat gain:

$$Q_s + Q_i + Q_l + Q_e + H_{des}(t_{ao} - t_{ai}) = \sum(m^*C_p)\frac{dt_{ai}}{dT} \quad (11.25)$$

The heating system factor, F_2, often close to unity, has been left out. In terms of a time constant and temperatures, this can be rewritten as

$$H_{des}(\Delta t_s + \Delta t_i + \Delta t_l + \Delta t_e + \Delta t_c) = \sum(m^*C_p)\frac{dt_{ai}}{dT} \quad (11.26)$$

where the Δt terms are the temperature rises due to the heat gains; Δt_c is the ventilation and fabric conductance term. The time constant, τ, is given by

$$\tau = \frac{\sum(m^*C_p)}{H_{des}}$$

which dictates how much heat and how quickly the heat is absorbed into the fabric. A tent will have an insignificant time constant (although ventilation and high heat gain or loss may keep the temperature close to ambient). But a castle will have a time constant of days, possibly up to a week, and a cave will have a virtually constant temperature throughout the year. There might be a slight variation in the cave temperature depending on the type of rock and the cracks in it.

As Fig. 11.16 shows, for a constant heat gain into the room, it will eventually raise the room temperature to Δt_f above the initial room temperature, $t_{ai\ in}$. The heat into the fabric is above the curved exponential line. The steady-state heat balance (when the heat input equals the heat loss) is when the temperature curve is horizontal – its slope and differential are zero.

At night, when the heat gains from the sun, etc., are zero and the outside temperature is below the inside temperature, the building will cool down and equation (11.25) becomes

$$H_{des}(t_{ai} - t_{ao}) = \sum(m^*C_p)\frac{dt_{ai}}{dT} \quad (11.27)$$

where the fabric loses its heat by ventilation and fabric conductance to the outside. These equations are demonstrated in Example 11.3.

Example 11.3
Consider a naturally ventilated building during a sequence of hot summer days and initially with an inside temperature of 20°C. To examine the effect purely of the heat gains (solar, equipment, etc.), losses and fabric storage, it

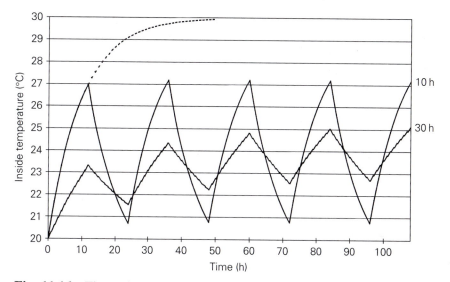

Fig. 11.16 Thermal response of a room with time constants of 10 h and 30 h.

is assumed that t_{ao} remains constant at 18°C. The heat gains are capable of raising the inside temperature to 12 K above the external temperature of 18°C, i.e. to 30°C. It is further assumed that the heat gains are present for 12 h during the day and negligible for the other 12 h.

(a) Calculate the building inside temperature through a number of days, assuming a single building time constant of (i) 30 h and (ii) 10 h.

(b) Consider the effect of night ventilation with the night ventilation rate at double the daytime rate, the daytime ventilation loss being equal to the fabric loss.

(c) Consider the outside air temperature varying sinusoidally during the day from 18°C to 28°C.

Solution

(a) We can use integrated equations from equations (11.26) and (11.27) which are similar to the heating and cooling equations in Chapter 9. The integrated version of equation (11.26) is:

$$t_{ai\ end} = t_{ai\ st} + (1 - \exp(-T/\tau))(\Delta t - (t_{ai\ st} - t_{ao})) \qquad (11.28)$$

where $t_{ai\ st}$ = the inside temperature at the start of the heating period due to the gains

$t_{ai\ end}$ = the inside temperature at the end of the heating period due to the gains

Δt = the temperature rise due to the heat gains (10 K here)

When the gains stop and the cooling starts, equation (9.20) can be adapted for integrating equation (11.27); this gives

$$t_{ai\ end\ c} = t_{ao} + (t_{ai\ st\ c} - t_{ao})\ \exp\ (-T/\tau) \tag{11.29}$$

where $t_{ai\ end\ c}$ = the inside temperature at the end of the cooling period
$t_{ai\ st\ c}$ = the inside temperature at the start of the cooling period
It is worth checking equations (11.28) and (11.29) for their steady-state values, i.e. when $T = \infty$, when the fabric has fully absorbed the heat (11.28) or fully discharged it (11.29). Equation (11.28) gives $t_{ai\ end} = t_{ao} + \Delta t$, which in our example means $t_{ai\ end} = 30°C$. Equation (11.29) gives $t_{ai\ end\ c} = t_{ao}$, which means $t_{ai\ end\ c} = 18°C$. Both of these values are consistent with the steady-state calculations. Figure 11.16 shows the temperature variation due to the heat gains and heat losses using equations (11.28) and (11.29); it was developed on a spreadsheet with hourly values. It also shows, for the shorter time constant, the check calculation of the heat gains continuing indefinitely and the inside temperature approaching 30°C.

The longer time constant reduces the peak internal temperature and the temperature swing quite substantially. After 4.5 days the peak internal temperature is 25°C, whereas for the 10 h time constant the peak is just over 27°C. These temperatures compare with the maximum potential temperature of 30°C which a lightweight building would achieve in a day and the 10 h building achieves in about 2 days with continuous heat gains. In Fig. 11.16 the heat stored in the fabric is represented by the area above the curve but bounded by the 30°C horizontal line; the heat loss through the fabric and ventilation to the outside is represented by the area under the curve, bounded by the 18°C horizontal. The 30 h building has a lower rise in temperature than the 10 h building as it can store more heat in its fabric.

Notice that the 10 h building has a consistent daily maximum of about 27°C whereas for the 30 h building the maximum inside temperature creeps up over the week, from just over 23°C to 25°C. So even though the 30 h building can mitigate the hot weather, the fabric cumulatively stores the heat and there is a slow rise in temperature. This is often experienced in winter heating as well. Monday morning is cold after the weekend, when the heating is off, but the temperature gradually rises through the week, so by Friday the building can be too hot.

(b) To consider the effect of the increased night ventilation, we can note that the ventilation term $NV/3$ is in the heat loss term H_{des}. This will therefore affect the time constant:

$$\tau = \frac{\sum m * C_p}{\sum (UA) + NV/3}$$

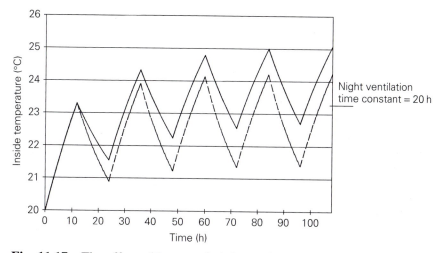

Fig. 11.17   The effect of increased night ventilation (day 30 h, night 20 h).

making it possible by controlling the ventilation to vary the time constant and hence the release or absorption of heat in the fabric. Now the question states that the night ventilation rate is double the daytime rate, the daytime ventilation loss being equal to the fabric loss. Put mathematically,

$$N_d V/3 = \Sigma(UA) = H_{des}/2 \quad \text{and} \quad N_n = 2N_d$$

where N_d = the daytime ventilation air change rate (h)
$\quad\quad N_n$ = the night-time ventilation air change rate (h)
Choosing the building with the longer time constant, the night-time time constant, τ_n, becomes 20 h with the daytime time constant remaining at 30 h. Figure 11.17 shows the reduction in the peak temperatures reached. The final peak temperature is almost 1 K below the 25°C peak without increased night ventilation.
(c) Now we have to consider the more realistic situation of the outside temperature varying sinusoidally throughout the day. Mathematically t_{ao} varies as

$$t_{ao} = 23 + 5 \sin 2\pi \left(\frac{T + 6}{24} \right)$$

This form can be used on the Excel spreadsheet, which calculates the sine of an angle in radians. The factor 6 is included to start t_{ao} at its minimum value of 18°C. Figure 11.18 shows the steady rise of the internal temperature so it is above the outside temperature for 3.5 of the 4.5 days. The increased night-time ventilation rate has been retained with the varying outside temperature.

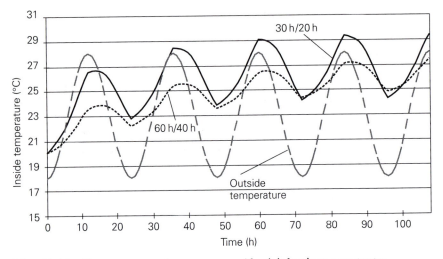

Fig. 11.18   Room temperature response (day/night-time constants: 30 h/20 h and 60 h/40 h) with sinusoidally varying outside temperature.

Also shown is a building with an even longer time constant of 60 h; this reduces to 40 h at night. The varying outside temperature substantially raises the internal temperature. Initially the 60 h building's peak temperature is about 3 K below the 30 h building's peak temperature, but after 4.5 days the difference is 2 K.

The mass of a naturally ventilated building is crucial in controlling the internal temperature, as Example 11.3 has shown. Although 28°C is a reasonable peak temperature for a UK summer day, it is unusual to have five days together with such a temperature [12]. So this is a severe test for the building. But it does emphasize that with long time constant, or heavy-weight, buildings the weather over a number of days does influence the inside temperature. It is often during periods of successive warm to hot days in moderate climates that occupants wish for air conditioning. To design a naturally ventilated building it is sensible to simulate the building on a computer against real design weather data for the site. Data is available in the CIBSE Weather and Solar Data Guide [12]. The model in Example 11.3 could be further developed to account for stack ventilation, which would vary with the inside–outside temperature difference and so vary the time constant with temperature difference. As the outside temperature drops quickly at night, but the inside temperature drops slightly, this would enhance the stack ventilation and make the night ventilation more effective.

11.8.3 The effect of suspended ceilings and carpets

The simple model in Example 11.3 needs to be expanded to show the effect of suspended ceilings and carpets that can reduce the heat storage capacity of the fabric. Also note that the comfort temperature, often called the dry resultant temperature, is the average of the air temperature and the mean surface temperature. In the simple first-order model used in Example 11.3, the relationship between the air temperature and the fabric temperature is not explicit. It depends on the heat transfer between the air and the fabric, and also the radiative heat transfer between surfaces. A more accurate model is

$$Q_s + Q_i + Q_l + Q_e - Q_{\text{inf+vent}} = m_1 C_{p1} \frac{dt_1}{dT} + m_2 C_{p2} \frac{dt_2}{dT} + m_3 C_{p3} \frac{dt_3}{dT} + \dots$$

(11.30)

or

$$H_{\text{des}} \Delta t - Q_{\text{inf+vent}} = m_1 C_{p1} \frac{dt_1}{dT} + m_2 C_{p2} \frac{dt_2}{dT} + m_3 C_{p3} \frac{dt_3}{dT} + \dots \quad (11.31)$$

where the subscripts 1, 2, 3, . . . refer to the different elements of heat storage in the building, primarily the fabric, and $Q_{\text{inf+vent}}$ is the infiltration and ventilation heat loss. There is no steady-state fabric heat loss term, as the external fabric has to absorb and transmit the heat to the outside surfaces before it can lose heat to the external environment. If the main heat storage capacity of the building is in its internal walls, floors and ceilings then for ease of analysis they are considered as similar structures. If the external fabric has relatively small storage capacity in comparison then the heat storage can be considered to reside in an internal fabric element with a fabric heat loss through the external elements such as the walls. The internal element can be split into layers so that finite difference techniques may be used to solve the differential equation numerically. Figure 11.19 shows the internal element split into five layers. For this equation (11.31) becomes

$$H_{\text{des}} \Delta t - H_{\text{des}}(t_{ai} - t_{ao}) = m_{\text{air}} C_{p\,\text{air}} \frac{dt_{ai}}{dT} + \frac{mC_p}{5} \frac{dt_{w1}}{dT} + \frac{mC_p}{5} \frac{dt_{w2}}{dT} \quad (11.32)$$

As this is an internal element of fabric, the inside temperatures on both sides of the element are probably equal and vary similarly with time. So the heat flows from both sides are equal and there is no heat flow across the central resistance between t_{w2} and t_{w3}. Equation (11.32) also has a term for the heat flow into the air. From examining the heat flows in the circuit, we can derive the relationship between t_{ai} and t_{w1}:

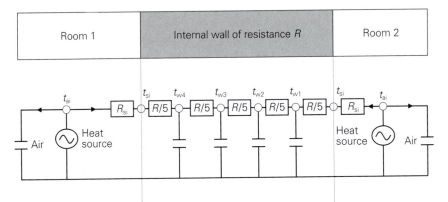

Fig. 11.19   Heat flow circuit of a building with its storage in internal elements.

$$\frac{dt_{w1}}{dT} = \frac{dt_{ai}}{dT}\left[1 + \frac{H_{des}}{A}\left(R_{si} + \frac{R}{5}\right)\right] \tag{11.33}$$

where A is the area of the internal surfaces storing heat (m²).
So considering just the first layer, equation (11.32) gives

$$\Delta t - (t_{ai} - t_{ao}) = \frac{dt_{ai}}{dT}\frac{\tau}{5}\left[1 + \frac{H_{des}}{A}\left(R_{si} + \frac{R}{5}\right)\right] + \tau_{air}\frac{dt_{ai}}{dT} \tag{11.34}$$

From equation (11.34) R_{si} has an influence on the time constant of the first element of fabric even if the thermal capacity, mC_p, is unaltered. If insulation, a suspended ceiling or carpeting are used in the building then this will also increase the time constant as the resistance R_{ins} is in series with R_{si}:

$$\Delta t - (t_{ai} - t_{ao}) = \frac{dt_{ai}}{dT}\frac{\tau}{5}\left[1 + \frac{H_{des}}{A}\left(R_{si} + R_{ins} + \frac{R}{5}\right)\right] + \tau_{air}\frac{dt_{ai}}{dT} \tag{11.35}$$

The fact that the thermal mass may not have been altered but that carpeting or a false ceiling may have been installed would increase the first time constant in equation (11.35) as R_{ins} increases. This would reduce the heat flow into the fabric within a given time. Although this shows the influence of the carpeting, etc., in R_{ins}, equation (11.35) gives a single time constant solution. A better description of the heat absorption process is given by an empirical equation with two time constants:

$$t_{ai\ end} - t_{ai\ st} = [\Delta t - (t_{ai\ st} - t_{ao})][A(1 - \exp(-T/\tau_1)) + B(1 - \exp(-T/\tau_2))] \tag{11.36}$$

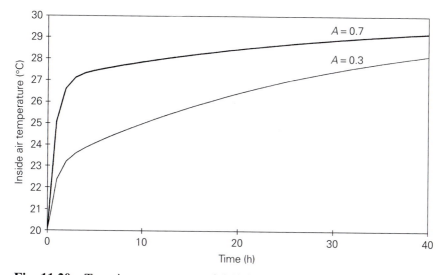

Fig. 11.20   Two time constant model (0.8 h and 30 h).

where A and B are constants such that $A + B = 1$ and τ_1 and τ_2 are long and short time constants; the long time constant is related to the fabric thermal storage and the shorter time constant is related to the air and surface layers of the building. If there is a lot of insulation on the inside surfaces of the building due to suspended ceilings, etc., then A will be high and the room air temperature will rise quickly (Fig. 11.20).

Although a building can have a large thermal mass, if it has false ceilings and extensive carpeting then the fabric thermal mass is effectively insulated from the internal heat gains. Consequently, it acts as a lightweight building.

11.8.3.1   The influence of surfaces with the Admittance Method

The effect of suspended ceilings, etc., is also shown by the admittance method (Chapter 7), where the approximate swing in the internal environmental temperature is given as follows [39]:

$$\tilde{Q} = \left(\Sigma AY + \frac{1}{3}NV \right) \tilde{t}_{ei} \qquad (11.37)$$

$$\tilde{t}_{ei} = \hat{t}_{ei} - \overline{t}_{ei}$$

where $\tilde{Q}$ is the alternating mean-to-peak heat gain (W) and $\tilde{t}_{ei}$ is the alternating mean-to-peak internal environmental temperature (°C). These values vary sinusoidally about the mean gain and temperature respectively. Y is the admittance value of the internal or external fabric element (W m^{-2} K^{-1}). All surfaces contribute in varying measures to absorbing the heat gains. These *Y*-values, hence the temperature swings, are influenced by the surface layers, as shown in Example 11.4.

Example 11.4

A room on an intermediate floor of a large office block has one external wall 4 m wide, 3 m high and 5 m deep. It has a double-glazed window of 4 m². The mean-to-peak temperature rise is 5 K with its standard construction, given below. What is the mean-to-peak temperature when (i) the ceiling is changed to a suspended ceiling, the floor is covered with a carpet and the partition walls are made of plasterboard, (ii) the room is made open plan? Assume that there is one air change per hour. The standard construction is as follows.

	Y-value (W m⁻² K⁻¹)
External wall, block, cavity, insulation, block	4.3
Double-glazed window	2.9
Partitions, 100 mm lightweight concrete block	2.0
PVC-covered cast concrete floor	5.0
Plastered concrete ceiling	2.0
Construction (i)	
Partitions, 25 mm plasterboard, 25 mm airgap, 25 mm plasterboard	1.4
Suspended ceiling	1.0
Carpet-covered concrete floor	2.5

Solution

We can use equation (11.37) to find the heat input initially. The *YA* values can be determined as follows.

	A	*Y*	*YA*
Double-glazed window	4	2.9	11.6
External wall, block, cavity, insulation, block	8	4.3	34.4
Partitions, 100 mm lightweight concrete block	42	2.0	84
PVC-covered cast concrete floor	20	5.0	100
Plastered concrete ceiling	20	2.0	40
Total			270

From equation (11.37) we get

$$\tilde{Q} = (270 + 20) \times 5$$

$$= 1450 \text{ W}$$

For construction (i), which is more lightweight, the *YA* values are given below.

	A	Y	YA
Double-glazed window	4	2.9	11.6
External wall, block, cavity, insulation, block	8	4.3	34.4
Partitions, 25 mm plasterboard, 25 mm airgap, 25 mm plasterboard	42	1.4	58.8
Carpet-covered cast concrete floor	20	2.5	50
Suspended ceiling	20	1.0	20
Total			174.8

From equation (11.37) we get

$$\tilde{t}_{ei} = 1350/(174.8 + 20)$$

$$= 7.4 \text{ K}$$

For the open-plan room there are no internal walls and the *YA* values are as follows.

	A	Y	YA
Double-glazed window	4	2.9	11.6
External wall, block, cavity, insulation, block	8	4.3	34.4
Carpet-covered cast concrete floor	20	2.5	40
Suspended ceiling	20	2.0	100
Total			116

From equation (11.37) we get

$$\tilde{t}_{ei} = 1350/(116 + 20)$$

$$= 10.7 \text{ K}$$

So if the mean temperature in the room were 20°C then the peak temperatures would be 25°C, 27.4°C and 30.7°C – quite a difference.

11.8.4   Control of night ventilation

Most modern, naturally ventilated (and mixed-mode) buildings use night cooling. This is where the appropriate windows and louvres are opened to capitalize on the lower night-time outside temperatures to ventilate the summer day's heat, stored in the building's fabric. The application of night ventilation can reduce the daytime temperature by up to 2–3 K. Essentially the control strategy should be based on a measurement of the heat stored in the building fabric. However, this is not always that easy, so the control strategies are very varied, as a study by BSRIA has shown [32–34]. The strategies can be based on the room (or zone) temperature or the fabric temperature, or both. The fabric temperature is measured by a **surface temperature sensor** or a sensor actually embedded in the fabric itself. The depth of embedding is important as the further into the fabric, the less the exchange of heat and the less the variation in temperature. From the **admittance theory**, the cyclic nature of the heat flow means that only the structure's first 75 mm, or thereabouts, is actively storing heat. However, the further into the structure the sensor is embedded, the greater the difference from room temperature and the lower its amplitude of variation.

In a strategy some basic conditions can be applied such as the night ventilation being started if

(1)   room air temperature > outside air temperature
(2)   room air temperature > heating setpoint
(3)   outside air temperature > 12°C

Condition 1 is so that the room is cooled and not heated. Condition 2 is so that heating or preheating is avoided. Condition 3 minimizes unnecessary cooling or the slight risk of condensation. BSRIA studied four mixed-mode buildings over the hot summer of 1995. The night ventilation strategies are considered here.

11.8.4.1   Fabric setpoint control

Fabric setpoint control is based on the Inland Revenue building, Durrington. The strategy, which was complicated, was based primarily on the fabric and room air temperature. If the fabric temperature was above its setpoint and the building was to be occupied the next day, the automatic windows and the atrium vents were opened. Mechanical ventilation could be applied during the night, off-peak tariff period if the fabric temperature was still too high by the start of the off-peak period. Essentially the cooling strategy was to calculate and adjust the fabric temperature setpoint so that the fabric temperature and the room temperature equalled the fabric temperature setpoint at the end of the occupancy period on the following day. The fabric setpoint was adjusted by an adaptive algorithm [33] which calculated three error values, Δt_1 Δt_2 and Δt_4:

$$\Delta t_1 = \text{room setpoint} - \text{fabric temperature at 1700} - \text{offset}$$

$$\Delta t_2 = \text{room setpoint} - \text{room temperature at 1700}$$

The offset differentiated between winter and summer operation. In summer the offset was applied to make the fabric temperature at the end of occupancy lower than the room setpoint. Then

$$\Delta t_3 = \Delta t_1 + \Delta t_2 \text{ current day's adjustment}$$

$$\Delta t_4 = \text{fabric temperature at 0700} - \text{fabric temperature setpoint}$$

$$\Delta t_3 \text{ reduced if } \Delta t_4 \text{ too high}$$

The new day's fabric temperature setpoint was then calculated as

$$\text{previous day's fabric setpoint} + \frac{\Delta t_{3\text{current}} + \Delta t_{3\text{previous}}}{2}$$

In fact, during the 1995 monitoring period, the fabric temperature setpoint was never achieved (in July it was 22°C); the variation was between 1.2 K and 7.8 K. Also the setpoint change was very limited, about 0.7 K. Consequently, the strategy resulted in very high night utilization even in the months of September and October [32].

11.8.4.2 *Average afternoon outside temperature control*

Average afternoon outside temperature control is based on the Inland Revenue building, Nottingham, which has the ventilation towers that were discussed earlier and are shown in Fig. 11.3. The night ventilation control is based on the minimum average outside air temperature and room or zone temperature. Night cooling was activated if

(1)   between 1200 and 1700 the average outside temperature, $t_{ao} > 18°C$
(2) $t_{ao} > 12°C$
(3) $t_{ai} > t_{ao}$
(4) $t_{ai} > 15.5°C$ with 3 K deadband centred on the setpoint

This control meant that the cooling was stopped when $t_{ai} = 14°C$ and not started again until the air had been heated to 17°C, by the fabric giving off stored heat. In fact, this was too low a setpoint as the minimum zone temperature in the monitored summer of 1995 was between 18.2°C and 20.7°C [32]. The vents stayed open until one of the four conditions no longer applied. If all these conditions applied then the tower roofs were kept fully open. The wind speed, rain, security and fire conditional control that applied to the daytime operation also applied to the night-time operation.

This was considered a successful strategy. The use of average afternoon outside temperature as a limit was more representative of the daytime conditions than the use of peak temperature.

11.8.4.3 *Peak inside air temperature and degree-hours control*

At the Ionica building in Cambridge, the control was based on the minimum zone temperature and degree-hours above a setpoint of the fabric temperature of 21°C. Unfortunately, it did not operate during the BSRIA monitoring period, which could not be explained. Night cooling occurred between 2100 and 0600 if the zone temperature exceeded 24°C for more than 1 h in the day. Here are the conditions for night cooling to occur:

Table 11.3 Self-learning schedule for night cooling [23]

Zone average – zone setpoint (K)	Day degree-hours heating : night degree hours cooling
−8	1 : 0.8
−4	1 : 0.9
0	1 : 1
4	1 : 1.1
8	1 : 1.2

(1) outside air temperature $> 7°C$
(2) zone air temperature $> 14°C$
(3) peak zone air temperature during occupancy $> 24°C$
(4) zone air temperature at beginning of night cool $> 19°C$
(5) zone air temperature $>$ outside air temperature
(6) night cooling continued until degree-hours limit (see below)

The control was such that the night cooling would continue until the night-time cooling degree-hours equalled the daytime degree-hours. (The daytime degree-hours were calculated for temperatures either side of the setpoint fabric temperature of 21°C, whereas the night-time degree-hours target to match was only based on temperatures above the setpoint.) The target night cooling degree-hours were increased if the average daytime zone temperature exceeded 21°C and reduced if the average daytime zone temperature was less than 21°C.

The daytime degree-hours were calculated including periods above and below 21°C. However, the night-time degree hours only included periods below 21°C. A self-learning algorithm adjusted the night cooling degree-days depending on how much the zone average temperature exceeded its setpoint (Table 11.3).

In the analysis of this strategy, BSRIA considered that the limit of 24°C for the zone temperature in condition 3 was too high. This was because once it had been exceeded, the night cooling would be unlikely to reduce the fabric temperature to the 21°C setpoint, let alone generate the required degree-hours.

Peak inside air temperature and degree-hours control is similar to the strategy at the Anglia Polytechnic Learning Resource Centre, except the room air temperature is monitored for the degree-hours with a night-time setpoint adjusted by an adaptive algorithm. Monitoring the surface temperature was dismissed as it was considered to vary with position and surface depth. Likewise the exhaust temperature was not considered as it might vary irrespective of the fabric temperature [1].

11.8.4.4 Average room temperature and peak outside air temperature control

Based on the PowerGen building, Coventry, average room temperature and peak outside air temperature was the simplest strategy and potentially the most successful [32]. Night cooling (the BEMS opens the top row of office windows and the atria windows) was commenced if

(1) the average room or zone temperature at the end of the day > 23°C, and
(2) the maximum external temperature during the day > 21°C

To stop overcooling, night ventilation ceased if

(3) the room or zone temperature ≤ 18°C

Cooling started again if

(4) the room or zone temperature ≥ 20°C

During the night the windows were closed if the wind speed exceeded 20 m s^{-1}; during the day the condition was reduced to 10 m s^{-1}.

11.8.4.5 Night-cooling strategy: conclusions

During July and August, the peak months of the monitored period, all the strategies invoked night cooling for the maximum amount of time. And once initiated the cooling would operate for the whole night. However, in the marginal months of May, June, September and October there were wide variations. For September and October, out of a maximum of 47 days for operation, Ionica (peak room + degree-hours) would not use night ventilation whereas IRO Durrington (fabric setpoint) would use all 47 days, the IRO Nottingham (average afternoon t_{ao}) would use 31 days, and PowerGen (average room + peak t_{ao}) would use 7 days. Apart from the Ionica strategy, with its overcautious initiation setpoint, all the strategies would have the room temperatures above a comfort temperature for a similar number of hours [35, 32], so on this basis they were equally successful.

BSRIA concluded that measurement of the fabric temperature gave a better indication of the heat stored in the fabric than the air temperature. This accords with the theory in Sections 11.8.2 and 11.8.3. However, the location of the fabric sensor could be a problem, depending on the position of heat sources and inlet vents. BSRIA found that the best correlation of peak internal temperature was with the fabric temperature two days following. This demonstrates the long time constant associated with fabric storage.

From this study, and other work, BSRIA has developed a night-cooling strategy [36] which uses the temperature of the dominant exposed fabric mass in conjunction with the room or zone air temperature. This strategy initiates night cooling when

(1)   the average afternoon outside air temperature $> 20°C$

AND EITHER

(2)   the average daytime zone temperature (any zone) $> 22°C$

OR

(3)   the temperature of an exposed concrete ceiling slab $> 23°C$

Night cooling continues if the following conditions are maintained:

(4)   the zone temperature $>$ the outside air temperature
(+2 K if using mechanical ventilation to allow for fan pickup)

AND

(5)   the zone temperature (any zone) $>$ the heating setpoint (19°C is quoted)

AND

(6)   the outside temperature $> 12°C$

It is suggested that night cooling should be available for seven days a week, as hot conditions over the weekend can cause problems at the start of occupancy on a Monday.

Further advice is given that the natural, or passive, ventilation should be stopped and a BEMS alarm raised if

(7)   the average wind speed > 20 m s^{-1}
(averaged over 3 min with minimum sampling time of 15 s)

OR

(8)   rain is sensed by the roof-mounted rain intensity sensor

Under conditions 7 and 8 mechanical ventilation, if available, should be brought on.

11.8.5   The limit of ventilation cooling

So far the heat exchange due to the ventilation has been assumed to vary linearly with the number of air change rates per hour, N:

$$\frac{1}{3}NV(t_{ao} - t_{ai}) \tag{11.38}$$

But this is for low values of N as there is a limit to the heat exchange possible, determined by the heat transfer between the air and the fabric and also the mixing of the inlet and resident air. This is hinted at in the BSRIA night-cooling study [35], which found that mechanical ventilation was no more successful than natural ventilation. If the outside air was to rush through

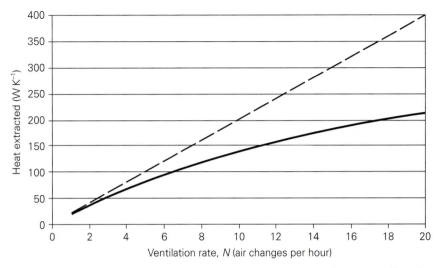

Fig. 11.21 The limitation of natural ventilation heat loss: (---) *NV/3* and (—) equation (11.39).

the building then it would hardly touch or interact with the fabric, where most of the heat is stored. Section A8 of the CIBSE Guide [37, 25] recommends that when the air change rate goes above $2\,h^{-1}$ (equivalent to a heat exchange of $0.6\,\mathrm{W\,m^{-3}\,K^{-1}}$), a more accurate equation for the heat loss is

$$\frac{1}{C_v} = \frac{1}{0.33NV} + \frac{1}{4.8\Sigma A} \qquad (11.39)$$

where ΣA = total area of the surfaces bounding the enclosure (m^2)

$\qquad C_v$ = the ventilation loss ($\mathrm{W\,K^{-1}}$) and the heat loss is $C_v(t_{ao} - t_{ai})$

As N increases in equation (11.39), the first term reduces but the second term stays constant. As $N \to \infty$ then $C_v \to 4.8\Sigma A$. This is shown in Fig. 11.21 for a small room $5\,m \times 4\,m \times 3\,m$ high.

Natural ventilation may also have to be limited for comfort reasons as ISO 7730 recommends that for the comfort of a sedentary worker mean air velocities should be less than $0.25\,\mathrm{m\,s^{-1}}$ in summer and less than $0.15\,\mathrm{m\,s^{-1}}$ in winter [19]. Air velocities above $0.8\,\mathrm{m\,s^{-1}}$ are likely to disturb papers [38].

The reduction in the heat exchange as the ventilation rate increases can also be appreciated by considering the room as a heat exchanger with the cold outside air entering with temperature t_{ao} at one end of the room and exiting at the other side with temperature t_{ae}. The steady-state heat balance equation then becomes

$$\frac{1}{3}NV(t_{ae} - t_{ao}) = CUA\left[\frac{\Delta t_{max} - \Delta t_{min}}{\ln(\Delta t_{max}/\Delta t_{min})}\right] \qquad (11.40)$$

where C = a correction factor for a non-perfect heat exchanger (dimensionless)
    U = the thermal transmission value of the room (heat exchanger) surfaces (W m^{-2} K^{-1})
    A = the area of all the internal room surfaces (m^2)
    $\Delta t_{max} = t_f - t_{ao}$, $\Delta t_{min} = t_f - t_{ae}$ where t_f is the fabric, or surface, temperature

It is assumed that all the fabric is at the same surface temperature, t_f, inside the building. Equation (11.40) can be simplified to

$$\frac{1}{3}NV(t_{ae} - t_{ao}) = CUA\left[\frac{t_{ae} - t_{ao}}{\ln(\Delta t_{max}/\Delta t_{min})}\right]$$

$$\ln\frac{\Delta t_{max}}{\Delta t_{min}} = \frac{3CUA}{NV} \tag{11.41}$$

Taking the exponential of both sides and substituting for Δt_{max} and Δt_{min}:

$$t_{ae} = t_f - (t_f - t_{ao})\exp\left(-\frac{3CUA}{NV}\right)$$

so, using the left-hand side of equation (11.40), the heat loss due to ventilation now becomes

$$\frac{1}{3}NV(t_f - t_{ao})\left[1 - \exp\left(-\frac{3CUA}{NV}\right)\right] \tag{11.42}$$

The exponential term can be expanded using Maclaurin's series, or the exponential series:

$$\exp x = 1 + x + \frac{x^2}{2!} + \frac{x^3}{3!} + \ldots + \frac{x^n}{n!} + \ldots$$

Using the first two terms of this expansion we have $\exp(-x) = 1/(1 + x)$ and substituting into equation (11.42) gives

$$(t_f - t_{ao})\left[\frac{CUA(NV/3)}{NV/3 + CUA}\right]$$

which can be rearranged to

$$(t_f - t_{ao})\left[\frac{1}{0.33NV} + \frac{1}{CUA}\right]^{-1} \tag{11.43}$$

Here the ventilation loss term, C_v, is the same as in the CIBSE equation (11.39). You can see how they are equivalent by taking CU in (11.43) and replacing it with 4.8 and by taking A in (11.43) and replacing it with ΣA. If t_f equals t_{ai}, and the CIBSE Guide assumes they are equal, then equation (11.43) equals the standard ventilation heat loss with equation (11.39) for C_v.

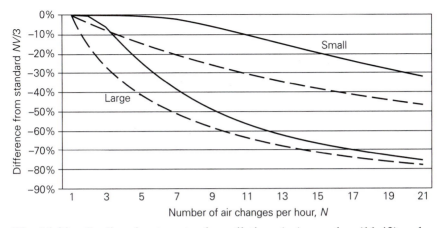

Fig. 11.22 Cooling due to natural ventilation: (——) equation (11.42) and (---) equation (11.43).

Figure 11.22 shows the ventilation loss term C_v given by the exact equation (11.42) and the approximate equation (11.43) compared to the standard CIBSE linear form (11.38). Results are shown for a small room (5 m long, 4 m wide and 3 m high) and for a large room such as an atrium (30 m long, 10 m wide and 16 m high). The solid lines are from equation (11.42) and the dashed lines are from equation (11.43). The effect is quite significant for the large room. At 8 air changes per hour, a 50% reduction in the heat loss would be expected from equation (11.42) compared to the standard linear equation (11.38), and a 60% reduction would be expected when comparing (11.43) with (11.38). This shows that care is required in calculating the cooling due to ventilation, especially for night ventilation when the stack-driven component will be high due to the cool night air.

References

[1] CIBSE (1997) Application Manual AM10, Natural ventilation in non-domestic buildings, Chartered Institution of Building Services Engineers, London.

[2] Wilson, S. and Hedge, A. (1986) *Office Environment Survey*, Building Use Studies, London.

[3] Raw, G. J. (1995) Air conditioning, friend or foe? *Building Services Journal*, Sept.

[4] Jones, P. J., Vaughan, N. D., Grajewski, T., Jenkins, H. G., O'Sullivan, P. E., Hillier, W., Young, A. and Patel, A. (1995) New guidelines for the design of healthy office environments. Project Report GR/H38645 to ESPRC/DTI.

[5] Twinn, C. A. (1995) A buoyant line of thinking. *Building Services Journal*, Jan.

[6] Brister, A. (1994) *Building Services Journal*, May.

[7] Cook, M. (1993) *Architect's Journal*, 16 June.

[8] Bunn, R. (1993) *Building Services Journal*, Oct.

[9] CIBSE (1986) CIBSE Guide, Section A4, Air infiltration and natural ventilation, Chartered Institution of Building Services Engineers, London.

[10] BS 5925: 1991 (1995) Code of practice for ventilation principles and designing for natural ventilation, British Standards Institution, London.

[11] Liddament, M. W. (1996) A guide to energy efficient ventilation. Air Infiltration and Ventilation Centre, Warwick, UK.

[12] CIBSE (2000) CIBSE Weather and Solar Data Guide, Chartered Institution of Building Services Engineers, London.

[13] Warren, P. R. and Webb, B. C. (1980) The relationship between tracer gas and pressurisation techniques in dwellings. In *Proceedings of the 1st AIC Conference on Air Infiltration Instrumentation and Measuring Techniques*, 6–8 October 1980, Windsor UK.

[14] Sherman, M. H. and Grimsrud, D. T. (1980) Infiltration-pressurization correlation: simplified physical modelling. *ASHRAE Trans.*, **86**(2).

[15] Awbi, H. B. (1991) *Ventilation of Buildings*, Spon, London.

[16] Liddament, M. W. (1986) Air infiltration calculation techniques – an applications guide. Air Infiltration and Ventilation Centre, Coventry, UK.

[17] Deaves, D. M. and Lines, I. G. (1997) On the fitting of low mean wind speed data to the Weibull distribution. *Journal of Wind Engineering and Industrial Aerodynamics*, **66**.

[18] Eastop, T. D. and Watson, W. E. (1992) *Mechanical Services for Buildings*, Longman, Harlow.

[19] BS EN ISO 7730 (1994) Moderate thermal environments – determination of the PMV and PPD indices and specification of the conditions for thermal comfort, International Organization for Standardization, Geneva.

[20] Fanger, P. O. (1970) *Thermal Comfort*, Danish Technical Press, Copenhagen.

[21] McCartney, K. and Nicol, F. (1998) Comfort in office buildings: results from field studies and presentation of the revised adaptive control algorithm. Paper presented at the CIBSE National Conference, Bournemouth, UK.

[22] Nicol, F. and Humphreys, M. A. (1972) Thermal comfort as part of a self-regulating system. In *Proceedings of the CIB Symposium on Thermal Comfort*, Building Research Establishment, Watford, UK, 1972.

[23] Humphreys, M. A. (1995) What causes thermal discomfort? Workplace Comfort Forum, RIBA, London.

[24] Brager, G. S. and de Dear R. J. (1998) Thermal adaptation in the built environment: a literature review. *Energy and Buildings*, **27**.

[25] CIBSE (1999) CIBSE Guide A, Environmental design, Chartered Institution of Building Services Engineers, London.

[26] Cooper, V., Wright, A. J. and Levermore G. J. (1996) Thermal analysis of mixed mode office buildings. Paper presented at the CIBSE/ASHRAE Conference, Harrogate, UK.

[27] Dubrul, C. (1988) Inhabitants' behaviour with regard to ventilation. AIVC Technical Note 23, Air Infiltration and Ventilation Centre, Coventry, UK.

[28] Monodraught Ltd (1998) Windcatcher natural ventilation systems, High Wycombe, UK.

[29] Bunn, R. (1993) *Building Services Journal*, Oct.

[30] BRESCU (1995) Avoiding or minimising the use of air-conditioning – a research report for the EnREI Programme. Building Research Establishment, Watford, UK.

[31] BRESCU (1998) Passive refurbishment at the Open University: achieving staff comfort through improved natural ventilation. BRESCU General Information Report 48, Building Research Establishment, Watford, UK.

[32] Fletcher, J. (1996) Night cooling control strategies – site monitoring results. BSRIA Report 11621/2, May.

[33] Fletcher, J. (1996) Night cooling control strategies – dynamic thermal simulation results. BSRIA Report 11621/3, May.

[34] Martin, A. and Fletcher, J. (1996) Night cooling control strategies – final report. BSRIA Report 11621/4, May.

[35] Fletcher, J. and Martin A. J. (1996) Night cooling control strategies. BSRIA Technical Appraisal 14/96, BSRIA, Bracknell, UK.

[36] Martin, A. J. and Banyard, C. P. (1998) Library of system control strategies. BSRIA Application Guide AG7/98, BSRIA, Bracknell, UK.

[37] CIBSE (1986) CIBSE Guide, Section A8, Summertime temperatures in buildings, Chartered Institution of Building Services Engineers, London.

[38] CIBSE (1986) CIBSE Guide, Volume B, Installation and equipment data, Chartered Institution of Building Services Engineers, London.

[39] Chadderton, D. V. (1997) *Air Conditioning, A Practical Introduction*, 2nd edn, E & FN Spon, London.

12
Low-energy air conditioning and lighting control

12.1 Displacement ventilation

Originally used for industrial buildings for ventilation and to remove air-borne contaminants from the occupied zone by natural convection, displacement ventilation is now applied to office buildings. Compared to conventional air conditioning, it reduces the concentration of polluting particles and gases generated by structural materials, furnishings and the occupants themselves. Conventional air conditioning relies on mixing the room air with the cold supply air, typically from ceiling diffusers (giving between 5 and 15 air changes per hour). Displacement ventilation uses about 2.5 to 3 air changes per hour [1] and is designed to minimize mixing in the occupied zone. Arguably this lower mixing can reduce sick building syndrome [2], but local discomfort due to draft and the vertical temperature gradient may be critical [3].

Displacement ventilation is achieved by the supply of conditioned air, with a temperature slightly lower (1 K to 3 K) than the desired room air temperature in the occupied zone. The occupied zone is often taken as up to 1.8 m above the floor level [4], but some displacement manufacturers consider it as up to 1.1 m or 1.2 m above the floor [5]. The supply outlets are at or near floor level and supply air at low velocities (typically $\leqslant 0.5 \text{ m s}^{-1}$, compared to 2–6 m s^{-1} for conventional mixed-flow systems) to form a shallow layer of cool, clean air with an average speed in the occupied zone of less than 0.1 m s^{-1} and a turbulence factor of less than 5%. The turbulence factor (%) is

$$Tu = s/v$$

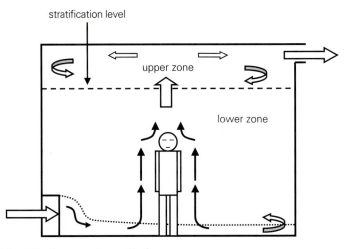

Fig. 12.1 Displacement ventilation.

where v is the average measured velocity of the air (m s^{-1}) at the point of measurement, 1.2 m above the floor, and s is the standard deviation of the air speed. The comparable values for standard mixed-flow air conditioning are $v = 0.15–0.20$ m s^{-1} and Tu = 25–35%.

The sea of cool supply air rises in **plumes** of warm air from heat sources, e.g. people and machines (Fig. 12.1).

12.1.1 Zones

The plume around a person achieves a velocity of 0.25 m s^{-1} at head height [6] and entrains a small amount of surrounding air to produce a plume of expanding warm air rising out of the occupied zone. This is then extracted at high level from the **upper zone** above the **stratification level** or boundary. The upper zone is characterized by recirculation of air into the plume, whereas in the **lower zone** there is entrainment of surrounding air into the rising plume. One could also consider a separate **pool zone** just above the floor, where the pool of cool air gathers before being entrained. To understand fully the air movement in a room, one would have to model it with **computational fluid dynamics** (CFD). However, the concept of zones does emphasize the stable, vertically stratified layers that are essential for the displacement ventilation to function effectively.

It is reported that an opened door can upset the plumes [7], and one could imagine other disturbances such as people's movement in the room also having an influence on the plumes. The control of displacement ventilation is to ensure that the pool zone does not occupy too much of the occupied zone or that the upper zone does not penetrate the occupied zone.

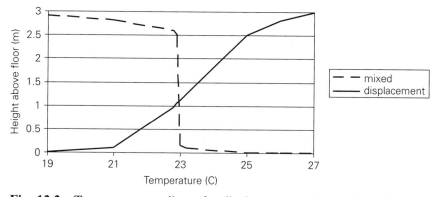

Fig. 12.2 Temperature gradients for displacement and mixed ventilation.

With displacement ventilation and its different zones it is not surprising that the vertical temperature gradient is markedly different from a conventional mixed-flow temperature profile (Fig. 12.2).

12.1.2 Mixing compared to displacement and temperature gradients

For mixed-flow ventilation the room air temperature is fairly constant throughout the room (23°C in Fig. 12.2), but for displacement ventilation the room temperature changes with height, so a design room temperature has to be defined with reference to height. Often this is the temperature at 1.1 m from the floor, $t_{1.1}$ (typically neck to head height for a seated person). Some designers use $t_{1.2}$ as the design room temperature. These heights are also sometimes used to define the height above floor level of the occupancy zone.

For comfort it is suggested [8] that the vertical air temperature gradient in the occupied zone should be less than or equal to 3 K m^{-1}, which corresponds to an air temperature difference of 3 K between the 1.1 m reference height and the ankle (about 0.1 m from the floor). Other sources recommend a lower value of 2 K m^{-1} (in Germany and Switzerland, for instance [5]), although 3 K m^{-1} has recently been confirmed as satisfactory for displacement ventilation [9]. It is also recommended that there are minimum values of $t_{0.1}$ between 19°C and 21°C in winter and between 21°C and 22°C in summer [5].

Besides air temperature, air movement is important for comfort, and the velocity at people's ankles should not be more than 0.15 m s^{-1}. Near the displacement diffusers there is an **adjacent zone** (defined as the length and width of the envelope at ankle level containing air moving at greater than 0.2 m s^{-1}) where discomfort will probably be experienced. This means that occupants' working positions should be away from floor-standing diffusers near walls or away from diffusers in the floor itself. The recommended upper limit on the air thermal gradient has implications for the design and control of displacement ventilation as Example 12.1 shows.

Example 12.1
Consider a room 3 m high. What is the likely sensible heat extraction by displacement ventilation? How does this compare with conventional mixed-flow air conditioning?

Solution
Assuming the maximum temperature gradient of 3 K m^{-1} throughout the height of the room, then for a room of height 3 m this produces a temperature difference of 9 K between supply and extract (assuming the extract is at the top of the room). The maximum air change rate [1] is between 2.5 h^{-1} and 3 h^{-1}. So for 3 air changes per hour the air supply rate for 1 m^2 of floor area is 9 m^3 h^{-1} (0.0025 m^3 s^{-1} or 2.5 l s^{-1}). This gives a sensible cooling capacity of

$$(\dot{V}\rho)C_p\Delta t \tag{12.1}$$

where $\dot{V}$ = volume flow rate (m^3 s^{-1})
ρ = density of air (1.2 kg m^{-3})
C_p = specific heat capacity of the air (1.02 kJ kg^{-1} K^{-1})
Δt = supply to extract air temperature difference (K)
So the cooling capacity is

$$0.0025 \times 1.2 \times 1020 \times 9 = 27.5 \text{ W m}^{-2}$$

This figure is not unreasonable, as it has been suggested [10] that the maximum convective load which can be dealt with by displacement ventilation in offices is limited to 25 W m^{-2}. If a floor-mounted induction unit is used, inducing room air 1 m above floor height, then the primary air can be reduced depending on the induction ratio. If an equal volume of room air is induced, then the primary air can be reduced to about 7 K below the room temperature at 1.1 m and the cooling capacity of the system raised to 50 W m^{-2}.

For mixed-flow air conditioning there can be up to 15 air changes per hour. For 15 air changes per hour the air supply rate for 1 m^2 of floor area is 45 m^3 h^{-1} (0.0125 m^3 h^{-1} or 12.5 l s^{-1}). With a typical temperature difference of 8 K between the supply air and the extract air (the same as the room temperature with good mixing), this gives a sensible cooling capacity of

$$(\dot{V}\rho)C_p\Delta t = 0.0125 \times 1.2 \times 1020 \times 8$$

$$= 122.4 \text{ W m}^{-2}$$

This is a little high; a more practical range would be 90–100 W m^{-2} [6].

Table 12.1 Temperatures with a selection of displacement ventilation systems

	t_{supply}	$t_{0.1}$	$t_{1.2}$	$t_{extract}$
Non-induction	16	18	22.2	24.2
Induction	16	19.8	22.1	24.1
Floor outlet	16	19.5	22.5	24.1

12.2 Control of displacement ventilation

To control a displacement ventilation system the occupant temperature has to be identified; it is usually taken as the temperature 1 m, 1.1 m or 1.2 m (t_1, $t_{1.1}$, $t_{1.2}$) above the floor. The normal criterion is that for comfort the supply air temperature should not be more than 3 K between the head and the ankle. There is also a heuristic rule that the supply temperature should not go below 16°C for sedentary workers, although it has been suggested that it should be between 18°C and 20°C [11]. With some floor-mounted diffusers discharging horizontally along the floor, the supply air can pick up heat and raise its temperature by about 1 K. So a **floor temperature**, t_{floor}, or ankle temperature, $t_{0.1}$, can be used to differentiate it from the supply temperature. Table 12.1 gives some typical temperatures at different heights in a 3 m high room.

With the criterion of maintaining a supply temperature that will not cause discomfort, the simplest form of control is proportional-only or PI control of the supply temperature. This produces open-loop control with no feedback from the room or zone. In fact, one manufacturer recommends that a displacement ventilation system should be operated throughout the year with a supply air temperature between 18°C and 20°C, although it concedes that the system can be operated economically as a VAV system [5]. With the room temperature gradient in displacement systems, and the consequent variation of temperature with height, location of the room temperature sensor can be difficult, unlike a mixed conventional system where the extract temperature is the same as the room air temperature in the occupied zone. Feedback control of displacement ventilation is considered below.

Remember that a displacement ventilation system can only cope with a small load (27 W m^{-2} in Example 12.1) and this will probably be a reasonably constant load. It could not cope with a large solar load. Also the fresh air load will be a large part of the supply air. With the recommended [11] fresh air supply per person of 8 l s^{-1} and assuming 10 m^2 floor area per person, this equates to an air change rate of 0.96 h^{-1} for a 3 m high room. The maximum air change rate is likely to be up to 3 h^{-1}, although 3.9 h^{-1} was used in an experimental room [9]. However, slight changes in occupant density could mean that most of the supply air is fresh air, so volume flow control would not be very flexible. Any air control is mostly to ensure

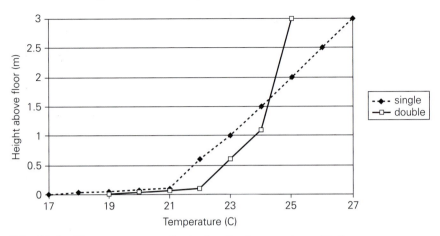

Fig. 12.3   Temperature gradients for displacement ventilation.

sufficient air is supplied and that the diffusers are balanced. Floor diffusers are commonly supplied by a floor plenum under the raised floor. In terms of network analysis (Chapter 13) this plenum has low resistance, so there is little pressure difference between the floor outlets and they are effectively balanced. For an extensive plenum, ducts may be used to help distribute the air within the plenum, with **regulators** at the ends of the ducts as they discharge into the plenum. The regulators are counterweighted dampers which can be set to the required flow rates.

Variable air volume (VAV) boxes can be used with displacement diffusers to control the volume flow (Chapter 10). These boxes can either maintain a constant flow to the room or zone at a constant supply temperature (the temperature control is at the central plant), or vary the volume flow from a room temperature signal. Volume flow variation requires care to maintain an adequate fresh air supply.

An alternative is to maintain a constant volume supply and to vary the supply temperature using the standard reset type of schedule. The reset signal can come from room air temperature sensors (with the usual problems of where to site them, especially in open-plan offices, and how many to have) or from an extract temperature sensor. To relate the extract temperature to the occupancy temperature, use an approximation of the temperature distribution through the room. This is a dog-leg shape (Fig. 12.3) [12, 13]. The diagram shows a double dog-leg and a single dog-leg, from the two references. In the double dog-leg the supply temperature is $t_s = 19°C$ and the extract temperature is $t_e = 25°C$; in the single dog-leg $t_s = 17°C$ and $t_e = 27°C$. Both dog-legs can be used in the design of displacement ventilation systems, using the **comfort method**. An alternative is the **air quality method** [11] using Skaret's equations for the airflow in a displacement plume [14].

The single dog-leg has been verified as a reasonable approximation in a study of three sites [12]. But it is a simplification and cold surfaces (e.g. external walls and windows) and infiltration or exfiltration can influence the dog-leg significantly [12]. In Fig. 12.3 the single dog-leg gradient changes at 0.1 m above the floor – ankle height. Empirically the temperature difference $(t_{0.1} - t_s)$ is between 0.3 and 0.5 times the temperature difference $(t_e - t_s)$ for a typical 3 m high room [11]. In Fig. 12.3 it is taken as the typical design value of 0.4:

$$t_{0.1} - t_s = 0.4(t_e - t_s) \tag{12.2}$$

The important temperature gradient $(t_{1.1} - t_{0.1})/(1.1 - 0.1)$ has the maximum value of 3 K m^{-1}, but when using the comfort method, 2 K m^{-1} is often recommended for sedentary workers. Kruhne has determined approximate values for the ratio $(t_{1.1} - t_{0.1})/(t_e - t_s)$ for various heat source positions and room usages [15]. For a museum with lighting near the ceiling it is 0.16; for an office area it is nearer to 0.25. For an office with a high thermal load and a chilled ceiling it is closer to 0.4 [5].

Example 12.2
A displacement ventilation system for an office is designed to provide a cooling load of 25 W m^{-2} with an occupancy design air temperature of $t_{1.1} = 23°C$. The design temperature gradient $(t_{1.1} - t_{0.1})$ is 2 K m^{-1}. If the cooling load reduces to 20 W m^{-2}, to what value would the supply temperature have to be raised in order for $t_{1.1}$ to be maintained at 23°C?

Solution
The straight-line gradient from 0.1 m to the ceiling height of 3 m gives

$$2 = \frac{t_e - t_{0.1}}{2.9} \quad \text{K m}^{-1} \tag{12.3}$$

Putting the design conditions into equation (12.2):

$$t_{0.1} - t_s = 0.4(t_e - t_s)$$

Rearranging gives

$$t_{0.1} - 0.6t_s = 0.4t_e \tag{12.4}$$

and given

$$2 = t_{1.1} - t_{0.1}$$

but $t_{1.1} = 23°C$ so

$$2 = 23 - t_{0.1}$$

$$t_{0.1} = 21°C \tag{12.5}$$

From equations (12.3) and (12.5) we see that $t_e = 26.8°C$ and from equation (12.4) we see that $t_s = 17.1°C$. For the removal of 25 W m^{-2} we require

$$\dot{m}C_p(t_e - t_s) = 25$$

$$\dot{m}C_p = 2.58 \qquad (12.6)$$

When the heat gains are reduced to 20 W m^{-2} we have

$$t_e - t_s = 20/2.58$$

$$= 7.75 \text{ K}$$

To keep $t_{1.1}$ at 23°C and assuming the dog-leg relationship still holds, equation (12.4) gives

$$t_{0.1} - t_s = 0.4(7.75)$$

$$= 3.1 \text{ K} \qquad (12.7)$$

For the upper line

$$\frac{t_{1.1} - t_{0.1}}{1} = \frac{t_e - t_{1.1}}{1.9} \qquad (12.8)$$

Also

$$t_e - t_{0.1} = 0.6 \times 7.75$$

$$t_e = 4.65 + t_{0.1} \qquad (12.9)$$

Substituting equation (12.9) into equation (12.8):

$$23 - t_{0.1} = \{(4.65 - t_{0.1}) - 23\}/1.9$$

$$43.7 - 1.9t_{0.1} = t_{0.1} - 18.35$$

$$t_{0.1} = 21.4°C$$

And from equation (12.7) we have

$$t_s = 18.3°C \quad \text{and} \quad t_e = 26.05 \text{ K}$$

So the changes in internal gains can be controlled by changes in the supply air temperature of the displacement ventilation. The feedback control signal would come from sensing the return air temperature. The ankle-to-head gradient reduces so comfort can be maintained.

Example 12.2 can be generalized to derive a control equation relating t_s to t_e but keeping $t_{1.1}$ constant. The heat load (kW) to be absorbed by the displacement ventilation is

$$Q = \dot{m}C_p(t_e - t_s)$$

or

$$(t_e - t_s) = Q/\dot{m}C_p \tag{12.10}$$

For a room 3 m high, the gradient of the lower line is

$$t_{0.1} - t_s = 0.4(Q/\dot{m}C_p) \tag{12.11}$$

and the gradient of the upper line is

$$\frac{t_{1.1} - t_{0.1}}{1} = \frac{t_e - t_{1.1}}{1.9}$$

Rearranging gives

$$t_{0.1} = t_{1.1} - \left(\frac{t_e - t_{1.1}}{1.9}\right) \tag{12.12}$$

From equations (12.10), (12.11) and (12.12) we can get t_s in terms of t_e and $t_{1.1}$:

$$t_e - t_s = \frac{1}{0.4}\left[t_{1.1} - \left(\frac{t_e - t_{1.1}}{1.9}\right) - t_s\right]$$

giving

$$t_s = 2.54t_{1.1} - 1.54t_e$$

To check this with Example 12.2, when $t_e = 26.8°C$ then t_s should be 17.1°C:

$$t_s = 2.54 \times 23 - 1.54 \times 26.8$$
$$= 17.1°C$$

The control of t_s is reset control from t_e, such that when $t_e = 26.8°C$ then $t_s = 17.1°C$ and the cooler coil is at a maximum, producing 25 W m^{-2}. The cooler coil output is zero when t_s is 23°C and t_e is also 23°C.

Although temperature control can be used with displacement ventilation, care has to be taken that it does not vary too much and cause discomfort. The supply temperature should generally be between 18°C and 20°C and never below 16°C for sedentary occupants. It has been shown that up to half the temperature rise between the supply and extract air can take place in the layer 0.1 m above the floor [16].

12.2.1 Energy saving

Compared to conventional air conditioning, supplying air at 13°C, displacement ventilation, supplying air at 17°C and upwards, will save on chiller energy. In fact, the free cooling by using cool fresh air will contribute a considerable amount to the cooling, especially in the spring and autumn, so the chiller is not required except during hot outside conditions.

Only summer ventilation and removal of heat loads have been considered so far. For heating, the displacement ventilation will act rather like convector heaters. With high ceilings it may be necessary to provide ceiling destratification fans to stop too much heat accumulating at the top of the room. However, Twinn considers that displacement ventilation is not suitable for winter heating and advocates a perimeter system [7].

Note that traditional air conditioning typically maintains the room at an air temperature of 21°C whereas displacement ventilation has a typical occupancy temperature of 23°C, so this also saves thermal energy, although it does reduce comfort. But the cooling differentials also have to be considered. For conventional cooling the differential is $21 - 13 = 8$ K. For the displacement systems in Fig. 12.3 the differentials are $25 - 19 = 6$ K and $27 - 17 = 10$ K. For the same heat load, the displacement system differentials of 6 K and 10 K will respectively require higher and lower flow rates than the conventional system. Displacement systems cannot cope with more than about 25 W m^{-2} whereas a conventional system can cope with up to 100 W m^{-2}, so the displacement ventilation will use less energy.

12.3   Chilled ceilings

The displacement ventilation systems discussed above can absorb about 25 W m^{-2} of heat load. This demands stringent design of a building to eliminate solar and other gains for displacement ventilation alone to work satisfactorily. However, chilled ceilings are often employed with displacement ventilation systems to increase the cooling capacity. In this section the ceilings alone will be considered, and their use with displacement systems will be discussed in Section 12.4. As no fans are involved, chilled ceilings are often known as **static cooling**, along with chilled beams (cooling convectors often placed at ceiling level).

Air conditioning has traditionally used ceiling diffusers to supply the air, injecting the tempered air with a velocity of 2–6 m s^{-1} to create a good mixing zone. The tempered air then diffuses into the comfort zone of the occupants. But it is cheaper and quieter to use water to transport heat and coolth (Chapter 13). The thermal power transmitted to the motive power of transmission is

$$\frac{\Delta P}{\rho C_p (t_f - t_r)} \tag{13.16}$$

For a typical water system transmitting 1 kW of thermal power, 7.14×10^{-3} W kW^{-1} m^{-1} was required to transmit it. For a typical air system it needed 0.1 W kW^{-1} m^{-1} about 14 times more transmission power.

This shows there is a power advantage in using cool water in chilled ceilings rather than cooled air. The early chilled ceiling schemes used pipes

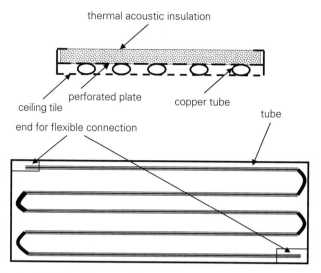

Fig. 12.4 A chilled ceiling section.

embedded in the actual floor or ceiling slab, hence they had a high capital cost. Modern systems use pipes clipped to perforated acoustic ceiling tiles made of aluminium or steel (Fig. 12.4), typically at 60 mm separation suspended from the concrete ceiling slab. The perforated tile encourages convection around the chilled pipes. A layer of thermal (and acoustic) insulation is placed above the tiles and pipes; this directs the heat transfer down into the room. These modern systems are cheaper than the embedded pipe systems and are proving more popular. However, good control of the flow water temperature is required to prevent condensation on the pipes and tiles, but there are fewer worries now about leaks and the flexibility of the modern systems. An auxiliary air supply system is essential to deliver cooled, dehumidified air to the occupied space. Often this is by a displacement ventilation system (Section 12.2). It is recommended that variations in sensible gain are dealt with by the ceiling, and the constant sensible heat gain and all the latent heat gains are dealt with by the auxiliary air supply system [17]. This implies control of the chilled flow water and/or its flow temperature.

For offices the chilled ceiling typically has a mean water temperature of approximately 17°C (16°C flow and 18°C return) and a room air temperature of 24–26°C. The flow can go down to 14°C to increase the output. With these temperatures the occupant perceives the room temperature (dry resultant temperature) as being 1 K to 1.5 K lower than the air temperature. Note that the average surface temperature of the panel depends on the spacing of its pipes and the insulation behind the pipes. From data for an aluminium ceiling panel with pipes spaced at 300 mm centres [6], the typical temperature difference between the panel surface temperature and the average water temperature is

$$t_s - t_w = 0.107q$$

where q is the cooling capacity of the ceiling (W m^{-2}).

The panels cool primarily by radiation, so the occupants perceive the room temperature as 1 K to 1.5 K less than the air temperature. The amount of radiant cooling is limited by condensation considerations as the panels and pipes may approach the dew point of the room air.

With a chilled ceiling using 8 mm copper tubes spaced 60 mm apart on aluminium or steel perforated ceiling panels, a 17°C mean water temperature in the pipes and a room air temperature of 27°C, then according to one manufacturer, the ceiling will provide 70–80 W m^{-2} cooling. If there are ceiling diffusers then the air movement can increase the panel's cooling effect by an extra 10%. Care should be taken to avoid laminar water flow in the pipes above the panels as this greatly reduces the heat transfer compared to turbulent flow. Small-bore polypropylene pipework (1.9 mm internal bore) is also used to form a lattice of cooled pipes in the ceiling panel. These are reported to produce 70 W m^{-2} [18].

Example 12.3
Consider a chilled ceiling with a mean water temperature of 17°C and a room air temperature of 27°C, as quoted above. Assume a mean radiant temperature of 25°C. Does theory confirm the output given?

Solution
Radiation heat loss was discussed in Chapter 7. There the radiation heat exchange between two surfaces, 1 and 2, was given as

$$Q_{12} = A_1 \sigma F_a F_e (T_1^4 - T_2^4) \qquad (12.13)$$

where σ is the Stefan–Boltzmann constant (5.67×10^{-8} W m^{-2} K^{-4}), F_a is a form factor for the relative geometries of the two surfaces and F_e is a factor accounting for the emissivities and absorptivities of the two surfaces. ASHRAE [19] suggests that Hottel's equation can be used for heating and cooling panels in a boxlike room with the panel at a uniform temperature, the other surfaces at another temperature and all surfaces perfectly diffusing:

$$F_a F_c = \cfrac{1}{\cfrac{1}{F_{12}} + \left(\cfrac{1}{\varepsilon_1} - 1\right) + \cfrac{A_1}{A_2}\left(\cfrac{1}{\varepsilon_2} - 1\right)} \qquad (12.14)$$

where F_{12} = view factor = 1.0
$\varepsilon_1, \varepsilon_2$ = emissivities of the surfaces
A_1, A_2 = areas of the surfaces (m^2)

The ASHRAE Handbook goes on to say that in practice the emissivity of non-metallic or painted metal non-reflecting surfaces is about 0.9 and that when used in equation (12.14) $F_a F_c$ is about 0.87 for most rooms. Substituting this into equation (12.13) gives

$$q = 4.93\left(\left[\frac{T_r}{100}\right]^4 - \left[\frac{T_p}{100}\right]^4\right) \tag{12.15}$$

where q = heat transferred from the room to the panel per unit area of panel (W m^{-2})

T_r = area-weighted surface temperature of the room's surfaces excluding the ceiling (K)

T_p = panel temperature (K)

The temperatures are divided by 100 to account for the 10^{-8} index of the Stefan–Boltzmann constant. The area-weighted surface temperature is taken as a practical approximation to the mean radiant temperature, as the radiant temperature varies with position in the room and is difficult to measure, whereas the surface temperature does not vary with position and is easier to measure [20, 11].

Using equation (12.15) with the mean water temperature as an approximation to the panel surface temperature gives the ceiling panels' output as

$$4.93\left(\left[\frac{298}{100}\right]^4 - \left[\frac{290}{100}\right]^4\right) = 40 \text{ W m}^{-2} \tag{12.16}$$

There is also natural convection from the ceiling panels [18], for which the heat absorption is

$$q = 2.18(t_{ai} - t_p)^{1.31} \text{ W m}^{-2} \tag{12.17}$$

where t_p = panel temperature (°C)

t_{ai} = room air temperature (°C)

In this example the output is

$$q = 2.18(27 - 17)^{1.31}$$

$$= 44 \text{ W m}^{-2}$$

This gives a total cooling output of 84 W m^{-2}, a little above the manufacturer's values. But remember that the average water temperature was used and this will be slightly lower than the panel surface temperature, giving an overestimate of the heat transfer. The room air temperature and mean radiant temperature are rather high for comfort. Combining these two values to give the dry resultant comfort temperature, t_c:

$$t_c = \frac{1}{2}t_{ai} + \frac{1}{2}t_r$$

$$= 26°C$$

This would be a little uncomfortable and a better temperature would be between 23°C and 25°C. A temperature of 25°C is the upper limit for assessing the comfort of natural ventilation before cooling is required [21]. A rule of thumb for determining the output of chilled ceilings is that the cooling effect is 10 W for each 1 m^{-2} of ceiling for each 1 K of temperature difference between the ceiling and the room.

12.3.1   Control of chilled ceilings

Compared to hot water radiators for heating, chilled ceilings have a greater surface area but a lower surface-to-room temperature difference. The driving force of the heat transfer for the ceiling is less than for a radiator, hence there is more self-regulation with the chilled ceiling and less need for control.

Basic control of chilled ceilings is either by varying the water inlet temperature or by varying the water flow rate with a constant inlet temperature. On/off control is also used. Water temperature and flow rate are varied using a three-port mixing valve and secondary pump, as with compensator control (Chapter 8), or by a diverting valve circuit. The variable temperature/constant volume mixing circuit will give a better distribution of chilled water through the panel, whereas a diverting constant temperature/variable volume circuit will have a larger temperature difference from flow to return and hence a more variable panel temperature. Two-port valves can be used instead of three-port mixing and diverting valves but the control will not be as good. Figure 12.5 shows a two-port variable temperature circuit but two-port variable flow circuits can also be used [18].

If the ceiling panels are supplied with chilled water from a primary chilled water circuit, perhaps supplying a cooling coil, then the flow temperature is likely to be about 6°C and an **injection circuit** will be required. As explained in Chapter 13, the injection circuit enables the secondary chilled ceiling circuit to have a higher flow water temperature, about 16°C.

The feedback control from the room is often just for on/off control. Although this is a rather basic form of control, even this is not very easy as the location of the room temperature sensor is sometimes difficult, especially in open-plan offices. The sensor should measure both air temperature and radiant temperature, although the radiant temperature will vary around the room. The air temperature varies more quickly than the radiant temperature as the radiant temperature relates to the fabric temperature, which can vary slowly. Hence a more responsive control comes from air temperature control. In one installation it was considered that a black bulb temperature sensor would give the best control signal.

Reset, proportional-only or PI control can be used for the feedback control instead of on/off, but the crucial element of chilled ceiling control is the **dew point control**. This is vital to stop any condensation forming on the chilled ceiling and falling as 'rain' on the occupants. For a typical zone at 22°C dry

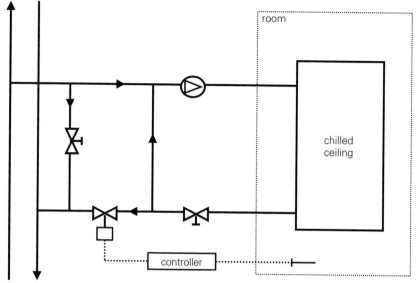

primary circuit

Fig. 12.5   Variable temperature control of a chilled ceiling panel with two-port valves.

bulb room air temperature and 50% RH the room dew point temperature is just above 11°C. As a safety measure, due to the variation in panel temperature between flow and return, 13°C would be considered a sensible lower limit for a chilled ceiling [6]. The chilled ceiling can only deal with sensible cooling loads. The latent loads in the zone have to be dealt with by the air supply, which is often a displacement ventilation system. A moisture sensor on the flow water pipe to the panel can be used for controlling the condensation. Although the room air temperature may not cause condensation on the ceiling, often the windows are openable and this may allow more humid air to enter and cause problems. Zone controls that close down the water supply to the relevant panels can alleviate this problem [18].

Hydronic control and balancing will often be necessary as the individual panels may be independently controlled. Network analysis will be necessary to design for adequate flows. There is also the possibility of using free cooling techniques to produce the chilled water, such as evaporative cooling towers [22]. Cooling towers can operate for a substantial period in moderate European climates and so save on mechanical cooling.

12.3.1.1   Self-balancing control

As with hot water radiators, to an extent, there is self-balancing with chilled ceilings. When the cooling output of the ceiling panel increases, the room temperature drops and reduces the temperature difference between room and panel, typically 25°C – 17°C = 8 K. The temperature difference is the driving force of heat transfer, so as it gets smaller the cooling effect is reduced. Ultimately, with little cooling load, the panel would reduce the room temperature to 16°C, the panel's flow water temperature. The ceiling panel is considered more self-balancing than a radiator, with a mean radiator-to-room temperature difference of 75°C – 20°C = 55 K. The radiator, with a small heat load, would drive the room temperature up to 75°C, hence it needs good control.

Example 12.4
A chilled ceiling system has been installed with a polypropylene pipework lattice behind the ceiling tiles. The output of the tiles is given by the equation in DIN 4715 [18]:

$$q = C(t_{mw} - t_{ai})^n \qquad (12.18)$$

For this system, $C = 5.56$ and $n = 1.105$; t_{mw} is the mean water temperature in the pipes and q is the heat absorbed by the panel (W m^{-2}). The flow water temperature is designed to be 14°C and the return 17°C with a maximum air temperature at high level of 25.5°C. If the flow water temperature is controlled at 14°C but the heat gain falls to half the design value, what does the room air temperature become if there is no control on the panel? Assume the heat gain is equally radiant and convective.

Solution
From the design conditions and the DIN 4715 equation, the design heat gain is

$$q = 5.56(15.5 - 25.5)^{1.105}$$

$$= 70.8 \text{ W m}^{-2}$$

and the heat extracted by the chilled water is

$$70.8A = \dot{m}C_p(t_r - t_f) \qquad (12.19)$$

where $\dot{m}$ is the mass flow of chilled water, C_p is the specific heat capacity of the water, t_r is the return water temperature, t_f is the flow water temperature and A is the chilled ceiling area. From the design conditions we have

$$\frac{\dot{m}}{A}C_p = 23.6 \text{ W m}^{-2} \text{ K}^{-1}$$

The only control equation we have is that the flow water temperature, t_f, is controlled at a constant value of 14°C. So when the load halves to 35.4 W m^{-2} then from equations (12.18) and (12.19) we have

$$35.4 = 5.56 \left(\frac{t_r + 14}{2} - t_{ai} \right)^{1.105} \tag{12.18}$$

$$35.4 = 23.6(t_r - 14) \tag{12.19}$$

From equation (12.19) $t_r = 15.5°C$ and substituting into equation (12.18) gives

$$6.37 = (14.75 - t_{ai})^{1.105}$$

$$5.34 = 14.75 - t_{ai}$$

$$t_{ai} = 20.09°C$$

There is a degree of self-control in that this temperature is not uncomfortable, but cooling is unlikely to be required continuously at this temperature, so further control from the room is required, preferably to modulate the ceiling temperature rather than simple on/off control.

12.4 Chilled ceiling and displacement ventilation

As displacement ventilation systems, through floor or side wall mounted units, can only cope with about 25 W m^{-2} of heat gain, chilled ceilings are often used with them to increase the cooling capacity to about 90 W m^{-2} with room temperatures of about 24–26°C. Correspondingly, chilled ceilings cannot cope with latent heat gains and need a separate air supply. So the two systems are complementary.

However, the displacement ventilation system uses less fan power than a conventional system and relies on the stack effect to move the air through the room. The chilled ceiling has a detrimental effect upon the displacement ventilation flow, suppressing the stratified boundary layer at ceiling temperatures of 18–21°C and destroying displacement flow altogether at ceiling temperatures of 14–16°C [23]. This work was done in a laboratory test room. The room had a ceiling with six circuits in parallel, providing chilled water to the ceiling panels. The circuits could be individually activated, each circuit comprising four or five panels connected in series. Control of the system was not discussed. To assess the effect of the chilled ceiling on the displacement ventilation, the vertical temperature profiles were measured for different ceiling temperatures between 14°C and 21°C and with the ceiling turned off.

A constant heat load of 62 W m^{-2} was used and the air change rate was fixed at 3.9 h^{-1}. The air supply temperature for the displacement system was 19°C. With just the displacement ventilation, the ceiling turned off, the profile was approximately a dog-leg with a floor temperature (0.1 m) of about 23°C and a 2.5 m high temperature of about 28.5°C. As the ceiling temperature reduced, the profile got steeper; the 14°C ceiling had a floor temperature of about 21°C and a 2.5 m high temperature of 21.5°C. These findings are in general agreement with those of Alamdari and Eagles [24]

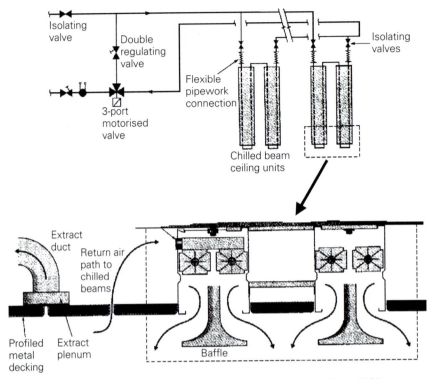

Fig. 12.6   A chilled beam. (Reprinted, with permission, from [28])

that the displacement ventilation, as identified by its temperature gradient, is suppressed by increasing output from the chilled ceiling.

There is a danger that the ventilation effectiveness can be greatly impaired by the chilled ceiling. It is therefore recommended [5] in the design of combined chilled ceiling and displacement ventilation systems that the ventilation system should cater for at least 30% of the total cooling load, Q_T, i.e. $Q_D \geqslant 0.3Q_T$. This is because there is a possibility of designing the combined system so that displacement ventilation deals with just the fresh air requirements for the occupants, which equates to $8\,\mathrm{l\,s^{-1}}$ per person [25, 11]. For an occupancy density of 1 person per $10\,\mathrm{m^2}$ this is $0.8\,\mathrm{l\,s^{-1}\,m^{-2}}$. For a temperature difference between the supply air and extract air of 10 K this yields

$$\dot{V}\rho C_p \times 10 = 0.0008 \times 1.2 \times 1.02 \times 10$$

$$= 9.8\ \mathrm{W\ m^{-2}}$$

A small cooling load and therefore a small displacement flow is dominated by the ceiling cooling. Effectively there would be a mixed ventilation system with the displacement stratification destroyed by the ceiling.

Swirl-type diffusers can also reduce the stratification of displacement ventilation as they introduce the air at higher velocities than standard displacement ventilation. This can promote mixing in the occupied zone and so reduce the buoyancy-driven plumes of the displacement ventilation [7].

12.4.1   Chilled beams

Passive chilled beams, effectively inverted finned-tube convectors sited in ceiling recesses, are also used in conjunction with displacement ventilation systems. Active beams combine the air supply in the ceiling unit, so they do not require a floor displacement system. The downward plumes from the passive beams hit the floor and spread out [26] with a 'filling-box' regime [27] forming a layer of cool air which gradually expands upwards from the floor. The beam plume upsets the buoyancy for the displacement system over time and the combined system can be regarded as a mixing system [26]. A chilled beam is shown in Fig. 12.6 with its diverting control valve and reverse return pipework to the beams to aid balancing [28].

12.5   Mixed mode

Many buildings have natural ventilation from windows that the occupants can open as well as mechanical ventilation and/or mechanical cooling to help maintain comfort. They are called mixed-mode air-conditioned buildings or simply mixed-mode buildings. The combination of systems and control strategies for mixed-mode systems is quite enormous but some refer to **zonal mixed mode**, where certain zones of the building are naturally ventilated and others are mechanically air conditioned. Another version is **seasonal mixed mode**, where the mechanical ventilation or air conditioning can be switched on when required, primarily in the summer and probably with the windows shut.

An example of a simple mixed-mode building is where fan coil units are installed as well as openable windows. Trip switches on the windows simply shut off the fan coils when the window is open. However, there are more sophisticated buildings, some of which are discussed below.

12.5.1   The Elizabeth Fry Building, University of East Anglia

The Elizabeth Fry Building at the University of East Anglia uses Termodeck integrated heating and ventilating, where the supply air is passed through the hollow cores of the concrete floor slabs (Fig. 12.7) [29]. Termodeck was devised in the 1970s by the Swedish engineer Loa Andersson. It aims to achieve a higher heat transfer efficiency into the building fabric due to the air supplied to the rooms or zones flowing within the core of the slab.

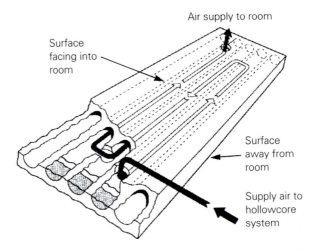

Fig. 12.7   Ventilation through a Termodeck slab. (Courtesy Termodeck)

Table 12.2   Thermal values for the Elizabeth Fry Building

Element	U-value $(W\,m^{-2}\,K^{-1})$
Walls	0.2
Roof	0.13
Windows	1.3

Generally the slabs are 1.2 m wide and 0.3 m deep and they have five smooth-faced cores, three of them normally used for the air [30]. The slabs can absorb 10–40 W m^{-2} of internal gains, assuming that 5–9 m^3 h^{-1} m^{-2}, (1.4–2.5 l s^{-1} m^{-2}) of air flows through the hollow cores of the slabs. With these low loads, good insulation and reduced solar gains are essential. Table 12.2 shows the insulation values for the Elizabeth Fry Building.

The storage efficiency [31] of a slab is given by

$$\frac{t_{in} - t_{out}}{t_{in} - t_{slab}}$$

where t_{in} is the inlet air temperature, t_{out} is the outlet air temperature and t_{slab} is the slab temperature. The slab accepts 75% of potential heat available from the air. The effective volume of the slab is 0.36 m^3 about one-third of the slab's volume of 1 m^3. The ventilation through the hollow cores is by fans; the fan power varies as the third power of the flow rate but the cooling varies directly with the flow rate. Off-peak cooling will be an advantage.

The Elizabeth Fry Building has four storeys [32], the top two having 50 cellular offices for about 70 staff. The lower ground and ground floors contain lecture theatres and seminar rooms. It is of narrow plan with the offices and meeting rooms being less than 6 m deep. The occupants can open the windows. The building was highly insulated, as Table 12.2 shows, with triple glazing and an integral sunblind. Infiltration was kept to a minimum with the building being specified and tested to have less than one air change per hour at the infiltration test pressure of 50 Pa. There is a small atrium, containing a staircase to all floors, with a glass roof and louvres to control solar gain.

Air for heating and ventilation comes from the hollow cores in the slabs through small circular ceiling diffusers. During occupancy there is a continuous supply of fresh air. In the main lecture theatres the ceiling air is ducted down to wall-mounted displacement ventilation units. As the cores can only supply one-third of the design maximum air to the lecture theatre, the rest is supplied from under the floor. The lecture theatre ventilation fans were of variable speed, controlled with a signal from a CO_2 level sensor. These maintained the CO_2 level at 800–1000 ppm, rising to 1300 ppm during periods of high occupancy.

Near the end of construction the BEMS was omitted to save money but the management team found that it needed data from the BEMS to run the building efficiently, so a BEMS was installed one year after the building's completion.

Summer night ventilation is possible with the fans blowing cool night air through hollow slabs. Extensive monitoring was conducted and a number of meetings were held between the designers and the client to ensure the building operated as intended. From the monitoring it was found that the internal temperature was maintained at a fairly steady value (Fig. 12.8). When the

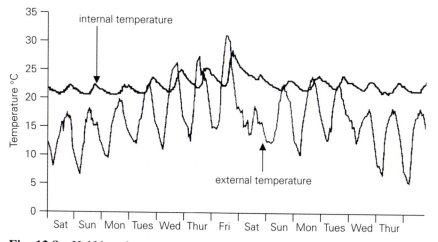

Fig. 12.8 Half-hourly temperatures in a south-facing office of the Elizabeth Fry Building for the period 1 June to 13 June 1996. (Adapted, with permission, from [29])

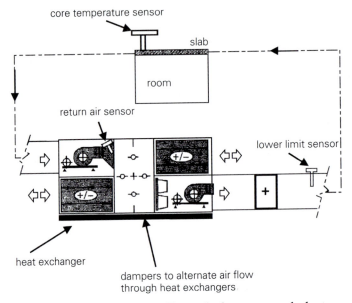

Fig. 12.9 Elizabeth Fry Building: Termodeck system and plant. (Reprinted, with permission, from Termodeck (UK) Ltd)

outside temperature rose to over 30°C on the Friday, the inside temperature did rise to 27°C but it occurred at 7 pm. Notice that the internal temperature peaks are significantly shifted compared to the external peaks just after midday.

Figure 12.9 shows the slab system and plant for the building. The plant shown is one of the four air-handling units, which have high-efficiency heat recovery units. These units use reversing regenerator heat exchangers with metal packs to absorb the heat from the exhaust airstream. The airflow between the fresh air duct and the extract duct is mechanically reversed once per minute to exchange the heat [32].

The control strategy is to maintain the core temperature, as measured by a sensor near the outlet of a hollow core, at 22°C during winter and summer. Initially the core sensor (an air temperature sensor suspended in the core) was placed near the air entry point to the core instead of near the exit. It was later repositioned and the control improved.

The heating strategy is very simple:

If $t_{core} < 21.5$°C then heating comes on
If $t_{core} > 22$°C then heating switched off

During occupied hours the system uses full fresh air, but during the unoccupied time after 2200 in the summer the cooling strategy is also simple:

If $t_{core} > 23°C$ and $t_{ao} < t_{core} + 2$ then the fans are switched on
If $t_{core} < 22°C$ then the fans are switched off

It is interesting to compare this simple control strategy with the more complex strategy for the Weidmuller building in West Malling, Kent [33]. This is a similar Termodeck system to the Elizabeth Fry Building, except there is an evaporative cooler before the air-handling unit; the air-handling unit contains the heat exchanger and dampers as before. The control strategy is designed to maintain the slab and room temperatures within the range 20–22°C. The night-time (off-peak) control schedule is as follows:

If $t_{ao} > 20°C$ and room or slab $> 23°C$
    then evaporative cooling until both $< 20°C$
If $t_{ao} > 10°C$ and room or slab $> 22°C$
    then free cooling until both $< 20°C$
If $t_{ao} < 6°C$ and room or slab $< 20°C$
    then heating until both $> 22°C$

And here is the daytime control schedule, 0700 to 1800:

If $t_{ao} > 10°C$, until $t_{ao} < 6°C$ and extract $> 23°C$
    then evaporative cooling until extract $< 22°C$
If $t_{ao} > 10°C$, until $t_{ao} < 6°C$ and extract $> 22°C$
    then direct ventilation until extract $< 22°C$ or supply $< 12°C$
If $t_{ao} > 10°C$, until $t_{ao} < 6°C$ and extract $< 22°C$ then heat recovery
If $t_{ao} < 6°C$ until $t_{ao} > 10°C$ and extract $> 22°C$ and supply $> 15°C$
    then direct ventilation until extract $< 22°C$ or supply $< 12°C$
If $t_{ao} < 6°C$ until $t_{ao} > 10°C$ and extract $< 22°C$ then heat recovery

There were some flaws in the initial control of the plant with this more complex strategy [33]. For a large part of the monitored period, the evaporative cooled return air was actually heating the supply air rather than cooling it. The control strategy failed to differentiate between weekdays and weekends.

In the Elizabeth Fry Building the artificial lighting was measured to be 310 lx. The total electrical consumption in 1997 was 61 kWh m^{-2} y^{-1} compared to a good practice figure for academic buildings of 75 kWh m^{-2} y^{-1} [34]. The total normalized gas consumption for 1997 was 37 kWh m^{-2} y^{-1}, lower than the previous two years and reflecting the fine tuning of the system.

An occupant survey of just over half the staff revealed high levels of satisfaction with the internal environment, suggesting that not only was the building energy efficient but also comfortable [32]. Further details on this building and fabric thermal storage can be found in the literature [35, 36].

12.5.2 Other mixed-mode control

12.5.2.1 *Inland Revenue, Nottingham*

The Inland Revenue building, Nottingham, has occupant-openable windows, as well as four-speed fans to supply fresh air in summer or heated air in winter [37]. The fans are under the floor and provide displacement ventilation but without any mechanical cooling. The air is extracted via towers (Fig. 11.3). The basic control of the towers to maintain a comfortable room temperature has been discussed in Chapter 11, but a look-up table of wind speed and rain is used for additional control conditions, e.g. for worsening rain or increasing wind speed. Wind direction and outside temperature are also used to modify the control of a tower opening.

12.5.2.2 *Inland Revenue, Durrington*

The Inland Revenue building, Durrington, has a two-speed mechanical ventilation system providing 1.5 or 4 air changes per hour, and there is natural ventilation from automatically controlled casement vents fitted above manually openable windows. Figure 12.10 shows the daytime control schedule [37] for controlling the mechanical supply and heating.

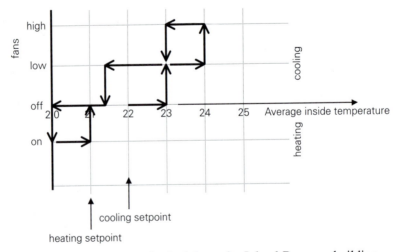

Fig. 12.10 Daytime control schedule at the Inland Revenue building, Durrington.

12.5.2.3   PowerGen, Coventry

The PowerGen building has a central atrium with an open-plan office built around it. There are three rows of windows in the perimeter, two of which can be manually operated by the occupants. The top row is controlled by a BEMS. During the occupied period there is continuous ventilation via a mechanical displacement system providing 1.5 or 3 air changes per hour with air extraction by fans in the atrium roof or via windows under automatic control at the top of the atrium. There are also blinds for the windows. When the inside temperature is greater than 23°C the automatically controlled perimeter windows are opened as well as the atrium windows on the opposite side of the building to provide crossflow ventilation. However, when the external temperature exceeds the internal temperature, the BEMS-controlled windows are made to shut.

12.5.2.4   Interlocks and occupant guidance

Interlocks are advisable on mixed-mode buildings, to prevent competition between mechanical cooling and natural cooling. The Refuge building, Wilmslow, has underfloor, four-pipe, three-speed fan coils and openable windows. During hot weather the occupants are uncertain whether to open the windows or rely on the air conditioning [38]. The natural tendency is to keep windows open on hot days, when it may well be hotter outside than in.

An interlock system was considered at the design stage but rejected on the grounds of cost and reliability. It was also felt that the style of the windows did not allow the occupants to fine-tune the ventilation.

However, at the BSI headquarters in Chiswick, the fan coils are shut off by window trip switches when the windows are opened [39]. At the Ionica building, Cambridge, which also has a Termodeck system and evaporative coolers, the design team gave seminars and distributed airline-style instruction sheets to help occupants use the window blinds and openable windows effectively. The occupants are expected to take part in controlling their environment. Note that blinds can be left down and the lights left switched on, although this is more common in older buildings where there are solar gain problems.

12.6   Lighting control

Lighting in many buildings can be a significant part of the total energy consumption, cost and CO_2 production. For an office building the installed power density can vary between 12 and 20 W m^{-2}; each year this consumes between 14 and 60 kWh m^{-2} and produces between 2 and 8.5 kg of carbon per square metre of treated floor area [40]. In cost terms this is between 15% and 31% of the total office energy costs, hence light switching, daylight and window size are important aspects of energy-efficient building design.

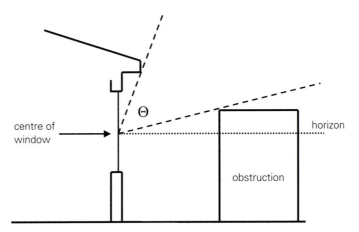

Fig. 12.11   Angle of visible sky.

Solar radiation not only gives daylight, it also generates a heat gain inside the building. The relationship between radiation power (W) and luminous flux (lm) is

$$\Phi = K \int P(\lambda)V(\lambda) \, d\lambda$$

where $P(\lambda)$ = the radiation power at wavelength λ (W nm^{-1})
$\qquad V(\lambda)$ = the visibility curve
$\qquad \Phi$ = the luminous flux or visible light (lm)
$\qquad \lambda$ = the wavelength of the radiation (nm)
$\qquad K$ = 683 lm W^{-1}

The visibility curve, $V(\lambda)$, is the eye response to different wavelengths of light (Fig. 12.11). For instance, the eye sees yellow and green light, at the centre of the visibility curve, much better than red and blue light, at the extremes. Effectively the unit of light, the lumen, is the 'eye-corrected' watt.

The **efficacy** is the ratio of the light output of a lamp to the electrical power input. Typical values are 10 lm W^{-1} for a tungsten lamp, 95 lm W^{-1} for a high-frequency fluorescent lamp, and 145 lm W^{-1} for a monochromatic low-pressure sodium lamp. Daylight varies in its efficacy but a typical value is 100 lm W^{-1}.

12.6.1   Daylight factor

With daylight levels of up to 100 000 lx, (1 lx = 1 lm m^{-2}) on a clear summer's day, there is a great potential for reducing artificial lighting. How much depends on the **daylight factor** (DF) of the room and window. This is the ratio of the daylight at a point in the room to the daylight outside the building in an unobstructed position, excluding direct sunlight. Often the

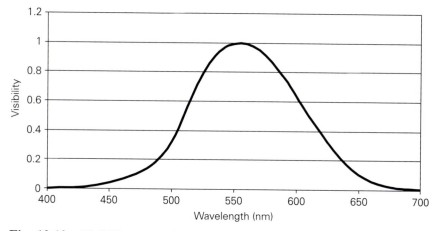

Fig. 12.12   Visibility curve for photopic vision.

average daylight factor is used; this is the average throughout the room rather than at one particular point. Here is the equation:

$$\overline{DF} = \frac{A_g \theta T}{A(1 - R^2)}$$

where $\overline{DF}$ = the average daylight factor (%)

A_g = glazed areas of windows (m²)

θ = angle of visible sky (Fig. 12.12)

T = transmittance of glazing to diffuse light, including the effect of dirt

A = total area of the room surfaces including the windows (m²)

R = mean reflectance of room surfaces (typical values: ceiling 0.7, light-coloured walls 0.5, floor 0.2, windows 0.1)

The diffuse radiation from an overcast sky, without direct sunlight, is used to calculate the daylight factor, and it can be used to assess the room appearance and the likelihood of saving energy [41]:

- $\overline{DF} < 2\%$
  (1)   Electric lighting is often needed during the day
  (2)   Room appears gloomy with only daylight
  (3)   Few savings from daylight
- $2\% \leqslant \overline{DF} \leqslant 5\%$
  (1)   Room is often adequately lit by daylight
  (2)   Good savings possible by using daylight
- $\overline{DF} > 5\%$
  (1)   Room appears bright in daylight
  (2)   Electric lighting rarely needed in the daytime

12.6.2   Luminaire switching

It is a requirement in UK buildings that a light switch should be no more than 8 m distant (in plan) from the luminaire it controls, or no more than three times the height of the luminaire above floor level if this is greater [42]. This will enable occupants to switch on only those luminaires that are necessary. Infrared, ultrasonic or microwave transmitters (often hand-held devices) and their receivers can also be used to switch or dim lights.

Time-operated controls, which switch off the lighting at lunchtime, at the end of occupancy and at the end of any early morning cleaning period are also good energy-saving systems. Often a **mains borne** signal is used for the switching, putting a short off pulse on the alternating mains sine wave at the zero-voltage switching point (i.e. when the voltage changes polarity). Effectively this extends the zero-voltage point by a short time without affecting the transmitted power significantly. This enables the **latching relays** in the luminaires to drop out of contact, disconnecting the luminaire's power supply. The latching relay can be reset to make contact, and so switch on the luminaire, by the occupant pulling a pull cord. Digital signals on the mains transmitted to receivers in the luminaires can also be used. In a narrow-plan, open-plan office in London, half the luminaires were fitted with latching relays and pull cords and the other half were left with the existing large panel of switches near the entrance door. Monitoring of both halves over a winter period showed electricity savings of 63% due to the pull-cord system.

Light switching is mostly through on/off signals, sometimes accompanied by a luminaire identity number. This simplicity explains why mains borne signalling is so common. Any noise or harmonics on the mains can easily be detected in the signals, which carry small amounts of information and for such a short time. For general BEMS use with HVAC plant, where perhaps temperatures and a range of control signals are constantly being sent, then the information being transmitted and the rate of transmission make mains borne signalling less attractive as it is more prone to noise and consequent errors. There is also a trend that where a bus system is used for transmission of control signals to luminaires, it is a separate bus to the HVAC BEMS as there are companies that specialize in lighting control and put in their own bus system with a rapid response and low information transmission.

12.6.3   Intelligent luminaires

Too much light switching may annoy the occupants and the energy savings may not reach expectation. However, **intelligent luminaires** employ high-frequency control circuits and lamps which can easily be dimmed sufficiently slowly that the occupant is not distracted or annoyed. The intelligent luminaire has a **light-dependent resistor** (LDR) pointing towards the working surface or the floor to sense the light reflected back. The output from the LDR is non-linear so a look-up table is used to linearize the output. A setpoint for the LDR can then be set. For instance, if 500 lx is required on the working plane, typically the plane at desk height in an office, and the desk has a reflectance of 20% then the LDR can be set at 100 lx. This is a very crude method and more exact calculations would be required to include the inverse square law of illumination, so in practice a light meter is placed on the desk and the LDR is set accordingly. The luminaire can then be controlled to this LDR setpoint. If daylight is present, the luminaire either dims to supplement the natural lighting or it switches off. High-frequency fluorescent lamps can typically be dimmed to 10%; below this they become unstable and may begin to flicker [43].

If a light sensor is used with just light switching then it is recommended that the luminaire is switched off when the combined daylight and electric lighting is at least three times the required task illuminance [40]. For most office tasks this will be between 300 lx and 500 lx, so the switching-off level will be 900 lx and 1500 lx respectively. A sensible switch-on level will be the task illuminance, i.e. 300 lx to 500 lx.

A dimming strategy is a little more complex than the simple on/off control. The control strategy for a popular intelligent luminaire varies the rate of change of light output from the luminaire in proportion to the error from the setpoint. If the control signal, u, varies from 0 V to 10 V then 10 V corresponds to the full light output of the luminaire, corresponding to an illuminance level on the working plane of E_{max}, and 0 V corresponds to a zero output (it may be offset to the practical dimming minimum of 10% output). Here the zero output will be assumed. If the setpoint illuminance is E_{set} then the error, e, is

$$e = E_{set} - E$$

and as a percentage it becomes

$$e' = \frac{E_{set} - E}{E_{set}} \times 100\%$$

Note that E is made up of the illuminance from the luminaire, E_l, and any daylight contribution, E_d. If the error is more than 50% away from the setpoint, i.e. $|e'| > 50\%$, then

$$\frac{du}{dT} = 1.0 \; \text{sign} \, e'$$

If $10\% \leqslant |e'| \leqslant 50\%$ then $\dfrac{du}{dT} = 0.2 \; \text{sign} \, e'$

If $|e'| < 10\%$ then $\dfrac{du}{dT} = 0.033 \; \text{sign} \, e'$

If e' is positive then sign $e' = 1$ and if e' is negative then sign $e' = -1$; if $e' = 0$ then sign $e' = 0$.

This is a form of integral control which effectively introduces a slowing down of the luminaire response. A proportional control would immediately change the control signal, u, which could give a noticeable change in lighting to the occupant. The sampling time of the control is typically 1 s with a dead time of 5 s before a change in the control schedule. An occupancy or presence detector can sense movement in the occupied space and switch off the luminaire if no movement is detected. Often there is a delay of about 10 min in case the occupant has been still and to avoid causing annoyance by switching off too soon.

Example 12.5
An intelligent luminaire can produce an illuminance of 800 lx on a person's desk. The desk has a reflectivity of 20%. The setpoint illuminance for the luminaire is 500 lx. The luminaire is providing a steady 300 lx on the desk when the window blind is down. The blind is then opened and daylight contributes a further 200 lx. How long does the luminaire take to adjust to this new condition with the above control schedule?

Solution
Here $E_{max} = 800$ lx which corresponds to $u_{max} = 10$ V and $E_{min} = 0$, i.e. $u_{min} = 0$. The luminaire is currently giving an illuminance of 300 lx and the daylight contribution through the blind is 200 lx. When the blind is opened there is an extra 200 lx from the daylight, so

$$e' = \frac{500 - 700}{500} \times 100\% = 40\%$$

so

$$\frac{du}{dT} = 0.2 \text{ V s}^{-1}$$

$$= 16 \text{ lx s}^{-1}$$

and when the luminaire is down to 150 lx (i.e. $E = 550$ lx, within 10% of the setpoint illuminance) then

$$\frac{du}{dT} = 0.033 \text{ V s}^{-1}$$

$$= 2.67 \text{ lx s}^{-1}$$

The time for the luminaire to reduce its output so that the desk illuminance is 500 lx is therefore

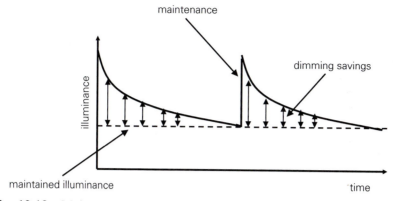

Fig. 12.13 Maintained illuminance and dimming savings.

$$\frac{150}{16} + \frac{50}{2.67} = 28.1\,\text{s}$$

This does not take account of the dead time that may be present to stop excessive changes, e.g. when clouds briefly alter the daylight contribution or an occupant passes under the sensor.

12.6.3.1 Maintained illuminance

An intelligent luminaire can make savings due to dimming when daylight is present and also from its presence detector. However, it can also save by not producing its full output and therefore going above the required lighting level. This is because lighting installations are designed to a **maintained illuminance**, i.e. the illuminance at which maintenance is required. So if the maintained illuminance is 500 lx then with various maintenance factors to account for the depreciation of the lamps, and the room and luminaire surfaces getting dirty, the initial illuminance level may well be about 700–800 lx from full output, undimmed lamps. Figure 12.13 shows that the savings may be very significant. Although 500 lx is the commonly accepted and recommended maintained illuminance for offices [44], many low-energy buildings are now using values around 400 lx or even 300 lx.

References

[1] Schultz, U. (1993) A new eye for indoor climate. *Heating and Air Conditioning*, 1 July.
[2] World Health Organization (1983) Indoor air pollutants, exposure and health effects, *European Reports and Studies*, **78**.

[3]   Melikov, A. and Nielsen, P. (1989) Local thermal discomfort due to draft and vertical temperature difference in rooms with displacement ventilation. *ASHRAE Trans.*, **95**(2).

[4]   Legg, R. C. (1991) *Air conditioning systems design, commissioning and maintenance*, Batsford, London.

[5]   Trox (1997) Displacement ventilation principles and design procedures S1.3/1/EN/1, Trox Technik, Norfolk.

[6]   Jones, W. P. (1997) *Air Conditioning Applications and Design*, 2nd edn, Edward Arnold, London.

[7]   Twinn, C. (1994) Displacement ventilation – fact or fiction? *Building Services Journal*, June.

[8]   ISO 7730 (1995) Moderate thermal environments – determination of the PMV and PPD indices and specification of the conditions for thermal comfort, International Organization for Standardization, Geneva.

[9]   Loveday, D. L., Parsons, K. C., Ahmed, H. T., Hodder, S. G. and Jeal, L. D. (1998) Designing for thermal comfort in combined chilled ceiling/displacement ventilation environments. *ASHRAE Trans.*, **104**, Pt 1.

[10] Sandberg, M. and Blomqvist, C. (1989) Displacement ventilation in office rooms. *ASHRAE Trans.*, **95**(2).

[11] CIBSE (1999) CIBSE Guide A, Environmental design. Chartered Institution of Building Services Engineers, London. This updates CIBSE Guide, Section A1, Environmental criteria for design, 1986.

[12] Jackman, P. J. (1990) Displacement ventilation. BSRIA Technical Memorandum 2/90, BSRIA, Bracknell, UK.

[13] Alamdari, F., Bennett, K. M. and Rose, P. M. (1993) Displacement ventilation performance – office space applications. BSRIA Technical Note TN3/93, BSRIA, Bracknell, UK.

[14] Skaret, E. (1986) Ventilation by displacement – characterisation and design implications. In *Ventilation '85*, Elsevier, Amsterdam.

[15] Kruhne, H. (1996) Experimentelle und theoretische Untersuchungen zur Quelluftstromung. Dissertation, Technical University, Berlin.

[16] Mundt, E. (1992) Convection flows in rooms with temperature gradients – theory and measurements. In *Proceedings of Roomvent '92*, Aalborg.

[17] Jamieson, H. C. and Calland, J. R. (1963) The mechanical services at Shell Centre. *JIHVE*, **31**(1).

[18] Thomas, D. (1994) First among Equalios: chilled ceilings – a design primer. *Building Services Journal*, June.

[19] ASHRAE (1993) *ASHRAE Handbook*, Chapter 7, Panel heating and cooling systems, American Society of Heating, Refrigerating and Air-conditioning Engineers, Atlanta GA.

[20] CIBSE (1979) CIBSE Guide, Section A5, Thermal response of buildings, Chartered Institution of Building Services Engineers, London.

[21] CIBSE (2000) CIBSE J Guide Weather and Solar Data, Chartered Institution of Building Services Engineers, London.

[22] Butler, D. J. G. (1998) Chilled ceilings – free cooling opportunity. Paper presented at the CIBSE National Conference, Bournemouth, UK.

[23] Loveday, D. L., Parsons, K. C., Ahmed, H. T., Hodder, S. G. and Jeal, L. D. (1998) Designing for thermal comfort in combined chilled ceiling/displacement ventilation environments. *ASHRAE Trans.*, **104**, Pt 1.

[24] Alamdari, F. and Eagles, N. (1996) Displacement ventilation and chilled ceilings. BSRIA Technical Note TN2/96, BSRIA, Bracknell, UK.

[25] CIBSE (1979) CIBSE Guide, Section A1, Environmental criteria for design, Chartered Institution of Building Services Engineers, London. Reprinted with amendments 1986.

[26] Davies, G. (1994) A model performance: chilled ceilings – mathematical modelling. *Building Services Journal*, June.

[27] Baines, W. and Turner, J. (1969) Turbulent buoyant convection from a source in a confined region. *Journal of Fluid Mechanics*, **37**.

[28] Bunn, R. (1994) Looking radiant. *Building Services Journal*, June.

[29] DETR (1998) New practice final report 106: the Elizabeth Fry Building, University of East Anglia – feedback for designers and clients. Department of the Environment, Transport and the Regions, London.

[30] Bunn, R. (1991) Termodeck: the thermal flywheel. *Building Services Journal*, May.

[31] Winwood, R. *et al.* (1994) Modelling the thermal flywheel. *Building Services Journal*, Oct.

[32] Anon (1998) Probe 14: Elizabeth Fry Building. *Building Services Journal*, April.

[33] Winwood, R. *et al.* (1997) Advanced fabric energy storage IV: experimental monitoring. *BSERT*, **18**(1).

[34] DETR (1994) Introduction to energy efficiency in further and higher education. DETR 6/94, Department of the Environment, Transport and the Regions, London.

[35] Bunn, R. (1994) Slab and trickle. *Building Services Journal*, Feb.

[36] Winwood, R. (1996) What's in store for Termodeck? *Building Services Journal*, June.

[37] Martin, A. and Fletcher, J. (1996) Night cooling control strategies – final report. BSRIA Report 11621/4, May.

[38] DETR (1995) Good practice case study 20: energy efficiency in offices. Department of the Environment, Transport and the Regions, London.

[39] CIBSE (1997) Application Manual AM10, Natural ventilation in non-domestic buildings, Chartered Institution of Building Services Engineers, London.

[40] DETR (1998) Energy consumption guide 19: energy use in offices. Department of the Environment, Transport and the Regions, London.

[41] DETR (1998) Good practice guide 245: desktop guide to daylighting – for architects. Department of the Environment, Transport and the Regions, London.

[42] Building Regulations Approved Document, Part 1 (1995). HMSO, London.

[43] DETR (1997) Good practice guide 160: electric lighting controls – a guide for designers, installers and users. Department of the Environment, Transport and the Regions, London.

[44] CIBSE (1994) CIBSE Code for Interior Lighting, Chartered Institution of Building Services Engineers, London.

13
Network analysis

Pipework and ductwork systems are typically and traditionally designed for full-load operation. But with the advent of cheaper variable-speed drives (e.g. static frequency inverters) [1] and the savings that can be made on reasonable size fans and pumps, good control is now required to change circuit flows in a stable manner. One has to understand how the network pressures vary with changes in flow rates and loads, and the relevant tool is network analysis. It also allows complex circuits to be reduced to one equivalent resistance, which gives the system characteristic.

13.1 Friction loss

In the network analysis of ductwork and pipework we are primarily interested in the pressure required from the fan or pump to overcome the frictional loss due to ducts, pipes and fittings. (Static and dynamic pressures will be considered later.) For network analysis of duct and pipe systems one can use D'Arcy's approximate equation for frictional pressure loss, traditionally expressed as as an equivalent pressure head (m):

$$h = 4flv^2/2g\rho d$$

where h = head (m)
f = friction factor
v = velocity (m s^{-1})
ρ = density (kg m^{-3})
g = acceleration due to gravity (m s^{-2})
l = length of pipe/duct (m)
d = diameter of the pipe or duct (m)

The pressure can be due to gravity and the height of the fluid above a datum point, hence the old measure of 'head' in metres. There is also the pressure due to the pump or fan to overcome the friction of the fluid movement over the pipe or duct surface. The normal form of D'Arcy's equation is in terms of the pressure drop (Pa), noting that the equivalent pressure drop is $\rho g h$:

$$\Delta P = 2flv^2/d \qquad (13.1)$$

When the diameter of the pipe or duct changes then the velocity of the fluid changes. Although this does not affect the **total pressure**, ΔP_{tot}, it does affect the **static pressure**, ΔP_{static} (the pressure pushing outwards on the pipe or duct surface), and the **velocity pressure**, $\Delta P_{velocity}$ (the pressure due to the motion of the fluid). The total pressure is the sum of the static pressure and the velocity pressure:

$$\Delta P_{tot} = \Delta P_{static} + \Delta P_{velocity}$$

where

$$\Delta P_{velocity} = \rho v^2/2$$

For air systems this velocity pressure is 15 Pa at 5 m s^{-1}, a significant pressure compared to a typical pressure drop in an air duct of 1 Pa m^{-1}. For a water system with a velocity of 1 m s^{-1} the velocity pressure of 500 Pa is not as significant as in air systems; this is because the typical pressure drop in a water pipe is 300 Pa m^{-1}. Hence velocity pressure is often ignored in water system pressure calculations but not in air pressure calculations. Figure 13.1

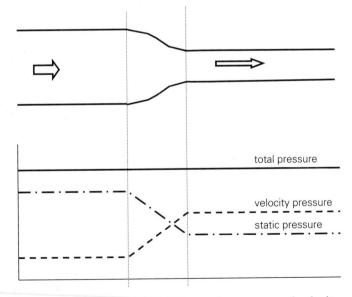

Fig. 13.1   Frictionless duct: total pressure, static pressure and velocity pressure.

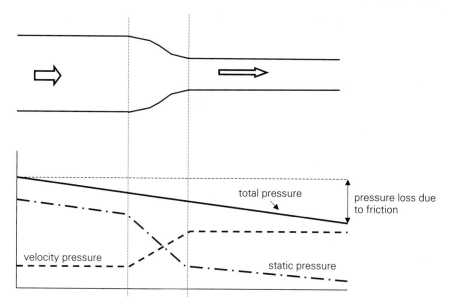

Fig. 13.2   Friction and pressure drop in a duct.

shows the static pressure and the velocity pressure at two duct diameters. The total pressure stays constant, assuming there is no friction loss in the ideal duct, but there is an interchange between static pressure and velocity pressure (similar to the interchange between kinetic energy and potential energy of a falling ball).

For a normal duct there is friction to overcome in moving the air through it. For a duct of uniform diameter (Fig. 13.2) the velocity at the start of the duct is the same as the velocity at the end. Assuming there is no compression of the air, the air volume going into the duct must equal the volume coming out. As the velocity is the same at the start and at the end, the velocity pressure is also the same. So the pressure drop due to overcoming the friction of the duct is seen as a reduction in static pressure. However, the pressure loss at fittings, such as dampers, is related to the velocity pressure by a pressure loss factor, ξ, so the pressure loss (Pa) is

$$\Delta P_{\text{fitting}} = \xi \rho v^2 / 2 \tag{13.2}$$

Equation (13.1) can also be expressed in terms of the volume flow rate V (m^3 s^{-1}) as

$$\Delta P = 32 f l v^2 / \pi^2 d^5 \tag{13.3}$$

The pressure drop varies dramatically with the pipe or duct diameter and a reduction in size can increase the pressure considerably. For a 25 mm medium grade steel pipe, 1 m long with a water flow of 0.46 kg s^{-1}, the pressure drop is 300 Pa. The pressure drop for the same flow in a pipe one size up (diameter 32 mm) is 75 Pa; in a pipe one size down (diameter 20 mm) it is 980 Pa. For an air duct with 0.35 m³ s⁻¹ airflow the pressure drop in a 300 mm duct is 1 Pa; for one size up (350 mm) it is about 0.5 Pa and for one size down (250 mm) it is about 2.6 Pa.

Equations (13.1) and (13.2) can be generalized to

$$\Delta P = RV^2 \tag{13.4}$$

But this is a simplification, as in D'Arcy's equation f varies with the fluid velocity; this is why designers use CIBSE or ASHRAE pipe loss tables and duct loss graphs. The Reynolds number (Re) represents the ratio of inertia to viscous forces on the fluid. And the value of f varies with velocity in a complex manner that involves Re:

$$\text{Re} = \frac{\rho v d}{\mu} = \frac{v d}{v} \tag{13.5}$$

Here μ is the coefficient of viscosity and v is the kinematic viscosity. Duct or pipe roughness is important in the fluid flow and is described by a relative roughness, k_s, the ratio of the mean height for the roughness of the duct or pipe wall (absolute roughness) divided by the internal duct or pipe diameter.

There are four flow regimes defined by the Reynolds number:

- Laminar, Re < 2000
- Critical zone, $2000 <$ Re < 3000
- Turbulent (transition zone), Re > 3000
- Fully turbulent

Poiseuille's formula holds for laminar flow:

$$f = 16/\text{Re}$$

The laminar flow becomes unstable in the critical zone and the friction factor is unpredictable. The Colebrook–White equation [2] holds in the turbulent zone:

$$f^{-1/2} = -4 \log_{10} \left[\frac{k_s}{3.7d} + \frac{1.255}{\text{Re}} f^{-1/2} \right] \tag{13.6}$$

This is a transcendental equation and needs numerical solution like the heat balance equation for the compensator (8.3). Newton's method of approximation could be used or Solver in Microsoft Excel [3]. However, Miller [4] provides an explicit version of the equation:

$$f = \frac{0.25}{4 \log_{10} \left(\dfrac{k_s}{3.7d} + \dfrac{5.74}{\text{Re}^{0.9}} \right)} \tag{13.7}$$

In the fully turbulent zone f becomes independent of Re and the $(k_s /3.7d)$ $\gg (5.74\mathrm{Re}^{-0.9})$.

CIBSE uses equations (13.2) and (13.6) for its pipe-sizing tables and duct-sizing chart. For air systems with circular ducts CIBSE uses $k_s = 0.15$ mm, and for steel piping it uses $k_s = 0.046$ mm (2.5 mm for rusty pipes).

13.2 Fluid resistances

If we assume that f is constant then R in equation (13.4) is a constant; R is the resistance of the pipe or duct and it makes network analysis much easier. This is an approximation but it is quite a good one. Suppose we consider a medium grade steel pipe of nominal diameter 25 mm with water at 75°C, the values taken from the CIBSE pipe-sizing tables for a range of pressures are given in Table 13.1. And the resistance (m^{-2} kg^{-1}) is calculated from

$$R = \frac{\Delta P}{M^2}$$

where M is the mass flow rate (kg s^{-1}). The average value of R is 1512 with a standard deviation of 12%. However, using the method of least squares, which may be set up on a spreadsheet program such as Microsoft Excel, it is possible to calculate a best-fit resistance of 1393. This resistance value can be used to estimate the pressure and the mass flow rate, allowing its accuracy to be checked. The errors are shown in the two right-hand columns, indicating that a resistance and the square law are reasonably trustworthy over this range. A similar exercise can be done for ducts, indicating the approximation of the square law. So for analysing a network the design values of pressure drop and flow rate can be used to calculate resistances.

Further accuracy can be obtained by estimating the index using the least-squares method, and a value of 1.864 instead of 2 is found to be better for

Table 13.1 Resistance values for a medium grade steel pipe[a]

Pressure drop (Pa m^{-1})	Flow rate (kg s^{-1})	Resistance per unit length (m^{-2} kg^{-1})	Error in pressure (%)	Error in flow rate (%)
10	0.073	1877	26	16
50	0.177	1596	13	7
100	0.258	1502	7	4
200	0.373	1438	3	2
300	0.462	1406	1	0
400	0.537	1387	0	0
500	0.602	1380	−1	0

[a] Nominal diameter 25 mm and water at 75°C. Values are taken from CIBSE pipe-sizing tables.

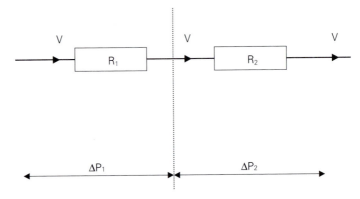

Fig. 13.3   Two resistances in series.

ductwork, with less variation in the resistance [5]. However, this can complicate network analysis as the fitting losses still have a square law.

We can now consider pipes or ducts, valves and dampers, etc., as resistances
– like an electrical circuit except that we have a square law not a linear law
(Ohm's law). The resistance for an element, such as a pipe or duct, will be
treated as lumped at the symbol.

13.2.1   Series resistances

Using the standard symbol for an electrical resistance, Fig. 13.3 shows two
resistances in series, R_1 and R_2. In network analysis we wish to simplify
circuits and here we can make R_1 and R_2 into one equivalent resistance, R_{eq}:

$$\Delta P_1 = R_1 V^2$$

$$\Delta P_2 = R_2 V^2$$

$$\Delta P_{eq} = R_{eq} V^2$$

since V flows through all three resistances. And $\Delta P_{eq} = \Delta P_1 + \Delta P_2$ so

$$R_1 + R_2 = R_{eq} \tag{13.8}$$

Series fluid resistances are added together just like electrical resistances.

13.2.2   Parallel resistances

Figure 13.4 shows resistances R_1 and R_2 in parallel. Here the pressure drop
across both of them is the same but the volume flow is different. This means

$$\Delta P = R_1 V_1^2 = R_2 V_2^2$$

$$= R_{eq} V^2$$

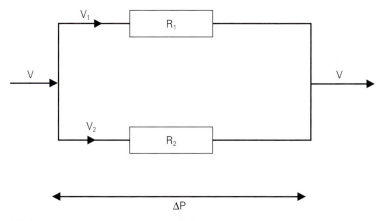

Fig. 13.4 Two resistances in parallel.

From $V = V_1 + V_2$ we get

$$\left(\frac{\Delta P}{R_{eq}}\right)^{1/2} = \left(\frac{\Delta P}{R_1}\right)^{1/2} + \left(\frac{\Delta P}{R_2}\right)^{1/2}$$

which gives

$$\frac{1}{\sqrt{R_{eq}}} = \frac{1}{\sqrt{R_1}} + \frac{1}{\sqrt{R_2}}$$

or

$$R_{eq} = \frac{R_1 R_2}{R_1 + R_2 + 2\sqrt{(R_1 R_2)}} \tag{13.9}$$

If $R_1 = R_2 = R$ the equivalent resistance is $R_{eq} = R/4$. This is different to electrical resistance, which obeys a linear law. Two equal electrical resistances, R, in parallel produce an equivalent electrical resistance of $R/2$. Two equal **hydronic** resistances, R, in parallel produce an equivalent resistance of $R/4$. With three hydronic resistances in parallel – boilers or other system elements, e.g. radiators or pipes – the resistance would be $R/9$. This has implications for boiler and chiller sequencing, and they will be discussed later on.

13.3 Water circuits

It is now possible to analyse water circuits for pressure losses due to friction. Consider this example.

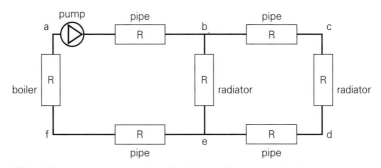

Fig. 13.5   Resistance network of boiler, radiators and pipes.

Example 13.1
Figure 13.5 shows a circuit containing a boiler, pump, radiators and pipework. For simplicity all the components have a resistance of R. (Bends have been included in the lumped resistances.) What is the equivalent resistance of the four right-hand resistances?

Solution
The three right-hand resistances, *bcde*, are in series and have an equivalent resistance of $3R$. This is in parallel with the resistance across *bc*. From equation (13.9) the equivalent resistance is

$$\frac{3R \times R}{3R + R + 2\sqrt{(3RR)}} = \frac{3R}{4 + 2\sqrt{3}}$$

$$= 0.4R$$

Example 13.2
If the radiators in Example 13.1 have the same heat output, they will need the same flow of water through them. (a) Will they get the same flow? (b) What will be the ratio between the two flows?

Solution
(a)   The last radiator, *cd*, has more resistance between it and the pump than the radiator across *be*. Consequently, more water will flow through the radiator across *be*.
(b)   The pressure drops across the circuits *bcde* and *be* are the same but their resistances are different. If V_{be} and V_{bcde} are the respective flows then

$$R(V_{be})^2 = 3R(V_{bcde})^2$$

giving a ratio of flows of $\sqrt{3}$ to 1 (or 1.73 to 1).

This is quite a difference in flow. If this system were left like this, there would not be the required volume flow rates through the radiators. The right-hand radiator would be colder than the left-hand radiator. The circuit needs to be balanced with **balancing valves**.

Example 13.3
(a) What are the resistances of the balancing valves needed in Example 13.2 to balance the circuit?
(b) What is the equivalent resistance of the whole balanced circuit?
(c) What is the pump pressure to circulate this water if its volume through the pump is $2V$?

Solution
(a) The circuit *abcdef* is the circuit with the greatest resistance. It is called the **index circuit**. The circuit *abef* has less resistance, but as the radiator across *be* needs the same flow, a balancing valve is required in *be*. If the balancing valve has a resistance denoted by R_{bv} then its value is determined from

$$(R + R_{bv})V^2 = 3RV^2$$

$$R_{bv} = 2R$$

(b) This makes the equivalent resistance of circuit *bcde* equal to $3R/4$, i.e. two resistances of $3R$ in parallel. This is now in series with the boiler and pipework of resistance $3R$, so the total circuit equivalent resistance is $15R/4$.
(c) The pump pressure required to circulate the water is therefore

$$\frac{15R}{4}(2V)^2 = 15RV^2$$

Note that if the circuit is balanced, and therefore the flows in the branches are at their known design values, then equation (13.9) can be simplified. At balance we have

$$R_1V_1^2 = R_2V_2^2$$

so

$$R_2 = R_1\left(\frac{V_1}{V_2}\right)^2$$

Putting this into equation (13.9) gives

$$R_{eq} = R_1\left(\frac{V_1}{V_1 + V_2}\right)^2 \qquad\qquad 13.10$$

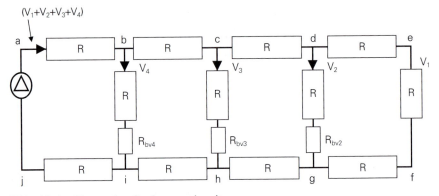

Fig. 13.6   Network of a larger circuit.

This can be used on larger circuits such as Fig. 13.6, which is designed to have equal volume flows, V, in the ladder resistances (*bi, ch, dg, ef*). Working from the right-hand side, *dg* needs a balancing valve of $2R$ and therefore the equivalent resistance of the balanced circuit *defgd* is $0.75R$. To work out the equivalent circuit for *cdefghc*, we need to know the value of the balancing valve at *ch*, which we will denote as R_{bv3}. This can be calculated from equating the parallel circuits *ch* and *cdgh*. Circuit *cdgh* includes the equivalent resistance for *defgd*, so it has a resistance of $2.75R$. Hence

$$(R_{bv3} + R)V^2 = 2.75R(2V)^2$$

and after some calculation this gives $R_{bv3} = 10R$. The equivalent resistance for circuit *cdefghc* can then be found, using equation (13.9), to be $1.22R$. The three balancing valves are $2R$ for *dg*, $10R$ for *ch* and $28R$ for *bi*.

   Rather than use this protracted method for determining the equivalent resistances, we can use equation (13.10). The equivalent resistance R_{defgd} for circuit *defgd* is

$$R_{defgd} = 3R\left(\frac{V}{2V}\right)^2$$

$$= \frac{3R}{4}$$

The next equivalent resistance is

$$R_{cdefghc} = \left[\frac{3}{4}R + R + R\right]\left(\frac{2V}{V + 2V}\right)^2 = \frac{11R}{9}$$

$$R_{bcdefghib} = \left[\frac{11}{9}R + R + R\right]\left(\frac{3V}{V + 4V}\right)^2 = \left(\frac{29}{9} \times \frac{9}{16}\right)R = \frac{29R}{16}$$

Thus the equivalent resistance for the whole circuit is

$$2R + 29R/16 = 61R/16$$

It is left as a fraction so the pump pressure is easy to calculate. The pump pressure is

$$\frac{61R}{16}(4V)^2 = 61RV^2$$

13.4   Pipe sizing

If we were designing the pipework for this circuit we would refer to the CIBSE or ASHRAE pipe-sizing tables, and select pipes on the equal friction method, also called the equal pressure drop method [6]. There are other methods but they are less popular nowadays [7]. The equal pressure drop method would typically have a $300 \, \text{Pa m}^{-1}$ pressure drop. Then the **index run** would be determined, which is the circuit with the greatest pressure drop across it. Working out the pressure drop for this circuit would establish the overall maximum pressure required from the pump. The pump could then be sized. But this is not the method of analysing circuits shown above, where the equivalent resistance of the circuit was determined.

Example 13.4
(a)   What is the index run in the circuit of Fig. 13.6?
(b)   What is the resistance of this index run? Find it without calculating a number of equivalent resistances for various parts of the system?

Solution
(a) The index run is the circuit with the greatest pressure drop across it. From inspection, and from a previous example this is likely to be *abcdefghij*. But we can check this. Traditionally the index circuit is used to determine the maximum pressure drop around the system to determine the pump size. In determining the pressure drop around the index run, it is assumed that the circuit is balanced and that the design water flow rates are established.
(b) When the circuit is balanced then the flow rates are known and network analysis is simplified. One simply has to go round the index run. From the previous discussion the circuit in Fig. 13.6 has equal design flows, V, through the resistances (*bi, ch, dg, ef*) so $V_1 = V_2 = V_3 = V_4 = V$. Hence the pressure drop around the balanced circuit *abcdefghij* is as follows, starting from *a* then going to *b*, *c*, and so on:

$$R(4V)^2 + R(3V)^2 + R(2V)^2 + 3RV^2 + R(2V)^2 + R(3V)^2 + R(4V)^2 = 61RV^2$$

which is the same as we obtained before. So we have calculated the index circuit pressure drop without using equivalent resistances. The resistance of

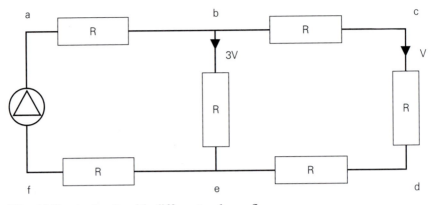

Fig. 13.7   A circuit with different volume flows.

the index circuit is its equivalent resistance, which is the equivalent resistance of the complete system. This is also a quick way of establishing the equivalent resistance of the complete system. The equivalent resistance of the index circuit is

$$R_{eq}(4V)^2 = 61RV^2$$

$$R_{eq} = 61R/16$$

It is interesting to note that the index circuit is often described as a circuit 'having the highest total pressure drop' across it [6]. Care is needed with this definition as all the circuits in a system which are in parallel have the same pressure drop across them. For instance, circuit *abija* in Fig. 13.6 has the same pressure drop as the index circuit. We found earlier that the equivalent resistance of *bcdefhib* is 29R/16, so the pressure drop across *abij* is

$$R(4V)^2 + \frac{29R}{16}(4V)^2 + R(4V)^2 = (16 + 29 + 16)RV^2$$

$$= 61RV^2$$

which is the same pressure drop as across the index circuit.

Another definition that could be misleading is for the longest circuit. Figure 13.7 is an interesting example in that $3V$ is required in *be* and V is required in *bcde*. It could be wrongly assumed that *abcdef* is the index circuit as it is the longest circuit. If it is assumed that a balancing valve, R_{bv}, is required in *be* then its value is calculated from

$$(R + R_{bv})(3V)^2 = 3RV^2$$

giving the unrealistic answer of $R_{bv} = -2R/3$. The balancing valve should be on *bcde* and of value 6R, so

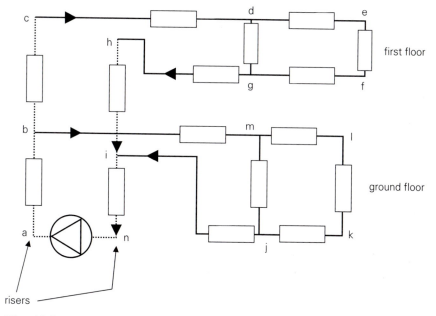

Fig. 13.8   A more complex circuit.

$$R(3V)^2 = (3R + R_{\mathrm{bv}})V^2$$

A better way to define the index circuit is as the circuit which does not require a balancing valve across it (or rather in its heat emitter branch). It is just a coincidence that this happens to be the longest circuit in many cases. It is worth remembering that the conventional pipe-sizing method assumes that all the circuits are balanced, so only the index circuit needs to be considered for the initial pipe and pump sizing.

13.5   Complex circuits

Figure 13.6 is a simple circuit. Figure 13.8 is a more complex circuit for a two-storey building with a ladder system on each floor coming from the risers. However, circuit analysis can also reduce this to a single equivalent resistance. Each floor's circuit has been shortened to simplify matters and the boiler resistance has been combined in the initial pipe resistances, again to simplify the circuit. But there are now two distinct circuits, *bcdefgh* and *bmlkji*. If each resistance is R and balancing valves are installed for equal volume flows through *ef*, *dg*, *lk* and *mj*, then by inspection *abcdefghin* is the index run. Redrawing this as Fig. 13.9 makes it easier to see the parallel and series resistances.

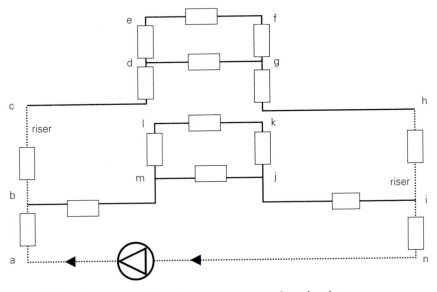

Fig. 13.9   Redrawn version of two-storey complex circuit.

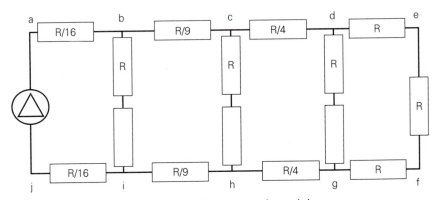

Fig. 13.10   Equal friction or equal pressure drop sizing.

13.6   Constant friction method

Consider the circuit of Fig. 13.6 sized by the equal pressure drop method. If the end part of the circuit, *defg*, is left at its original resistance then the upstream risers will be altered by the equal pressure drop method to produce equal resistances (Fig. 13.10). This assumes that the sections of the risers, *ab, bc, cd, de, fg, gh, hi, ij*, are of equal length and that they will therefore be of equal pressure drop, say 300 Pa m^{-1} times the length (m).

Example 13.5
What is the equivalent resistance of the circuit in Fig. 13.10?

Solution
The equal pressure drop method gives equal pressure drops in the riser sections *ab*, *bc*, *cd* and *de*:

$$\Delta P_{cd} = (R/4)(2V)^2 = RV^2$$

$$\Delta P_{bc} = (R/9)(3V)^2 = RV^2$$

and so on. All the riser sections have the same pressure drop of RV^2. Assuming that the circuit is balanced then simply going around *abcdegfghij* gives the total pressure drop around the circuit and the required pump pressure. The pressure drop across each of the riser sections is RV^2, so we have

$$\Delta P_{abcde} = 4RV^2 \quad \Delta P_{ef} = RV^2 \quad \Delta P_{fghij} = 4RV^2$$

giving a total pressure drop of $9RV^2$. And the equivalent resistance is

$$9RV^2/(4V^2) = 9R/4$$

This is a lot less than $61R/16$, the equivalent resistance of the original circuit (Fig. 13.6).

13.7 Pressure loss diagram

It is often useful to draw the pressure drops on a pressure loss diagram. For an equal friction design the riser pressure drop should have a constant gradient. Figure 13.11 shows the pressure loss diagram for Fig. 13.10. If the last leg of the riser were longer than the others – if it went to a more distant point – then the pressure drop would be higher and the overall pressure drop and

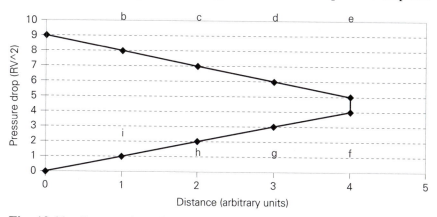

Fig. 13.11 Pressure loss diagram for the circuit in Fig. 13.8.

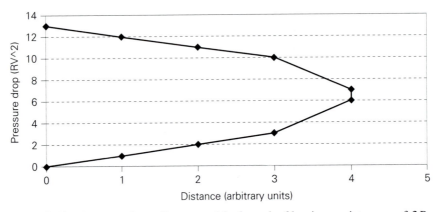

Fig. 13.12   Pressure loss diagram with *de* and *gf* having resistances of 3*R*.

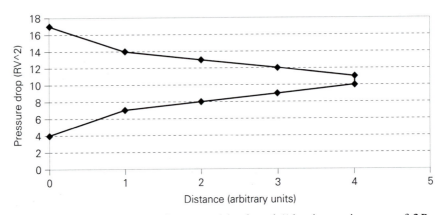

Fig. 13.13   Pressure loss diagram with *ab* and *ij* having resistances of 3*R*.

the pump size would be larger (Fig. 13.12). A pressure loss diagram like this may influence the designer to reduce its pressure loss, by using a larger pipe diameter to reduce the pump size. Similarly, a smaller pipe diameter at the beginning of the circuit could also 'distort' the pressure loss diagram, as shown in Fig. 13.13. Again the diagram may influence the designer to alter the design.

The static head due to a pressure vessel, or header tank, may be included in the pressure loss diagram (Fig. 13.14). Here the pressure vessel is connected to point *j* of the circuit, giving it a pressure of $4RV^2$ above atmospheric pressure. If it were connected to point *a*, just after the pump, then some of the circuit may be below atmospheric pressure, which would induce air into the pipes (air can often get in as it is a gas whereas water cannot get out as it is a liquid). The equivalent pressure due to the header tank is given

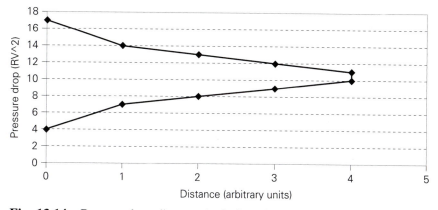

Fig. 13.14   Pressure loss diagram including the static head at point *j*.

in Fig. 13.14 in terms of *R* and *V*. If the header tank is *h* (m) above point *j* of the circuit, and assuming the surface of the water in the header tank is at atmospheric pressure, then point *j* will be ρ*gh* (Pa) above atmospheric pressure.

13.8   Pump characteristic

So far we have considered the resistance of the circuit and the consequent volume flows with the circuit pressure constant. But the pump has a characteristic curve and changes its pressure as the volume flow changes. Figure 13.15 is for a typical centrifugal pump. The operating point of the whole system is the intersection of the pump characteristic and the **system curve**, given by the equivalent resistance of the whole system, R_{eq}:

$$\Delta P = R_{eq} V^2$$

An equation that describes a number of centrifugal pumps, and fans, is the parabola

$$\Delta P = \Delta P_0 \left(1 - (V/V_0)^2 \right) \tag{13.11}$$

where ΔP_0 and V_0 are the points where the parabola hits the pressure axis, or y-axis, and the volume flow axis, or x-axis.

   A generalized polynomial may be used for fitting actual manufacturers' fan and pump data; *HVACSIM+* has a polynomial-fitting routine [8]. The polynomial can also be fitted by least squares (Chapter 14) or using the Solver routine in Microsoft Excel. Artificial neural networks (ANNs) may be used for complicated characteristics [9, 10].

13.9 Fan and pump laws

The pump and fan laws are useful for understanding how fans and pumps perform, especially when their speeds are varied. It is not practical to test every fan or pump, so the performance of geometrically similar fans and pumps (perhaps with a different impeller size) can be obtained by applying the fan and pump laws to the basic model's performance. There are a number of laws, but three will be quoted here. They relate to most centrifugal fans and pumps, but it is sensible to check with manufacturers' data when designing. The first law is

$$V \propto n \tag{13.12}$$

where V = volume flow rate ($m^3\ s^{-1}$)
$\qquad n$ = fractional fan or pump speed, $0 \leqslant n \leqslant 1$
$\qquad$ ($n = 1$ at full speed)
The second law is

$$\Delta P \propto n^2 \tag{13.13}$$

The third law relates to the fluid power, i.e. the power transmitted to the fluid:

$$\text{power in the fluid} = V \Delta P$$

so

$$V \Delta P \propto n^3 \tag{13.14}$$

This is the power in the fluid. The fan or pump motor power has not been considered and this will vary with the load. The fluid power divided by the electrical power into the motor gives the overall fan or pump efficiency. A typical value is about 80%. When inspecting fan and pump systems it is interesting to compare the motor power size with the fan or pump rating from its nameplate. This will reveal any oversizing of the motor.

When the pump speed is varied, the pump characteristic equation becomes

$$\Delta P = \Delta P_0 \left(n^2 - (V/V_0)^2 \right) \tag{13.15}$$

where n is the speed of the pump as a fraction of full speed, i.e. $n = 1$ at full speed and $n = 0$ at zero speed. ΔP_0 is the pressure drop at zero volume flow (Fig. 13.15) and ΔP is the general pressure drop. V_0 is the maximum volume flow rate at zero pressure drop. Equation (13.14) can be checked with the pump laws. At half-speed and $V = 0$ then $\Delta P = 0.25 \Delta P_0$ as per the pump laws. Also when $\Delta P = 0$ then

$$n^2 - (V/V_0)^2 = 0$$

and half-speed means $n = 0.5$, so

$$0.5^2 - (V/V_0)^2 = 0$$
$$V = 0.5V_0$$

again as per the pump laws.

13.10 Maximum power into fluid

It is not good for a pump or a fan to be operated too far outside its design range. If the pump operates at the extremes of its curve (Fig. 13.15) then insufficient power goes into the fluid. Instead it heats up the pump and its bearings. The maximum power in the fluid can be determined by differentiating the equation for fluid power, derived from the pump or fan characteristic equation, and equating to zero to find the maximum value. Start with

$$\text{power} = V\Delta P = V\big(1 - (V/V_0)^2\big)\Delta P_0$$

Differentiating and equating to zero:

$$0 = \big(1 - (3V^2/V_0^2)\big)\Delta P_0$$

which gives the maximum at $V = V_0/\sqrt{3}$, and substituting this back into the pump equation, we obtain $\Delta P = \frac{2}{3}\Delta P_0$. This is an indication of the most efficient point to operate a centrifugal pump or a centrifugal fan. It is only an indication because the motor driving the pump or fan will also alter its efficiency as the air or water load varies, and there is the friction of the

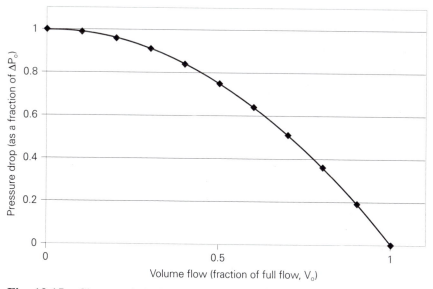

Fig. 13.15 Characteristic for a centrifugal pump.

connection between the motor and pump. However, for induction motors that invariably drive pumps and fans, there will not be a dramatic variation in efficiency with load.

13.11   Variable speed

Traditionally, three-port valves have been used for controlling water circuits, as either diverting valves or mixing valves (Chapter 8), and air-conditioning ductwork has been for constant volume flows. But this means at average to low thermal loads, the pump or fan is circulating more fluid than is required to meet demand, and using more energy than necessary in transmission and heat gains or losses. Hence, to save energy, there has been a trend of varying the pump or fan speed according to the thermal load, although it has been rather slow to take off in the United Kingdom. In water systems this often involves **two-port** valves and in air systems it involves **variable air volume** (VAV) terminal boxes.

The variable flow circuits have to have a method of controlling the pump or fan speed. There are many methods [11], including the inverter control for induction motors (common motors for fans and pumps). Inverters are becoming cheaper and hence more popular. In fact, some pumps now incorporate the variable-speed drive and controller in the pump.

The power into the water can be compared to the heat power transmitted. For instance, the heat power transmitted is

$$Q = \dot{m}C_p(t_f - t_r)$$

where $\dot{m}$ is the mass flow rate of the water (kg s^{-1}) or in terms of the volume flow rate, V (m^3 s^{-1}):

$$Q = V\rho C_p(t_f - t_r) \quad \text{W}$$

From the pump and fan laws, the power to transmit the fluid is $V\Delta P$, so we obtain the ratio of the thermal power transmitted to the motive power of transmission:

$$\frac{\Delta P}{\rho C_p(t_f - t_r)} \tag{13.16}$$

With water a typical pressure drop is 300 Pa m^{-1} and typically $t_f = 80°C$, $t_r = 70°C$; we also have $C_p = 4.2$ kJ kg^{-1} K^{-1}. So the power to transmit divided by the thermal power transmitted is

$$\frac{300}{1000 \times 4.2 \times 10} = 7.14 \times 10^{-3} \quad \text{W kW}^{-1} \text{ m}^{-1}$$

Bearing in mind that a small installation could have 100 m of pipework, with the consequent fitting losses, and also that overall pump efficiencies (motor efficiency combined with pump efficiency) will be of the order of

50% at maximum efficiency [12], the pumping power requirement can build up to about 3 W kW^{-1}.

For air and fan systems it is even more important. With a typical pressure drop of 1 Pa m^{-1} and a temperature difference of 8 K, equation (13.16) gives

$$\frac{1}{1.2 \times 1.02 \times 8} = 0.1 \quad W \, kW^{-1} \, m^{-1}$$

which is 14 times the value for a water system. This is one reason for using fan coil systems rather than all-air systems. As equation (13.16) depends on the density and specific heat capacity of the fluid, using latent heat can give a further reduction in the transmission power required for steam and refrigeration systems.

13.12 Two-port valves

There has been a recent tendency in the United Kingdom, although it is also quite common in the United States and Continental Europe, to modulate water flow by using two-port modulating valves instead of three-port valves. Domestic thermostatic radiator valves are common and modulating two-port valves are increasingly used on commercial installations and district heating systems serving non-storage heating calorifiers (heat exchangers) in blocks of flats. Three-port mixing valves maintain a constant flow rate in the pumped, unmodulated circuit and so use a high transmitting power. Three-port diverter valves also maintain a high pump transmitting power with load, as the undiverted circuit has a constant flow.

However, two-port modulating valves regulate the flow of water according to the load, so they allow the pump speed to be slowed at lower loads, saving power and energy. With the advent of cheaper static frequency inverters and other forms of motor control, this energy-saving method is increasing in popularity. A static frequency inverter is a variable-speed drive that consists of a full-wave rectifier, a d.c. link to smooth the rippled rectified a.c. and an inverter which converts the d.c. into variable-frequency power. A pulsewidth modulation inverter produces positive and negative pulses of variable width so their mean value forms a sinusoidal voltage of the desired frequency and amplitude [12, 13].

Extra control is required to vary the pump speed with two-port modulating valves. The two-port valves also need control to perform the modulation. As the load decreases in a circuit, the two-port valves modulate and the circuit resistance increases. The circuit resistance increases and the operating point of the system rides up the pump curve (Fig. 13.16). A better method is to control the speed of the pump using a controller, perhaps a PI controller. This maintains a constant pressure differential across the circuit, measured by two pressure sensors. Example 13.6 illustrates its operation.

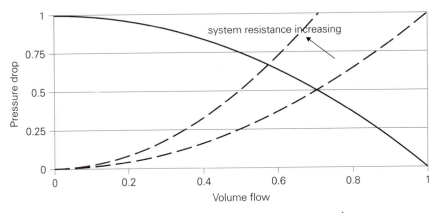

Fig. 13.16   Pump pressure and system curve as system resistance increases.

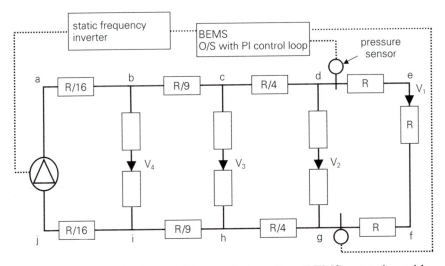

Fig. 13.17   Control of a variable-speed pump by a BEMS outstation with a PI loop.

Example 13.6

Consider the circuit in Fig. 13.17. Assume there is ideal control, i.e. sufficient pressure to maintain the required flow through the branches. Calculate the speed of the pump if all the four branches go to half the full design flow rate, i.e. $0.5V$ each. At full design flow, the pump operates at its point of maximum power into the water.

Solution

By going round the index run, the pump pressure at full design flow of $4V$ is $9RV^2$. This is the pressure above the datum level (atmospheric pressure plus the static pressure of the pressure vessel or feed and expansion tank); for ease of calculation we shall treat this as zero.

For good design the pump should be selected so its design full flow operating point is at the position of maximum power into the water on the pump characteristic, i.e.

$$\Delta P_0 = \frac{3}{2}(9RV^2)$$

$$= \frac{27}{2}RV^2$$

and such that

$$V_0/\sqrt{3} = 4V$$

$$V_0^2 = 48V^2$$

Then the pump equation is;

$$\Delta P = \frac{27}{2}RV^2\left(n^2 - \frac{V_{total}^2}{48V^2}\right)$$

where V_{total} is the sum of the flows through the four branches, i.e. $V_{total} = V_1 + V_2 + V_3 + V_4$.

The **control equation** for this **ideal control** is simply that the pump speed is sufficient to maintain the required flow rates through the branches and can be determined from the design full flow conditions to maintain ΔP_{dg} constant. This assumes that the PI control is perfect and maintains a constant value of pressure difference. Hence

$$\Delta P_{dg} = 3RV^2 = \text{constant}$$

But for a control equation this needs to relate to the pump speed, n. This can be done by substituting for ΔP_{dg} as follows:

$$3RV^2 = \Delta P_{pump} - 2\Delta P_{ab} - 2\Delta P_{bc} - 2\Delta P_{cd}$$

The term $2\Delta P_{ab}$ is for the pressure drop $\Delta P_{ab} + \Delta P_{ij}$. Putting in the pressures:

$$3RV^2 = \frac{27}{2}RV^2\left(n^2 - \frac{V_{total}^2}{48V^2}\right) - \frac{2R}{16}V_{total}^2$$

$$- \frac{2R}{9}(V_1 + V_2 + V_3)^2 - \frac{2R}{4}(V_1 + V_2)^2$$

This can be simplified by putting in the diversity factors, x, such that

$$x_1 = V_1/V \quad x_2 = V_2/V \quad x_3 = V_3/V \quad x_4 = V_4/V$$

giving

$$3 = \frac{27}{2}\left(n^2 - \frac{(x_1 + x_2 + x_3 + x_4)^2}{48}\right)$$

$$- \frac{1}{8}(x_1 + x_2 + x_3 + x_4)^2$$

$$- \frac{2}{9}(x_1 + x_2 + x_3)^2 - \frac{1}{2}(x_1 + x_2)^2 \tag{13.17}$$

With all the flows reducing to half, the control equation shows that the pump speed reduces to

$$3 = \frac{27}{2}\left(n^2 - \frac{2^2}{48}\right) - \frac{1}{8}2^2 - \frac{2}{9}\left(\frac{3}{2}\right)^2 - \frac{1}{2}1^2$$

$$= \frac{27}{2}\left(n^2 - \frac{1}{12}\right) - \frac{1}{2} - \frac{2}{9}\left(\frac{9}{4}\right) - \frac{1}{2}$$

$$\frac{9}{2}\left(\frac{2}{27}\right) = \left(n^2 - \frac{1}{12}\right)$$

$$n = \left(\frac{1}{3} + \frac{1}{12}\right)^{1/2} = 0.65$$

So the pump is controlled to 65% of its full speed to maintain ΔP_{dg} constant.

13.13   Diversity of load

In Example 13.6 the speed of the pump was worked out from the control equation. But the pressure may not be sufficient to supply all the branches with the required volume flow rates. To calculate the required system pressure to satisfy all the branches, we need the **system equation**. This is determined from the pressure drop over the whole circuit, which can be calculated from the index run, and the pressure developed by the pump:

$$\frac{27}{2}RV^2\left(n^2 - \frac{V_{total}^2}{48V^2}\right) = \frac{2R}{16}V_{total}^2 + \frac{2R}{9}(V_1 + V_2 + V_3)^2$$

$$+ \frac{2R}{4}(V_1 + V_2)^2 + 3RV_1^2$$

or in terms of diversities:

$$\frac{27}{2}\left(n^2 - \frac{(x_1 + x_2 + x_3 + x_4)^2}{48}\right) = \frac{1}{8}(x_1 + x_2 + x_3 + x_4)^2$$

$$+ \frac{2}{9}(x_1 + x_2 + x_3)^2$$

$$+ \frac{1}{2}(x_1 + x_2)^2 + 3x_1^2 \qquad (13.18)$$

For all branches on half flow this gives

$$\frac{27}{2}\left(n^2 - \frac{2^2}{48}\right) = \frac{1}{8}2^2 + \frac{2}{9}\left(\frac{3}{2}\right)^2 + \frac{1}{2}1^2 + 3\left(\frac{1}{2}\right)^2 = \frac{9}{4}$$

$$\left(n^2 - \frac{1}{12}\right) = \frac{9}{4}\left(\frac{2}{27}\right) = \frac{2}{12}$$

$$n = \left(\frac{3}{12}\right)^{1/2} = \left(\frac{1}{4}\right)^{1/2} = 0.5$$

This is lower than the speed at which the controller is running the pump, so there will be more than enough pressure to provide all branches with the required flow rates. In fact, the valves will be closed a little to establish the correct flow rates in their branches. This can be seen from the fact that when the controller is controlling the pump to 65% of its full speed ($n^2 = 5/12$) the pump pressure is

$$\frac{27}{2}RV^2\left(\frac{5}{12} - \frac{1}{12}\right) = \frac{9}{2}RV^2$$

$$= 4.5RV^2$$

and the total volume flow rate is $2V$. So the equivalent resistance of the circuit is $4.5RV^2/4V^2 = 1.125R$ which compares to the equivalent resistance at full flow of $9RV^2/16V^2 = 0.56R$. Note that when only the system equation is considered, the pump speed for half flow is 0.5, which agrees with the pump law $V \propto n$. But this is only when the whole system resistance stays constant and the control is ideal.

13.13.1   Starvation

Starvation occurs if a branch is getting insufficient flow due to poor balancing or due to the pump running too slow to satisfy the pressures for all the branches and their required flow rates. The controlled speed of the pump is below the system equation speed. The control speed, n_c, must always be greater than or equal to the speed required by the system, n_s. Rearranging the control equation (13.17) gives

$$\frac{27}{2}\left(n_c^2 - \frac{(x_1 + x_2 + x_3 + x_4)^2}{48}\right) = 3 + \frac{1}{8}(x_1 + x_2 + x_3 + x_4)^2$$

$$+ \frac{2}{9}(x_1 + x_2 + x_3)^2 + \frac{1}{2}(x_1 + x_2)^2$$

and rearranging the system equation (13.18) gives

$$\frac{27}{2}\left(n_s^2 - \frac{(x_1 + x_2 + x_3 + x_4)^2}{48}\right) = \frac{1}{8}(x_1 + x_2 + x_3 + x_4)^2$$

$$+ \frac{2}{9}(x_1 + x_2 + x_3)^2$$

$$+ \frac{1}{2}(x_1 + x_2)^2 + 3x_1^2$$

And for $n_c \geq n_s$ we have

$$3 + \frac{1}{8}(x_1 + x_2 + x_3 + x_4)^2 + \frac{2}{9}(x_1 + x_2 + x_3)^2 + \frac{1}{2}(x_1 + x_2)^2$$

$$\geq \frac{1}{8}(x_1 + x_2 + x_3 + x_4)^2 + \frac{2}{9}(x_1 + x_2 + x_3)^2 + \frac{1}{2}(x_1 + x_2)^2 + 3x_1^2$$

$$3 \geq 3x_1^2$$

$$x_1 \leq 1 \tag{13.19}$$

As $x_1 \leq 1$, under design conditions, there is never going to be starvation. However, if the control differential pressure is set below the design value to save more energy – and from Example 13.6 the control equation maintains the system pressure at too high a value for most of the volume flows – then, depending on the value of x_1, there could be starvation at high volume flow rates. Another source of starvation is due to the balancing valves not being used or being incorrectly set. This may be due to cutting costs. It may also be due to the belief that two-port modulating valves are self-balancing, and to an extent they are.

13.13.2   A two-branch circuit

Consider a simple circuit of two branches (Fig. 13.18) with balancing valves, modulating valves and radiator, or heat exchanger, resistance. With a total volume flow of $2V$ (V through each radiator) the total equivalent resistance is

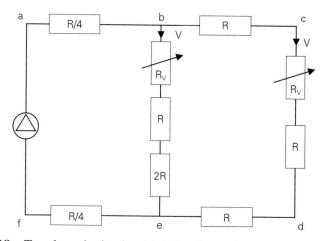

Fig. 13.18   Two-branch circuit with balancing and modulating valves.

$$R_{abcdef} = (10RV^2)/4V^2 = 5R/2$$

This assumes that the modulating valves, R_v, each have a resistance of $5R$ to give the valve in cd an authority of 0.5. (Authority is discussed in the next section.)

If the balancing valve, $2R$ in be, is **omitted** in the circuit there will be just the resistances of the radiators, each R, and the modulating valves, which when fully open are each of resistance $5R$. If the heating system is started on a cold morning then all the modulating valves will be open. The equivalent resistance of the full circuit is now reduced compared to the circuit with the balancing valves. To calculate the new equivalent resistance without the balancing valve in be and with the modulating valves fully open, we first have to calculate the equivalent resistance for $bcde$ using the parallel resistance formula (13.9):

$$R_{bcde} = (8 \times 6)R^2/(8 + 6 + 2\sqrt{48})R = 1.72R$$

So the equivalent resistance for the whole circuit is

$$R_{abcdef} = (1.72 + 0.5)R = 2.22R$$

To calculate the volume flow we need to know the pump characteristic. Based on the design flow of $2V$ and a pressure drop of $10RV^2$ it is

$$\Delta P = 15RV^2\left(1 - \frac{V^2_{total}}{12V^2}\right)$$

So the flow rate can be determined from the intersection of the pump characteristic and the pressure drop across the equivalent circuit, R_{abcdef}:

$$2.22RV_{total}^2 = 15RV^2\left(1 - \frac{V_{total}^2}{12V^2}\right)$$

$$(2.22 + 1.25)V_{total}^2 = 15V^2$$

$$V_{total} = 2.08V$$

The volume flow through each can be calculated from equating the pressure drops through each branch. These give $1.11V$ through *be* and $0.97V$ through *cd*, not too bad compared with the balanced circuit.

If the system has speed control on the pump and if the pressure sensors are just after *be*, then from the design conditions, the control pressure is $8RV^2$. But the pump will not be able to achieve this as the operating point of the pump at full speed and the lower equivalent resistance of the unbalanced circuit will be below the designed control pressure. This is discussed further in the following section on the larger circuit.

13.13.3   A 20-branch circuit

Consider a ladder circuit with n branches instead of just two. With the end circuit for the large ladder circuit remaining the same as *bcde* in Fig. 13.18, working around the index circuit, the balanced circuit will have a pump pressure of

$$\left(2(n - 1)R + 3R + R_{ve}\right)V^2$$

where R_{ve} = the modulating valve at the end of the ladder, branch n. This assumes that each horizontal arm to the branch is sized according to the equal pressure drop method so that its pressure drop will be RV^2. With an authority of 0.5 for R_v this equals the pressure drop across the rest of the circuit:

$$0.5 = \frac{R_{ve}V^2}{\left(2(n - 1)R + 3R + R_{ve}\right)V^2}$$

so

$$R_{ve} = 2(n - 1)R + 3$$

If no balancing is attempted on this circuit and the modulating valves are the same size for all the branches then there will be problems.

Consider 20 branches, n = 20. This means that the first horizontal arm to the branch nearest to the pump will have a resistance of $R/400$ by the method of equal pressure drop. The end modulating valve, R_{ve}, with an authority of 0.5, will have a resistance of

$$2(n - 1)R + 3R = 41R$$

and the balanced circuit will have a pump pressure of

$$2\left(2(n - 1)R + 3R\right)V^2 = 82RV^2$$

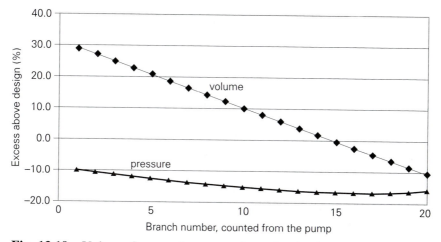

Fig. 13.19 Volume flows and pressure drops for the 20-branch unbalanced circuit.

Using equation (13.9) recursively in an Excel spreadsheet, the equivalent resistance of the whole circuit is $0.153R$. This can be approximately checked from the fact that 20 branches, each of resistance $41R$, in parallel with arms of zero resistance connecting them, would give an equivalent resistance of $41R/400 = 0.103R$. The unbalanced circuit's equivalent resistance of $0.153R$ is smaller than the balanced circuit's equivalent resistance of

$$\frac{2(2(n-1)R + 3R)V^2}{(20V)^2} = \frac{2 \times 41R}{400}$$

$$= 0.205R$$

Figure 13.19 shows the results in terms of the flow rates through each branch. They have been calculated using a spreadsheet. Notice that the modulating valves do have an element of self-regulation as the nearest branch takes almost 30% too much water and the end branch is starved by about 10%. Five of the end branches will be starved. But the pressures throughout the circuit are below the designed values, so for a variable-speed pump controlled by pressure, it will not be able to increase its speed as it will be on its full speed to maintain the calculated flows in the unbalanced circuit. This is because the unbalanced circuit has a lower equivalent resistance than the balanced circuit.

13.14   Sensor position

In the four-branch example (Fig. 13.17) the sensors are positioned near the end of the index run, across the section *defg*. Here the pressure drop is $3RV^2$

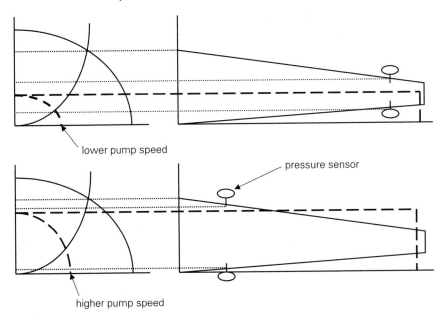

lower pump speed

pressure sensor

higher pump speed

Fig. 13.20   Effect of sensor position on pump speed.

where V is the full flow rate through a branch. If the sensors were moved closer to the pump then the pressure drop would have to be increased. Suppose the sensors are across bi then the pressure drop at full flow is

$$9RV^2 - \frac{2R}{16}(4V)^2 = 7RV^2$$

The speed reduction available for the pump is much reduced.

This can be seen by considering the flows to all branches reducing to zero flow, an unlikely situation but one that produces the ultimate energy saving. In this situation the pressure is still maintained at the same set value (assuming a perfect PI controller), but the system position on the pump characteristic is at zero flow and on the pressure axis (Fig. 13.20). Note that at the zero flow rate, the pressure loss diagram is a rectangle as the friction loss is now zero too. In fact, as the flow rate reduces, the pressure loss diagram tends to a shallower slope and a rectangular form.

Consider the situation with the sensors across de and the flow rate going to zero. Now the equation for the pump pressure gives

$$3RV^2 = \frac{27}{2}RV^2\left(n^2 - \frac{0^2}{48V^2}\right)$$

$$n = (6/27)^{1/2} = 0.47$$

The pump has reduced to 47% of full speed.

If the sensors are across *bi* then the setpoint control pressure there is $7RV^2$, so at zero flow we have

$$7RV^2 = \frac{27}{2}RV^2\left(n^2 - \frac{0^2}{48V^2}\right)$$

$$n = (14/27)^{1/2} = 0.72$$

This compares with 47% for the sensors across *de*. So the further the sensor from the pump, the greater the possible energy savings.

13.15   Authority

As discussed in Chapter 10, the heat outputs from most heat exchangers vary non-linearly with the flow of tempered water through them. There is a 'log mean temperature difference' (LMTD) to calculate the temperatures of the flows. Consequently, control valves are often designed to complement the heat emitters so that the overall response of the valve and emitter is a linear relationship between the control signal to the valve affecting its stem movement and the heat emitted. It is also important that the movement of the stem controlling the opening and closing of the valve produces significant changes in flow over its whole movement. In other words, it is not very useful to have a valve where it is only the last 10% of the stem travel that produces any real change in the flow of water. A slight change in this 10% region and there is too much or too little flow. The **authority** of a valve is a measure of its ability to regulate the flow evenly over its stem travel. As shown in Fig. 8.19, the authority, N, is

$$N = \frac{\Delta P_v}{\Delta P_v + \Delta P_s}$$

This can also be interpreted as the ratio of the pressure drop across the valve when fully open to the pressure drop when the valve is fully closed and the whole pump pressure is across it. A ladder circuit complicates this definition (see later). For the simple circuit of Fig. 8.19, with the valve and system in series with the pump, we have

$$N = \frac{R_{vopen}}{R_{vopen} + R_{sys}}$$

where R_{vopen} is the resistance of the valve when it is fully open and R_{sys} is the resistance of the rest of the simple series circuit. For an equal percentage valve, suitable for non-linear heat exchangers (e.g. coils and radiators), the relationship between the volume flow and the stem position [16] is

$$V = V_{max} e^{a(s-1)} \qquad (13.20)$$

where V_{max} = the maximum flow through the valve when fully open

s = valve stem position, $0 \leqslant s \leqslant 1$

a = a constant, given as 3.22 in the literature [16]

In terms of resistance this yields

$$R = R_{open} \exp\left(-2a(s - 1)\right) \qquad (13.21)$$

where R_{open} is the resistance of the valve when fully open.

Equation (13.20) can be determined when there is a constant pressure across the valve for all flows. But it is not valid for low flow rates; equation (13.20) shows that the flow does not reduce to zero when the stem is zero, i.e. when the valve is closed. Likewise the resistance in equation (13.21) does not reach infinity when the valve is closed. For a valve following this equation the flow will be uncontrollable at low flow rates. The rangeability defines the ratio of the maximum flow to the minimum controllable flow and for this equation it is 25.

A better equation may be derived from the valve characteristic being complementary to the heat exchanger for a linear response [13]:

$$V = V_{max}\left(\frac{s\Phi}{1 - s(1 - \Phi)}\right) \qquad (13.22)$$

where Φ is the thermal effectiveness of the heat exchanger; further details can be found in Petitjean [16]. With $\Phi = 0.25$, a typical value for a heater battery, equation (13.22) becomes

$$V = V_{max}\left(\frac{0.25s}{1 - 0.75s}\right)$$

and

$$R = R_{open}\left(\frac{0.25s}{1 - 0.75s}\right)^{-2}$$

The influence of the valve authority, N, can be seen by incorporating N in the pressure drop around the circuit. With the valve at an intermediate resistance, R_v, between fully open and fully closed, we have

$$\Delta P = (R_{sys} + R_v)V^2$$

$$= \left[\left(\frac{1 - N}{N}\right)R_{open} + R_v\right]V^2$$

and with the valve fully open we have

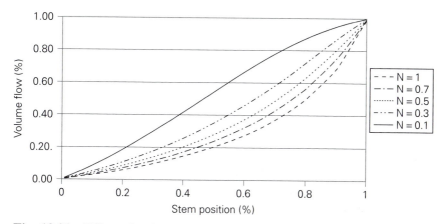

Fig. 13.21 Effect of valve authority on the volume flow rate of a valve.

$$\Delta P = \left[\left(\frac{1-N}{N} \right) R_{\text{open}} + R_{\text{open}} \right] V_{\text{max}}^2$$

Assume that the pump pressure in the circuit is constant. This is an ideal situation but it simplifies the equation and it is the same assumption as used when measuring the valve characteristic. We have

$$\frac{V^2}{V_{\text{max}}^2} = \frac{\left(\dfrac{1-N}{N} \right) R_{\text{open}} + R_{\text{open}}}{\left(\dfrac{1-N}{N} \right) R_{\text{open}} + R_{\text{v}}}$$

$$\frac{V^2}{V_{\text{max}}^2} = \frac{1}{1 - N + N \left(\dfrac{0.25s}{1 - 0.75s} \right)^{-2}}$$

Figure 13.21 shows the effect of the valve authority on the characterized valve. Notice that the curves are reasonably close together from $N = 0.5$ to $N = 0.1$. The complementary curve to most heat exchangers is between 0.3 and 0.5, hence the heuristic rule that the authority for a control valve should be between 0.3 and 0.5.

The circuit for the valve considered above has been Fig. 8.19, the two-port valve in series with a system resistance and a pump. In this circuit it has been assumed that the design flow rate occurs when the valve is fully open. However, the pump pressure can vary and the flow could be above or below the design flow rate. To distinguish between the design flow condition and the non-design flow condition, Petitjean suggests that two authorities should be used: N' when the design flow is achieved with the valve fully open and

N when the design flow is not achieved with the valve fully open. So N' is defined as

$$N' = \frac{\text{pressure drop in the control valve fully open with design flow}}{\text{pressure drop when the valve is shut}}$$

Petitjean [16] implies that in a ladder circuit like Fig. 13.17 the pressure drop across the branch, bi, ch, etc., is considered for the authority. For the whole circuit, with the other valves possibly shutting down and the pressure across the branch approaching the pump pressure, a third authority is defined:

$$N'_{min} = \frac{\text{pressure drop in the control valve fully open with design flow}}{\text{pressure drop across the pump}}$$

Petitjean recommends a value of $N'_{min} \geqslant 0.25$, i.e. the design pressure drop must be at least 25% of the pump pressure.

Such authorities dramatically change the pump power required for a circuit. Consider the modulating valves in Fig. 13.18. The modulating valve in cd, with an authority of 0.5, is $5R$. Hence the total pressure required from the pump for full design volume flow is $10RV^2$ (compared to $5RV^2$ without the modulating valve). Lower authorities may be considered to reduce the pump pressure, but in a variable flow circuit, with pump speed control, the modulating valves, with authorities of 0.5 do reduce the short-circuiting and starvation to an extent. The diversity of the loads and the consequent movement of the modulating valves, and whether balancing valves are included, both need to be considered in relation to the authorities employed.

13.16   Modulating two-port valves

Care is required when designing with two-port valves, especially where minimum or constant flow rates through boilers or chillers are required. Suppose two-port modulating valves are used with a constant-speed pump, as in a domestic heating system with thermostatic radiator valves (TRVs). As some of the TRVs close down, the pressure across the remaining open modulating valves increases significantly. It might be such that the actuator cannot easily move the valve. Rather than using a more powerful actuator, it requires a bypass pipe at the end of the index run, preferably with a pressure relief valve. The pressure relief valve can be a regulating valve set to provide a high resistance. This will ensure that it does not bypass water through it when the rest of the circuit requires flow.

Alternatively a pressure differential control valve can be used (Fig. 13.22). As the loads reduce, the pressure across the pump rises, riding the pump curve. The sensors of the pressure differential valve sense this, alter the pressure across the actuator diaphragm and open up the bypass valve. This

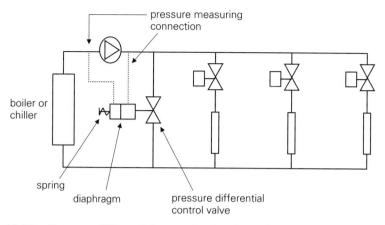

Fig. 13.22   Pressure differential control valve for maintaining minimum flow.

will maintain a constant flow of water through the boiler or chiller. Another alternative is to put a three-port diverting valve on the last load so that water flow is maintained around the index circuit [14].

13.17   Boiler sequencing

Multiple boilers are often piped in parallel rather than series to reduce the overall resistance of the boiler circuit. It also allows boilers to be valved out and even removed for maintenance. Unvalved boilers that allow hot return water through them act as heat exchangers connected to the outside by a flue. Older boilers are especially prone to losing heat this way. So the best solution is to valve out older unfired boilers. The possible snag is the thermal shock when the boilers are fired up again. However, some modern high-efficiency boilers are available with full-load efficiencies in the 90% range and less than 1% zero-load losses [16], i.e. at no heat load, so valving out of unfired boilers is not necessary, although it is sensible to refer to manufacturers' details to confirm the high efficiencies.

Connecting the boilers in series would make sequencing quite easy. The first boiler would take the main heat load and the subsequent boilers would top up the heat required. However, this requires a lot of pump power, so most boilers are connected in parallel.

To sequence the boilers, a signal is required to indicate the heat load. Figure 13.23 shows three boilers in parallel with flow and return **headers**, a common but rather poor arrangement as will be shown later. A header is a pipe of larger diameter, a few pipe sizes up on the pipes connected to it. The header, having a large diameter compared to the other pipes, has a much lower resistance and hence a lower pressures drop across it. So if a valve

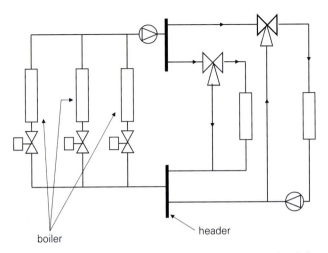

boiler

header

Fig. 13.23   Three boilers with two headers, diverting and mixing circuits.

shuts in the boiler circuit or the pump pressure changes, the consequent pressure drop across the header is small and so is the pressure change in the circuits downstream of the header. The header stops **interaction** between circuits, especially those which have pumps in them. This is a benefit for control systems and valves downstream of the header, as interaction from another pumped circuit will interfere with the control of that circuit.

The boilers in Fig. 13.23 serve a heating circuit with a mixing valve (under compensator control), and a domestic hot water (DHW) circuit with a diverting valve and control from a cylinder temperature, or secondary flow water pipe temperature sensor. Mixing valves give better control of circuits but they require pumps in their circuits, which is expensive. Diverter valves do not require pumps in their circuits and they are often used for the sake of cheapness, except in radiator circuits where the better control can be justified.

Traditionally both heating and DHW circuits are served from the same boilers. The boiler sequencer requires a signal of the heat load to switch in the correct number of boilers. A temperature sensor on the return water pipe in the boiler circuit is ideal. But the heating circuit with its compensator reduces its flow temperature as the heating load decreases and consequently the return water temperature also reduces. However, the DHW control increases its return water temperature as the DHW load reduces by diverting hot flow water straight into the return water pipe. With this circuit there is no clear relationship between the combined load (DHW and heating) and the return water temperature.

With the boilers on the **ring circuit** shown in Fig. 13.24 this problem is overcome. As the mixing valve changes, e.g. as it opens to allow some

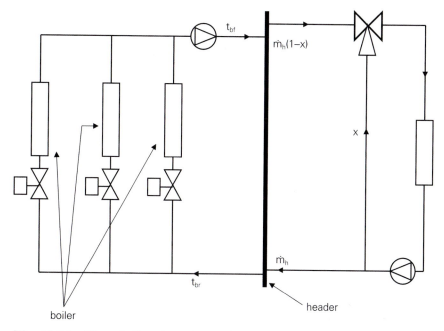

Fig. 13.24 Three boilers in a ring main with mixing circuit.

recirculation and closes to the boiler water from its previous full flow, the recirculation fraction, x, increases. The hot water from the boiler ring is reduced correspondingly by $\dot{m}_h(1 - x)$. This water instead goes through the ring's header. The amount through the header equals the recirculation in the heating circuit, $x\dot{m}_h$. The circuit from the boiler ring to the DHW diverting circuit shown in Figure 13.23 but not in Figure 13.24, is a constant volume circuit, so $\dot{m}_{dhw}$ will not change and it will not affect the flow through the ring header. The example below shows this.

Example 13.7
Calculate the return water temperature in the ring when the radiator heating load is half the design value. Assume that the DHW load is at full load and that the DHW flow water is a proportion y of the heating water flow $(0 < y < 1)$. The design conditions are as follows:

Boiler ring flow water temperature, $t_{bf} = 80°C$
Boiler ring return water temperature, $t_{br} = 70°C$
Heating circuit flow water temperature, $t_{fh} = 80°C$
Heating circuit return water temperature, $t_{rh} = 70°C$
Domestic hot water circuit flow water temperature, $t_{fdhw} = 80°C$
Domestic hot water circuit return water temperature, $t_{rdhw} = 70°C$

Solution

The heating system is a radiator system so the heat output is a 1.3 index relationship (Chapter 8) and its heat output is

$$B(t_m - t_{ai})^{1.3}$$

where t_m = mean radiator temperature

$$= (t_f + t_r)$$

t_{ai} = room air temperature

At half load:

$$0.5B(t_m - 20)^{1.3} = B(75 - 20)^{1.3}$$

and

$$t_f - t_r = 5 \text{ K}$$

giving

$$t_m = 52.3°C \quad t_f = 54.8°C \quad t_r = 49.8°C$$

This assumes that the room temperature stays constant at 20°C.

At the mixing valve with xm_h being recirculated (m_h is the water flowing in the heating circuit, kg s^{-1}):

$$49.8x\dot{m}_h + 80x\dot{m}_h(1 - x) = 54.8\dot{m}_h$$

$$x = 0.83$$

At the return point of the header we get

$$(\dot{m}_h + \dot{m}_{dhw})t_{br} = \dot{m}_h x 80 + \dot{m}_h(1 - x)49.8 + \dot{m}_{dhw}70$$

If the ratio between the DHW flow and the heating flow is y then

$$(1 + y)t_{br} = 80x + 49.8(1 - x) + 70y$$

$$t_{br} = (74.9 + 70y)/(1 + y)$$

As $0 \leqslant y \leqslant 1$ then $72.4°C \leqslant t_{br} \leqslant 74.9°C$. So, with the heating load at half its full value, the boiler ring return temperature increases. With separate headers the boiler ring return temperature would decrease due to the heating load contrary to the DHW return temperature.

Sometimes the header is made into a **buffer tank** so that the water velocity is reduced and dirt in suspension deposits in the bottom of the tank rather than clogging up the pipework and systems. The buffer vessel also increases the amount of water in the system, which increases the thermal inertia of the system and hence increases the cycle time of the on/off control.

This can be seen by examining an ideal heating system in which the boiler, the pipework and the radiator system do not have a significant storage capacity for heat. The system water is the major store of heat. From Chapter 5 the system equation is

$$Q_{in}(1 - \Phi) = MC_p dt/dT$$

where Φ = the heat load $(0 \leqslant \Phi \leqslant 1)$

       Q_{in} = the heat input from the boiler

In this example M is the mass of the water in the system, C_p its specific heat capacity, t its temperature and T the time. As the boiler in this example has an insignificant thermal capacity, it raises the water flowing through it immediately by its flow and return differential, $\Delta t_{f/r}$. With a flow rate through the boiler of $\dot{m}$ (kg s^{-1}) then

$$\dot{m} C_p \Delta t_{f/r}(1 - \Phi) = MC_p dt/dT$$

which yields a cycle time (Chapter 5) of

$$T_{cycle} = \frac{\dot{m}\Delta t_{diff}}{M\Delta t_{f/r}} \left(\frac{1}{\Phi(1 - \Phi)} \right)$$

Here Δt_{diff} is the differential of the on/off control. The addition of a buffer vessel will increase the mass of water in the system, M, so T_{cycle} will increase. However, **double circulation** – where the primary flow from the chillers or boilers simply flows back into the primary return and the secondary return flows back into the secondary flow – needs to be avoided. This would simply isolate the flow between the primary and secondary circuits. It is more common with chiller circuits where the natural convection to mix the primary and secondary water is less. Offset pipes and baffle plates can reduce or eliminate double circulation [16].

The buffer is also important for chiller circuits where the chilled water volumes are not as large as heating systems. Buffer tanks are especially important in older systems where the chiller modulation is not too efficient. Often the chiller just supplies water to a cooling coil in the plant room, whereas a heating system often has radiators and heat exchangers throughout the building. For chillers there are often limits on the amount of switching of the compressors and their motors. The motors can have large inrush currents on start-up when the windings are cool and of low resistance, but they heat up quickly. Motors can also produce inductive spikes and significant harmonics that can be transmitted through the electrical supply system. Good motor starters can reduce these problems but there is also the thermal shock for the refrigerator system with much starting and stopping. Chillers using cylinder unloading to vary output controlled from the chilled water flow temperature will also induce frequent cycling when they are on the last step of cooling, and this is likely to burn out the motors [15].

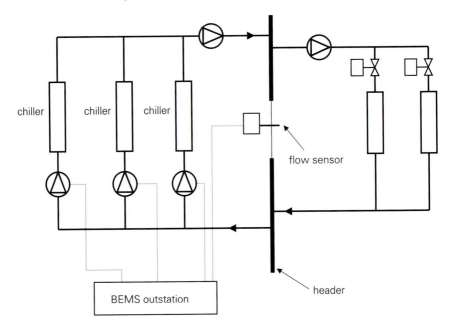

Fig. 13.25   Chiller circuit for sequencing.

13.18   Chillers

Chillers provide cool air to an air-conditioning system. In a DX system the refrigerant flows directly through the cooling coil. In other systems the refrigerant is used to cool water and the cooled water then passes through the cooling coil. This involves water flowing through one side of the evaporator with refrigerant through the other side. Chilled water circuits operate typically with a flow of 6°C and a return of 11°C. Therefore a reduction in water flow without a change in load could lower the chilled flow water to near freezing. It is therefore important to maintain adequate water flow through chillers to avoid freezing [16]. Often a pump is installed in series with each chiller coil (evaporator), as shown in Fig. 13.25. This can also be adopted with boiler sequencing but is less common as boilers are more tolerant of flow changes. As with multiple boilers, multiple chillers are connected in a ring main.

As the flow through the chillers is so important, their sequencing is often controlled on the flow rate with a flow sensor rather than using temperature. Hence in Fig. 13.25 there is a flow sensor to sequence the chillers. This monitors the demand for chilled water in the circuits that it serves, which in this case is a variable flow circuit. Although not shown, a pressure differential valve should be installed in the secondary circuit in Fig. 13.25 to protect

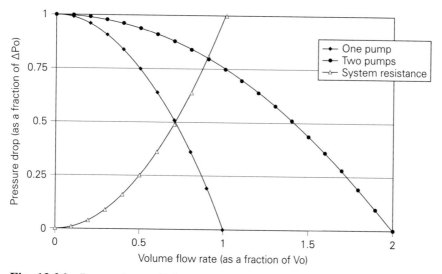

Fig. 13.26 Pumps in parallel.

against low flow conditions, as discussed earlier. The header is split in two and connected by a pipe of smaller diameter to get a reasonable water velocity for the flow sensor to collect accurate values.

With both boiler and chiller primary ring circuits there can be a danger of some reverse flow through the header, i.e. a powerful secondary pump drawing return water into the secondary flow pipe. This would lower the flow temperature in a heating system and raise it in a cooling system. Reverse flows would then cause problems in both primary and secondary circuits, with the consequent poor sequencing of the plant [17].

Circuit analysis is required to determine the flow rates in the primary and secondary circuits under part-load conditions to ensure there is no reverse flow, etc. This involves analysis of pumps in parallel. As with parallel resistances, the pressure drops across the pumps will be the same. So two identical pumps with pump curves given by

$$\Delta P = \Delta P_0 \left(1 - (V/V_0)^2\right) \tag{13.11}$$

combine to give a curve

$$\Delta P = \Delta P_0 \left(1 - (V/2V_0)^2\right)$$

shown in Fig. 13.26. The operating point (where the system resistance curve cuts the pump curve) for one pump is at $V = 0.7V_0$, $\Delta P = 0.5\Delta P_0$ and for two pumps it is at $V = 0.9V_0$, $\Delta P = 0.8\Delta P_0$.

A pump throttled by a valve in series, with resistance R, can be regarded as a pump with the characteristic

$$\Delta P = \Delta P_0 \left(1 - (V/V_0)^2\right) - RV^2$$

Note that chilled water requires about five times more water flow than hot water for the same thermal power output (Chapter 10). Therefore pump control and efficiency is even more important for chilled water.

13.19   Injection circuits

All the circuits we have so far considered operate over similar temperatures and pressures. For heating circuits there is typically a flow temperature of 80°C and a return of 70°C. For chilled water circuits, a flow typically of 6°C and a return of 11°C. However, for large heating schemes, office blocks and group or district heating schemes serving separate blocks of flats, it is economic to increase the temperature of the flow to reduce the mass flow rate of water. This reduces the diameter of the main pipes and so reduces the capital expenditure. These systems are pressurized so the water temperature can be raised above the normal atmospheric boiling temperature of 100°C, making them either medium- or high-temperature circuits. However, the secondary circuits are often radiator heating circuits, where the temperatures are the standard low temperatures of 80°C and 70°C, or underfloor heating systems with even lower temperatures. The primary main circuit has both a greater mass flow rate and a higher potential temperature drop than the secondary circuit, i.e. it has greater potential thermal power:

$$\dot{m}_p C_{pp}(t_{pf} - t_{pr}) \gg \dot{m}_s C_{ps}(t_{sf} - t_{sr})$$

where the subscripts refer to the primary and secondary circuits.

The ideal design is to have the circuits isolated from each other by using a heat exchanger, or non-storage calorifier, because if there is a failure of a valve on the primary or secondary side of the heat exchanger then the two circuits are still isolated. The medium-temperature water would be dangerous in a low-temperature circuit, flashing to steam where the pressure dropped to ordinary atmospheric.

However, injection circuits have been used here. They are also useful for chilled ceilings. The main chilled water circuit will vary between 6°C and 11°C for the cooling coil. However, the chilled ceiling supply and return temperatures are typically 16°C and 18°C respectively.

An injection circuit uses a three-port valve or a two-port valve. But whereas a compensator circuit, which uses a three-port valve, can vary from fully open (to let through the boiler flow water at about 80°C), an injection circuit needs extra pipework and controls to inject just a fraction of the primary water. Figure 13.27 shows one form of injection circuit due to Samson Controls. The balance pipe stops interaction between the primary and secondary circuits and allows surplus water to bypass the three-port valve when it is

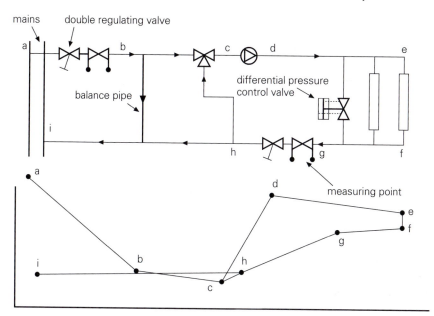

Fig. 13.27   An injection circuit and pressures. (Courtesy Samson Controls)

reducing its input from the primary circuit. The second bypass has a differential pressure control valve to regulate the pressure as the loads, with two-port valves, reduce.

13.20   Reverse return systems

Reverse return systems are sometimes used for self-balancing. Figure 13.28 shows a reverse return system for three branches, each with two resistances, and flows V_1, V_2 and V_3. This method can be used on radiator pipework, but the extra pipework involved and its expense means that it is rarely used, although it is sometimes used on boiler and chiller circuits. The resistances in Fig. 13.28 are all R for ease of calculation. However, branches *bc*, *de* and *fg* each represent a boiler and valve. The system is assumed to be self-balancing as the resistance through any branch is $4R$, hence it is assumed that the flows are equal. This is not the case, as network analysis shows. (Note that the branches *bc*, *de*, and *fg* are not in parallel. We have to consider the parallel circuits as being between *b* and *g*.) In Fig. 13.28 if we wrongly assume that the flows are equal, i.e. $V_1 = V_2 = V_3 = V$ then the pressure drops across *bceg* and *bdeg* are

$$\Delta P_{bceg} = 3RV^2 + R(2V)^2 = 7RV^2$$

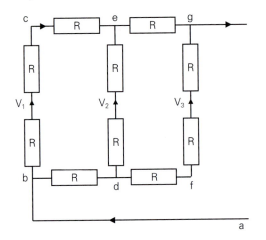

Fig. 13.28   A reverse return system.

and

$$\Delta P_{bdeg} = R(2V)^2 + 2RV^2 + R(2V)^2 = 10RV^2$$

This cannot be true, as the pressure drop between the same points must be the same, even via different routes. As the circuit is symmetrical about *de*, the flows through *bc* and *fg* are equal, call them V_{outer}; and the flow through *de*, call it V_m, will satisfy

$$2V_{outer} + V_m = 3V$$

Then as

$$\Delta P_{bceg} = \Delta P_{bdeg}$$

we have

$$3RV_{outer}^2 + R(V_{outer} + V_m)^2 = R(V_{outer} + V_m)^2 + 2RV_m^2 + R(V_{outer} + V_m)^2$$

$$3V_{outer}^2 = 2V_m^2 + (V_{outer} + V_m)^2$$

$$3V_m^2 + 2V_{outer}V_m - 2V_{outer}^2 = 0 \qquad (13.21)$$

We can relate the volume flows using

$$2V_{outer} + V_m = 3V \qquad (13.22)$$

where V is the volume flow we would like through each branch. Substituting this in equation (13.21) yields

$$3V_m^2 + 12VV_m - 9V^2 = 0$$

Solving this quadratic gives $V_m = 0.65V$ (the unrealistic root is $-4.65V$). Equation (13.22) then gives $V_{outer} = 1.175V$.

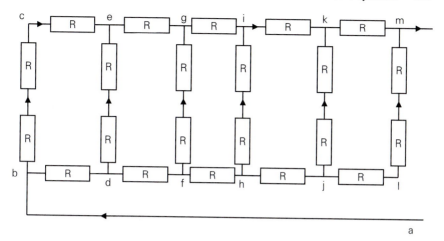

Fig. 13.29 A six-branch reverse return circuit.

Table 13.2 Flows in a six-branch reverse return network

Branches 1 and 6	Branches 2 and 5	Branches 3 and 4
2.24	0.683	0.077

Bigger reverse return networks can be analysed by using these two principles:

- The pressure drop between two points is the same whatever path is taken.
- Water is conserved at a node.

The first point is illustrated above and may be extended so that in Fig. 13.28 $\Delta P_{bceg} = \Delta P_{bdeg} = \Delta P_{bdfg}$ and $\Delta P_{bce} = \Delta P_{bde}$. The second point is simply water conservation: the water flowing into a node equals the water flowing out. Equations can then be set up and solved on Microsoft's Excel spreadsheet using the Solver routine. Alternatively, numerical techniques can be used, such as the Hardy–Cross method [18]. Using the spreadsheet method for the six-branch network in Fig. 13.29, a reverse return system with branch resistances $2R$ and connections R (as for the three-branch network earlier), the flows are as shown in Table 13.2. They are quite different!

But it would be wrong to regard reverse return networks as unbalanced, although the flows are symmetrical about the middle branch or branches (depending on whether there is an odd or even number of branches). If in Fig. 13.28 the resistances in *bd* and *eg* are reduced to $0.25R$, to produce an equal pressure drop in the horizontal connections to the branches, then the reverse return system is balanced and all the flows through the branches

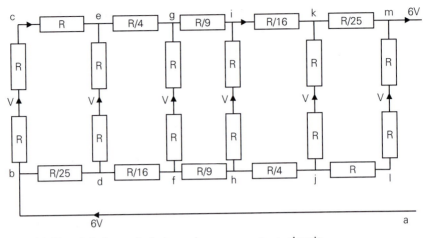

Fig. 13.30   A balanced six-branch reverse return circuit.

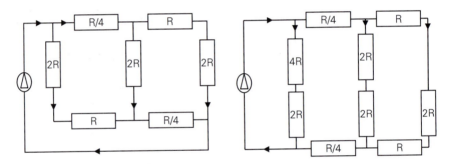

Fig. 13.31   Comparing a balanced ladder circuit (right) with a reverse return circuit.

are equal. Likewise with the six-branch reverse return circuit, the flows are balanced if the connections between the branches vary from $R/25$ to R (Fig. 13.30).

Comparing the reverse return circuit resistance to that of a normal ladder circuit, the equivalent resistance of the balanced reverse return circuit, R_{eqrr}, in Fig. 13.31 (each connection between the branches producing a pressure drop of RV^2) is

$$R_{eqrr}(3V)^2 = 4RV^2$$

so

$$R_{eqrr} = 4R/9$$
$$= 0.44R$$

For a normal ladder circuit of three branches, with balancing valves, also shown in Fig. 13.31, the equivalent resistance, R_{eqladder}, is

$$R_{\text{ladder}}(3V)^2 = 6RV^2$$

giving

$$R_{\text{ladder}} = 2R/3 = 0.67R$$

Hence the equivalent resistance and, as the circuits have the same volume flows, the pump power for the reverse return circuit are only 66% of their values in the normal ladder circuit and no balancing is required!

References

[1] BRECSU (1996) Variable flow control. BRECSU General Information Report 41, Building Research Establishment, Watford, UK.

[2] CIBSE (1977) CIBSE Guide, Section C4, Flow of fluids in pipes and ducts, Chartered Institution of Building Services Engineers, London.

[3] Liengme, B. V. (1997) *A Guide to Microsoft Excel for Scientists and Engineers*, Edward Arnold, London.

[4] Miller, D. S. (1990) *Internal Flow Systems*, 2nd edn, BHRA, Cranfield, UK.

[5] Shepherd, K. J. and Levermore, G. J. (1999) In *Proceedings of the 3rd International Symposium on Heating, Ventilation and Air Conditioning*, Shenzhen, China.

[6] McLaughlin, R. K., McLean, R. C. and Bonthron, W. J. (1981) *Heating Sevices Design*, Butterworths, London.

[7] Miller, L. M. (1976) *Students textbook of heating ventilating and air conditioning*, Technitrade Journals, London.

[8] Clark, D. R. (1985) *HVACSIM+ Building Systems and Equipment Simulation Program – User Guide*, US National Bureau of Standards, Gaithersburg MD.

[9] Mei, L. and Levermore, G. J. (1999) Modelling and simulation of a VAV test rig. In *Proceedings of the 3rd International Symposium on Heating, Ventilation and Air Conditioning*, Shenzhen, China.

[10] Underwood, C. (1999) *HVAC control system modelling analysis and design*, Spon, London.

[11] CIBSE (2000) CIBSE Applications Manual for Automatic Controls, Chartered Institution of Building Services Engineers, London.

[12] Grundfoss (1997) Frequency converter: design and functioning, Grundfoss literature FM-010/k.

[13] Gorman, S. M. (1996) Variable speed drives: practical guide to electrical engineering for HVAC&R engineers. Supplement to *ASHRAE Journal*, Nov.

[14] C. Parsloe, private communication.
[15] Jones, W. P. (1997) *Air Conditioning Applications and Design*, 2nd edn, Edward Arnold, London.
[16] Petitjean, R. (1994) *Total hydronic balancing: a handbook for design and troubleshooting of hydronic HVAC systems*, Tour & Andersson, Boras, Sweden.
[17] Avery, G. (1998) Controlling chillers in variable flow systems. *ASHRAE Journal*.
[18] Drake, J. M. K. (1983) *Essentials of Engineering Hydraulics*, 2nd edn, Macmillan, London.

14
Monitoring and targeting

Two of the great benefits of a BEMS are its ability to communicate and its ability to monitor plant (Chapter 1). The monitoring of plant performance often relates to logging sensor data for later transposing to graphs, the logging only being limited by the memory size of the outstations. Frequency of monitoring is in the hands of the operator and too much stored data can be as much of a handicap as lack of data.

However, this chapter is not primarily concerned with the logging of room temperature sensors, flow water temperature sensors, and so on, but with the overall performance of the plant and the building, for energy efficiency and maintenance.

14.1 Monitoring and alarms

The most basic part of monitoring relates to the standard alarms which, for instance, indicate that a sensor is reading high or low or is out of limits. Similar alarms can also apply to loops, drivers and digital inputs. But, in addition, drivers can also have maintenance interval alarms. A **logic hours run module** is required to facilitate this (Fig. 14.1). This configuration enables an alarm to be raised when an item of plant – here a pump with a binary driver – has gone beyond its period for maintenance. With this facility, **condition-based maintenance** can be implemented.

The logic hours run module counts the hours for which the source bit, S, is on and sends the output, O, to the hours run destination analogue address. This module performs a routine to check whether the hours run has exceeded the interval limit which has been set (input at I), and if so, it will set the

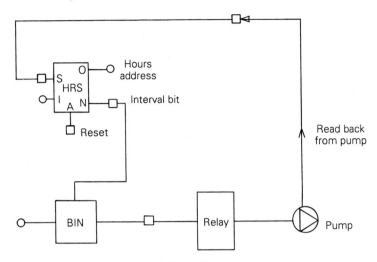

Fig. 14.1   Maintenance internal alarm for a pump.

interval bit, from output N, to on or 1. When the reset input node, connected to A, is set to on or 1, then the hours run total is set to zero.

With other logic modules, the logic hours run module can monitor the time during occupancy that the room temperature is above a certain limit, say 22°C, or below a limit, say 18°C. Both accumulated times are useful for monitoring the efficiency and performance of the heating system.

14.2   Energy audit

A BEMS can take monitoring a stage further than simply logging hours run of plant and alarms. Fuel and electricity meters can be interfaced serially to an outstation. Digital signals are sent from the meter counting the dial movements. This energy data can regularly and dynamically update a spreadsheet (see later) in the central station. Various charts and diagrams can be produced from the spreadsheet on the window-based BEMS software (Fig. 14.2).

Working to the axiom that a diagram is worth a page of writing and many pages of numbers, a Sankey diagram can be drawn from all the data collected by the BEMS [1, 2]. It has scaled arrows to indicate where the energy is going (Fig. 14.3).

14.3   Electricity monitoring

Electricity is comparatively easy to monitor. The main intake meter, belonging to the electricity company, will allow total consumption to be monitored if it has a port on it for a serial connection, e.g. by RS232-C, or other

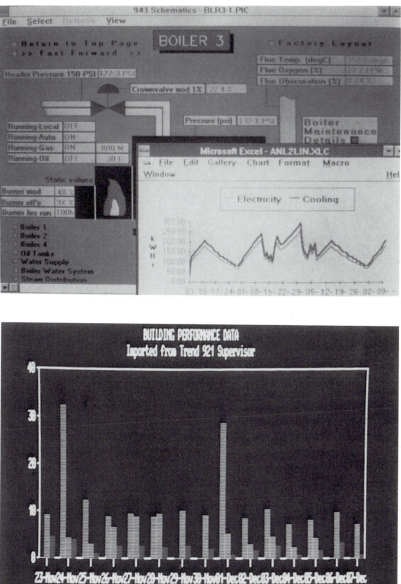

Fig. 14.2 Spreadsheet screens.

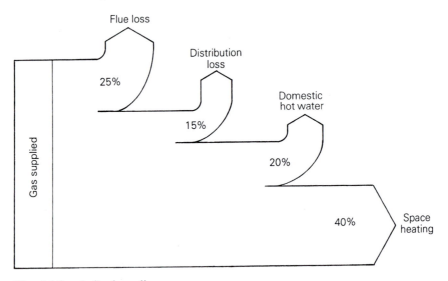

Fig. 14.3   A Sankey diagram.

suitable means, for pulse signals to be sent to a BEMS outstation. For the discerning energy manager, submeters can be installed around the building to determine where the electricity is used. Many BEMS manufacturers can supply serially interfaceable meters for this purpose.

Instead of fixed meters, portable monitoring equipment, not interfaced to the BEMS, can be put onto various items of equipment. This uses clamps which go around the live electrical cable forming the primary winding of a **current transformer**, the secondary winding being in the clamps connected to an ammeter device in the equipment. With the voltage of the supply being almost constant, simple computation allows the measurement of the electrical consumption, provided the power factor is known.

If the main meter is a maximum demand (MD) meter, then careful interfacing to the BEMS is required. This is because the MD meter consists of two parts: a unit measuring consumption (kWh) like an ordinary meter, and one which measures the kilowatt demand (some companies use KVA demand) over half-hour intervals. The BEMS outstation will require a pulse from the kWh meter and another from the demand meter, indicating the start of each half-hour interval, if it is to monitor the MD. If there is only one pulse output from the meter, the configuration in Fig. 14.4 can be used for monitoring the meter, although synchronization with the meters and MD will not be exact.

At the heart of the configuration is a **logic counter module** (CNTR) that counts pulses, at S, which are on or off for more than 15.6 ms. This pulse

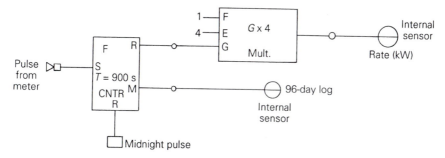

Fig. 14.4 Meter monitoring configuration.

time gives a limiting upper frequency response of 30 Hz. Notice that this module monitors the digital input channel at 15.6 ms intervals, much faster than the temperature sensors are monitored, as it is independent of the sequence, or program, cycle time which is much longer, typically 5 s. A **fast sequence cycle time** is available if prompt action is required once a pulse has been counted, to avoid the delay of the normal cycle time.

Outputs from the module give the cumulative total, from M, and the consumption rate, from R (the change in the cumulative consumption over a given period which is the average kilowatt demand). In the CNTR is a scaling factor, F, which relates each pulse to kWh, and a reschedule time, T, which defines the period (s) over which the rate, or demand, is measured. In Fig. 14.4 the CNTR is reset by a midnight pulse, at R, to count each day's consumption, which is logged at the internal sensor. For a typical log, this would give 96 days' data. The multiplication function module MULT determines the kW rate from the rate output, R. Here T is set to 900 s (15 min), so the MULT module multiplies this by four to give the kW rate.

The MD consumer is charged for the units consumed, as well as for the maximum demand recorded. Such meters are only used for large users, with supplies usually above 40 kVA [3].

The demand meter has a kW (or kVA) circular dial and a hand moves as the consumption in kWh is recorded. This hand pushes another hand. After a half-hour interval, the pushing hand is reset to zero, but the pushed hand stays at the maximum demand reading. The pushing hand then starts measuring the consumption for the next half-hour. The scale of the MD meter is in kW, and the pushed hand is read and reset by the meter reader each month.

The half-hourly resetting signal from the MD meter can be interfaced to the BEMS in a similar manner as in Fig. 14.4, using a CNTR module, so that the BEMS can monitor the demand from the meter. In Fig. 14.4 there is no separate half-hour pulse from the meter, so the BEMS demand may not be synchronized with the meter demand, and this could hinder load shedding.

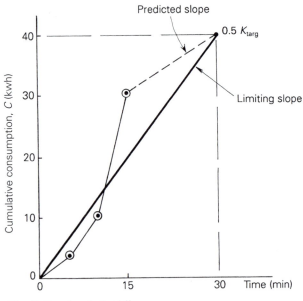

Fig. 14.5   Predictive load shedding.

14.3.1   Load shedding

If the kilowatt electrical demand goes too high in any half-hour during a month, then the user has to pay for this worst half-hour in the month. It is therefore worthwhile to keep the maximum demand at a low and steady level. A popular use of a BEMS is to monitor the demand and switch off plant on a priority and size basis. Two ways of configuring a BEMS to load shed are the predictive method and the offset method [4, 5].

14.3.2   Predictive method

The cumulative electrical consumption is shown in Fig. 14.5, as sampled by a BEMS every 5 min, for the first 15 min at a site. The aim is to keep the kilowatt demand to less than 80 kW, or a consumption of 40 kWh in half an hour. Already the consumption is at 30 kWh and a slope of consumption can be predicted for the demand not to exceed 80 kW.

The kilowatt demand can be determined from the rate of change of the consumption with time, that is

$$K = \frac{dC}{dT}$$

where K is demand (kW) and C is cumulative consumption (kWh).

If the BEMS monitors the consumption at regular intervals of δT minutes (5 min in Fig. 14.5) in each half-hour period, then at the ith interval the kilowatt demand, K_i, is

$$K_i = \frac{60 \ (C_i - C_{i-1})}{\delta T} \tag{14.1}$$

where C_i is the cumulative consumption at interval i from the start of the half-hour sampling period (kWh).

The kilowatt demand at the end of the half-hour period is

$$K_n = \frac{60 C_n}{n \ \delta T} = 2C_n$$

where n is the number of intervals in one half-hour.

Figure 14.5 shows the limiting slope, determined by the target kilowatt demand, K_{targ}, here 80 kW. If the consumption goes above this line then the consumption has to be reduced. The minimum electrical load that can be left switched on is given by the predicted slope:

$$\text{predicted slope} = \frac{60 \ (0.5 K_{targ} - C_i)}{(n - i)\delta T}$$

which might be

$$\frac{60 \ (0.5 \times 80 - 30)}{(6 - 3)5} = 40 \ \text{kW}$$

The consumption slope during the previous 5 min can be calculated from equation (14.1):

$$K_3 = \frac{60 \ (C_3 - C_2)}{5}$$

$$= 12(30 - 10)$$

$$= 240 \ \text{kW}$$

If this rate of consumption continued, 200 kW of equipment would have to be switched off. Conversely, if the rate dropped below 40 kW, equipment could be switched on to the level of 40 kW.

14.3.3   Offset method

With the offset method, the limiting slope is raised at the start of the demand interval by an offset (Fig. 14.6). Loads are only shed when the offset limiting slope is exceeded. This allows for larger kilowatt demands earlier in the

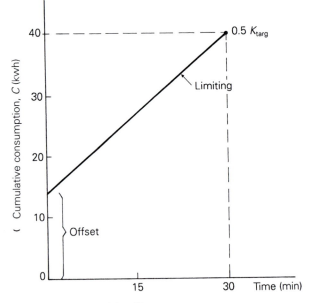

Fig. 14.6   Limiting slope with offset.

interval than the previous method. However, the offset limiting slope is less steep, so overall it has a lower kilowatt demand.

14.3.4   Shedding priority

Load shedding primarily saves maximum demand charges, although it may also save some electrical energy. But the decision has to be made as to which items of plant should be switched off, and in which order [4, 5]. But it is worth cautioning that careful thought should be given to what loads are shed, and indeed whether load shedding is carried out.

   Frequently, the decision is made to switch air-conditioning refrigerant compressors. These are not tolerant of too frequent switching and can easily be severely damaged. Induction motors also do not respond well to frequently being turned on and off, due to high inductive currents, although 'soft starters' greatly reduce these currents. The switching off of lights must not cause a safety hazard, so this is a further restriction.

14.4   Heat monitoring

Compared to electricity monitoring, heat monitoring is difficult. This is because both flow rate and temperature difference must be measured. Either

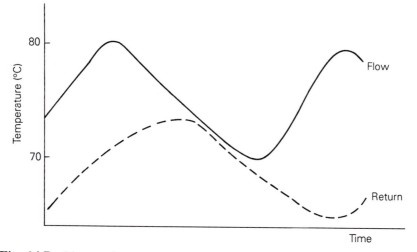

Fig. 14.7   Flow and return temperatures.

an interfaceable heat meter is used, which integrates the two measurements, of the BEMS directly reads the flow rate and temperature difference and uses its own configured software to calculate the heat flow.

For a hot water heating circuit, the temperature of the flow and return will most probably be measured in the plant room close to the BEMS outstation. Unfortunately, due to the transit time between the outgoing flow and the incoming return, the two will be out of synchronization (Fig. 14.7).

The instantaneous temperature difference varies between 10 K and 0 K, so the corresponding metered heat would vary in a similar way. Also a small error of temperature could correspond to a considerable error of heat output. For a heat supply of 100 kW with a 10 K temperature difference between flow and return, a 0.1°C error represents 1 kW.

Flow rate must also be measured and integrated with temperature difference to determine the amount of heat. There are several flow measuring devices for water systems. Many are based on differential pressure measurement around a constriction in the flow. The pressure differential is related by Bernoulli's equation to the fluid velocity [26, 27]. Such devices are the **orifice plate**, **venturi**, **Dall tube** and **nozzle**. The orifice plate is the most widely used and it is cheap.

Pressure differential has to be changed into an electrical signal for a BEMS to interface to one of the above meters, but an **axial flow turbine meter** directly produces an electrical signal. The turbine is a multi-bladed rotor suspended in the fluid stream with the axis of rotation parallel to the direction of flow. The rotor has between four and eight blades, each made of ferromagnetic material which forms a magnetic circuit with a permanent magnet and coil within the meter housing.

Turbine flowmeters and differential pressure meters can get clogged up if the water is not maintained in a clean condition. So there is a trend towards using non-intrusive flowmeters, which do not get clogged up. Examples are **electromagnetic**, **ultrasonic** and **cross-correlation** flowmeters.

The electromagnetic meter uses the Faraday effect, where a moving conductor in a magnetic field generates an electric current. The conductor in this case is the water in the pipe. In practice the water will not be pure, so it should have a small conductivity.

Bentley deals with flow measurement systems in detail, some of which have not been mentioned here [6]. Whatever the method of flow measurement, heat meters are expensive and from past experience they need to be well maintained and proven to be reliable. Interfacing of heat meter pulses will be very similar to the configuration for electricity meter pulses.

14.5   Spreadsheets and graphs

BEMS central stations can have spreadsheet programs and databases. Spreadsheets are tables with rows and columns of numbers. A spreadsheet program is like a word processing package except that numbers are manipulated rather than words [24], and it can perform much number crunching according to formulae entered by the user. From these numbers it is possible to draw graphs, histograms and pie charts, so a spreadsheet is a useful tool for a BEMS user, especially to manipulate energy data and temperatures. An increasing number of BEMS manufacturers are interfacing their central station software to spreadsheets, so the spreadsheet can automatically be updated with meter readings, temperatures from outstations, etc., when the central station is dialled up by the outstations to give their logged data. From the meter reading, the spreadsheet could calculate the latest consumption and readily display the latest consumption graph.

A distinct advantage of spreadsheets is that **what-if questions** can be analysed. If the cost of fuel rises, a spreadsheet of monthly consumptions can work through all the data and predict future monthly costs.

Common spreadsheets are SuperCalc, Microsoft Excell and Lotus 1-2-3. The earliest spreadsheet was VisiCalc (visible calculator) to run on Apple II PCs in the late 1970s. Lotus 1-2-3 was then developed to run on IBM-compatible machines.

The basic spreadsheet is a matrix of headed columns, lettered for identification, and rows, numbered for identification, as shown in Fig. 10.8(a). Each location is called a **cell**. The formulae for the spreadsheet calculations are entered on another sheet, and the user can view either the 'results' sheet or the formulae sheet. The two sheets are shown in Fig.10.8(b). The figure also shows a portion of output from a spreadsheet for oil consumption at a large boilerhouse.

Here the first column is headed Date and the first cell under it is 03-May-00. If this is called cell A8, then the 'stored' oil column is B and the B8 cell has 30 594 litres in it. Using this notation the formula for calculating some of the other cell values can be derived. The formula for Litres used, column F, is

$$= B(n) - B(n + 1) + C(n + 1) \text{ for cell } F(n + 1)$$

where n is a row number. We can check this by noting that the fuel used against 31-May-00, F9, is

$$30\ 594 - 55\ 301 + 54\ 532 = 29\ 825$$

as is F9. Also the price per litre is entered and the total cost calculated. The YRT columns refer to yearly running totals, explained shortly.

Generally spreadsheet programs are limited to the amount of data they can hold on their spreadsheet matrices. For instance, it can hold as much fuel data (Fig. 14.8) as there are rows; perhaps as many as a thousand rows. For details of a number of sites and large amounts of data for reference, a **database program** can be used. The information is held in files, rather like a card index, which can be quickly manipulated. Frequently outside the BEMS area, databases are used to hold mailing lists and account details. A commonly used database is Microsoft's Access. However, the art of good monitoring is to avoid 'data diarrhoea' and to transform columns of data into readily understood graphs and charts.

Often consumption data is presented as a histogram (Fig. 14.9); a spreadsheet program can produce histograms very quickly. But it is difficult to compare one year's consumption with another's. Figure 14.9 shows that in the current year, some months are higher and some lower than in the previous year.

A useful technique is to develop a trend graph, indicating whether the consumption is tending to increase or decrease. Such a graph is the yearly running total graph (YRT). This technique is also used for analysing company sales, especially where seasonal factors influence quantities sold, and a similar method applies to the determination of the retail prices index (RPI). Figure 14.10 shows a YRT from data used in a later example.

The YRT calculates the consumption for a complete year (all 12 months). The YRT for the 13th month is then calculated using the formula

$$YRT_{13} = YRT_{12} + M_{13} - M_1$$

where YRT_{13} = yearly running total for the 13th month
YRT_{12} = yearly running total for the previous 12 months

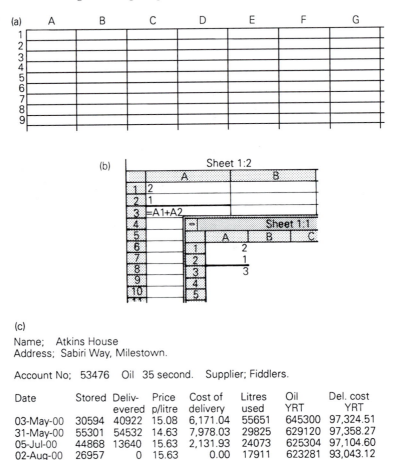

Fig. 14.8   Spreadsheets: (a) empty cells, (b, c) showing a site's consumption.

M_{13} = monthly consumption for the 13th month
M_1 = monthly consumption for the first month
In more general terms, for the nth month the formula is

$$YRT_n = YRT_{n-1} + M_n - M_{n-12} \tag{14.2}$$

If one month's consumption is above the same month's consumption a year ago, then the graph goes up slightly. If it is less then the graph slopes down. In this way, seasonal variations are smoothed out.

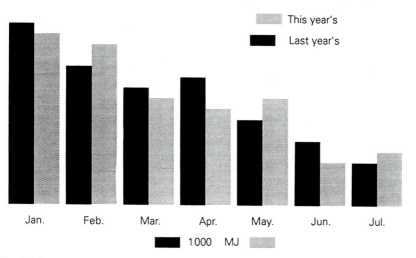

Fig. 14.9 A histogram of consumption.

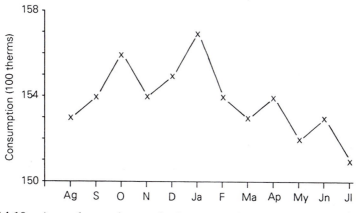

Fig. 14.10 A yearly running total of consumption.

Example 14.1
Calculate the yearly running totals for the following data.

Month	Consumption (100 therms)		
	1990/91	*1991/92*	*1992/93*
August	4	3	4
September	5	6	6
October	7	9	5
November	11	9	

Month	Consumption (100 therms)		
	1990/91	*1991/92*	*1992/93*
December	20	21	
January	25	27	
February	27	24	
March	18	19	
April	14	13	
May	8	6	
June	8	9	
July	7	5	
Total	154		

Solution

From equation (14.2) the YRT for August 1990 is

$$YRT_{Aug91} = YRT_{July91} + M_{Aug91} - M_{Aug90}$$

$$= 15\,400 + 300 - 400$$

$$= 15\,300$$

Similarly, for the following months one gets these values

Month	YRT (100 therms)	
	1991/92	*1992/93*
August	153	152
September	154	152
October	156	148
November	154	
December	155	
January	157	
February	154	
March	153	
April	154	
May	152	
June	153	
July	151	

Overall the trend is down, as is shown by the graph of the YRT in Fig. 14.10.

It is interesting to note that YRTs can be determined very easily from monthly meter readings. One simply takes last year's reading away from this year's to obtain the YRT.

The variations in consumption may be due to the weather, in which case the YRT of the monthly average temperature, $\overline{t}_{ao}$ will also show a similar variation.

cusum, the cumulative sum of savings is a similar method to the YRT but goes further and analyses the actual savings made [25].

14.6   Degree-days

The variation in the heating consumption is primarily due to the coldness of the weather. A common measure of weather coldness is the **degree-day**. Degree-days derive from the energy consumed in heating. From the integrated balance equation, the energy consumed is

$$\int \dot{m} C_p (t_f - t_r) \, dT = \int B(t_m - t_{ai})^{1.3} \, dT = \int A(t_{ai} - t_{ao}) \, dT$$

Here the storage effects in the fabric have initially been ignored and the time period assumed to be reasonably long, i.e. greater than a week. The above equation may be turned into a discrete equation:

$$\sum_{i=1}^{n} \dot{m} C_p (t_{fi} - t_{ri}) \Delta T = \sum_{i=1}^{n} (t_{mi} - t_{aii})^{1.3} \Delta T = \sum_{i=1}^{n} A(t_{aii} - t_{aoi}) \Delta T \quad (14.3)$$

where the time period is $n\Delta T$. This yields an equation with the average values over the same period:

$$\dot{m} C_p n \Delta T (\overline{t}_{fn} - \overline{t}_{rn}) = B n \Delta T (\overline{t}_{mn} - \overline{t}_{ain}) = A n \Delta T (\overline{t}_{ain} - \overline{t}_{aon}) \quad (14.4)$$

where $\overline{t}_{fn}$ is the average flow water temperature over the period $n\Delta T$; $\overline{t}_{rn}$, $\overline{t}_{mn}$, $\overline{t}_{ain}$ and $\overline{t}_{aon}$ are average values over the period $n\Delta T$. This is in fact the balance equation, although the average and the period n symbols are normally omitted, and $n\Delta T$ is cancelled out.

To determine the energy consumption over this period, $n\Delta T$, then the boiler output, the heat emitter output or the fabric heat loss parts of the equation can be used. But the first two need the average values $\overline{t}_{fn}$, $\overline{t}_{rn}$, $\overline{t}_{mn}$, which need to be determined from extensive monitoring. It is therefore easier to use the fabric heat loss part as this only relies on $\overline{t}_{aon}$ and $\overline{t}_{ain}$. Temperature $\overline{t}_{aon}$ need not even be monitored as it can be readily obtained from meteorological records, and $\overline{t}_{ain}$ should be maintained at a fairly constant level, at least during occupancy. Note that this average inside temperature is the inside temperature produced due to the heating alone. However, there will also be heating due to internal gains, such as lighting, equipment, people and sunshine. Although these gains are not considered in the heating system design calculation, if they are utilized, even unintentionally, they have to be considered in the energy consumption determination. This is done by redefining the components of the inside temperature as $t_{ai\ cont}$, the inside temperature at which the building is maintained by the controls, and t_g, a temperature due to the gains. So

$$\overline{t}_{ain} = t_{ai\,cont} - t_g$$

$\overline{t}_{ain}$ is often called the **balance point temperature** of the building, t_{bal}. The balance point temperature, t_{bal}, is so named because when the outside temperature, t_{ao}, is above t_{bal}, then no heating is required, and conversely when t_{ao} is below it, then heating is required.

From equation (14.3), for a continuously heated building, the energy consumed in maintaining a controlled inside temperature of $t_{ai\,cont}$ over N days is

$$24A \sum_{i=1}^{N} (t_{bal} - t_{aoi})$$

where $\Delta T = 24$ h and

$$t_{ai\,cont} = t_{bal} + t_g$$

The term

$$\sum_{i=1}^{N} (t_{bal} - t_{aoi}) \tag{14.5}$$

is the degree-days for N days, referred to the base temperature, t_{base}, of t_{bal}. Sometimes the 24 hours in the day is implied in the degree-day units but here, as equation (14.5) shows, degree-days are simply the sum of temperature differences and so the units are temperature difference (K).

Traditionally t_{base} has been taken as 15.5°C (60°F). This is derived from work done many years ago, primarily in the 1930s and 1940s [7], which showed there was a good correlation between monthly energy consumption and monthly degree-days with a base temperature of 15.5°C. For hospitals, with higher inside temperatures, a base temperature of 18.5°C is taken [8]. Monthly degree-day figures are published and available from the Department of the Environment Energy Efficiency Office and various journals and magazines.

Correction factors are available [9] to adjust a yearly degree-day total to a total referred to a different base temperature, t_{base}. This may be necessary as the base temperature should be equal to a building's balance temperature, t_{bal}. For modern, well-insulated buildings t_{bal} can be around 10°C, well below the standard base temperature of 15.5°C. The gains (solar, people and equipment) can supply sufficient heat to maintain the building at a comfortable temperature. However, these correction factors only apply to yearly degree-day totals and are for the determination of annual fuel consumption:

$$\sum_{i=1}^{N} (t_{base} - \Delta t_{base} - t_{aoi}) = \varphi \sum_{i=1}^{N} (t_{base} - t_{aoi}) \tag{14.6}$$

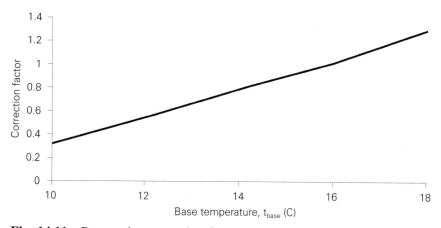

Fig. 14.11 Degree-day correction factors.

where φ is the base temperature correction factor, and Δt_{base} is a reduction in t_{base}, so that

$$t_{base} - \Delta t_{base} = t_{bal}$$

Values of φ can be calculated, but equation (14.6) is correct for only one value of N, the annual number of 365 days. Figure 14.11 shows a graph of φ for various base temperatures, derived from reference [8]. Alternatively, a formula derived by Hitchin [10] can be used for calculating the monthly degree-days to a different base temperature:

$$\text{degree days per month} = N_{month} \frac{(t_{base} - \overline{t}_{ao\ month})}{1 - \exp(-k)(t_{base} - \overline{t}_{ao\ month})} \quad (14.7)$$

where N_{month} = number of days in the month considered
$\overline{t}_{ao\ month}$ = average monthly outside air temperature (°C)
k = a constant which varies with site

Site	k
Heathrow	0.66
Manchester	0.7
Glasgow	0.74
Cardiff	0.78
Birmingham	0.66
Mean	0.71

Equation (14.6) is incorrect for general values of N different from 365, as can be seen in the following example.

Example 14.2
The average outside temperatures for three days are

$$\bar{t}_{ao1} = 10.5°C, \ \bar{t}_{ao2} = 12.5°C, \ \bar{t}_{ao3} = 6.5°C$$

and $t_{bal} = 13°C$. Calculate the degree-days to a base temperature of 13°C and $t_{base} = 15.5°C$ and determine the values of φ from equation (14.6).

Solution
For the first day, the degree-days are 5 K and from equation (14.6):

$$(15.5 - 2.5 - 10.5) = \varphi(15.5 - 10.5)$$

$$\varphi = 0.5$$

For the second day, the degree-days are 3 K, and for the first two days equation (14.6) gives

$$(2.5 + 0.5) = \varphi(5 + 3)$$

$$\varphi = 0.375$$

On the third day, the degree-days are 9 K, and for all three days equation (14.6) gives

$$(2.5 + 0.5 + 6.5) = \varphi(5 + 3 + 9)$$

$$\varphi = 0.56$$

Clearly, as Example 14.2 shows, the application of a correction factor is mathematically wrong for all but defined periods such as a year.

Degree-days can be measured with a BEMS, and the configuration for this is shown in Fig. 14.12. Loop L1 has its setpoint at the required base

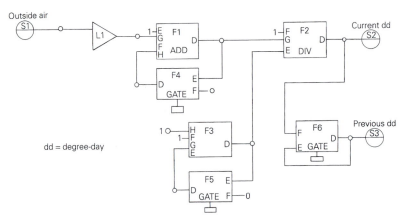

Fig. 14.12   Degree-day (dd) configuration.

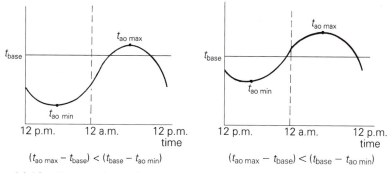

$(t_{ao\ max} - t_{base}) < (t_{base} - t_{ao\ min})$ $(t_{ao\ max} - t_{base}) < (t_{base} - t_{ao\ min})$

Fig. 14.13 Degree-day calculations.

temperature and its error signal output will be the temperature difference. ADD module F1 then adds this to the cumulative temperature difference (which is derived via gate F4), and the DIV module F2 divides by the number of readings outputted from the loop. ADD module F3 counts the readings outputted from the loop. With input H set at 1 and G connected to output D, the address node from D is incremented by 1 each time the module is serviced, at the same service step as the other modules and loop. Each gate will be pulsed at midnight to set the calculation to zero, except for GATE module F6, which will send the current degree-day reading on internal sensor S2 to internal sensor S3.

A simpler configuration, based on minimum and maximum function modules, would provide adequate though less accurate results. But this is the traditional way of measuring degree-days with an ordinary minimum–maximum thermometer, and the definition of the average daily temperature [25] is

$$\bar{t}_{ao} = \frac{t_{ao\ max} + t_{ao\ min}}{2}$$

where $t_{ao\ max}$ = maximum daily temperature (°C)
   $t_{ao\ min}$ = minimum daily outside temperature (°C)
This is satisfactory if $t_{ao\ max}$ is below t_{base} but when $t_{ao\ max}$ is just above t_{base} (Fig. 14.13), the degree-days for the day, DD, are

$$DD = \frac{(t_{base} - t_{ao\ min})}{2} - \frac{(t_{ao\ max} - t_{ao\ min})}{4}$$

If $t_{ao\ min}$ is just below t_{base}, then

$$DD = \frac{(t_{base} - t_{ao\ min})}{4}$$

These empirical equations can be appreciated if it is assumed that the daily temperature variation is sinusoidal, and numerical integration, such as Simpson's rule, is used to determine the average temperature under the t_{base} line.

The daily variation in temperature has been shown as sinusoidal in Fig. 14.13, as it is often regarded, although this is rather an approximation for the winter months [19].

14.6.1   Variable-base degree-days

Although degree-days can be useful, from earlier chapters we can see that the building internal temperature, t_{ai}, will vary with the heat load and the type of control. Fortuitous heat gains will also affect t_{ai}. It has been reported that in a reasonably well-insulated house t_{base} might be 16°C in December and 9°C in April. The difference is attributed to solar gains [9]. So to be accurate, the degree-day base for a building should be varied.

ASHRAE provides a **variable-base degree-day method** (VBDD) to reflect more accurately these gains [11]. Various factors are quoted to account for internal gains and solar gains. The monthly degree-days are calculated to variable base temperatures from the monthly average temperature by formulae similar to equation (14.6).

It is interesting that the CIBSE Energy Code [12, 28] does not use either standard degree-days or VBDD, but simply the average outside temperature. There is the justified comment that the degree-day method 'requires some measure of interpretive skill'.

14.7   Annual fuel consumption

The annual fuel consumption can be calculated from the annual degree-day total. The method is based primarily on work by Billington [13], who estimated the accuracy to be ±25%. These errors are borne out by work in the United States [11]. The method is outlined in the CIBSE Guide, Section B, Part 18 [9]. It is demonstrated in Example 14.3.

Example 14.3
Calculate the annual fuel consumption for the building with the following details:

Calculated heat loss from the building	1500 kW
Indoor design temperature	20°C
Outdoor design temperature	−1°C
London location with annual degree-days	2034 K
Heating system efficiency	0.6
Building occupied	5 days a week, 8 hours a day

The building is classified in the terminology of the guide as traditional with normal glazing, equipment and occupancy, and is regarded as heavyweight.

Solution

The CIBSE Guide (Table B18.7) [9] suggests that the average temperature rise maintained by miscellaneous gains alone is 4°C. In terms used earlier, $t_{ai\ cont}$, the inside temperature at which the building is maintained by the controls, t_g, a temperature due to the gains, and t_{bal}, the balance point temperature, are related as follows:

$$t_{bal} = t_{ai\ cont} - t_g$$

$$t_{bal} = 19 - 4$$

$$= 15°C$$

From Fig. 14.11 the correction factor for the degree-days to the base temperature of 15°C is 0.94. If the heating system were operated continuously throughout the day, and it exactly matched the design heat loss, then it would be on at full load for the **equivalent full load operating time**, E_q, of

$$E_q = \frac{24 \times 2034 \times 0.94}{20 - (-1)}$$

$$= 2185 \text{ h}$$

But the heating system is operated intermittently and the occupation is not continuous. The CIBSE Guide gives the following correction factors to account for this:

5-day week	0.85
Intermittent heating with system with	
a long time lag and a heavy building	0.95
8-hour day	1.00

So E_q becomes

$$2185 \times 0.85 \times 0.95 \times 1.00 = 1764 \text{ h}$$

It is interesting to note that, because of the building's thermal storage, the 5-day week only reduces the heat demand by 0.85, not by the direct ratio of $5/7 = 0.71$. From the equivalent hours, the annual heat requirement is

$$1764 \times 1500 = 2.64 \times 10^6 \text{ kWh}$$

$$= 9504 \text{ GJ}$$

The annual fuel requirement will therefore be

$$\frac{9504}{0.6} = 15\,840 \text{ GJ}$$

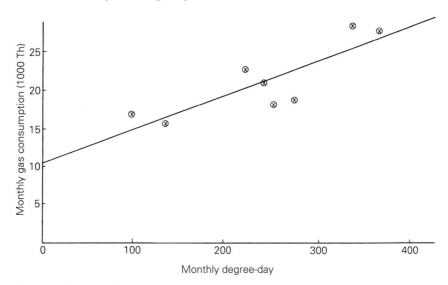

Fig. 14.14   A performance line.

14.8   The performance line

Besides using degree-days for determining the annual fuel consumption of a building, they can also be used for monthly targets.

From our earlier explanation of the theory of degree-days, the fuel consumption for heating a building should be related to the coldness of the weather, or degree-days. This has been demonstrated in a paper by McNair and Hitchin [14] which showed that the gas consumption in 96 dwellings in the UK was indeed related to degree-days. The following relationship was obtained:

$$C = 61 + 70 \times \text{DHL} \times \frac{\text{DD}}{2222} + 59 N_{\text{person}}$$

where C = annual gas consumption (therms yr^{-1})
     DHL = design heat loss (kW)
     N_{person} = number of persons in household
     DD = degree-days

The multiple correlation was $r = 0.74$. Hence a graph of monthly consumption against monthly degree-days should show a relationship similar to that in Fig. 14.14. The line through the points is the performance line, and it can be determined statistically by regression analysis. The slope, s, is given by

$$s = \frac{N \sum x_i y_i - (\sum x_i)(\sum y_i)}{N \sum x_i^2 - (\sum x_i)^2} \tag{14.8}$$

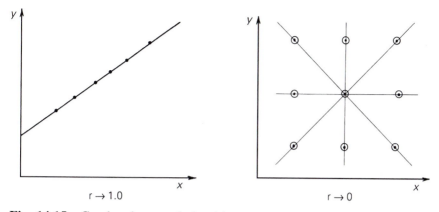

Fig. 14.15 Good and poor relationships.

where N = number of months considered
x_i = degree-days, along x-axis, for ith month
y_i = fuel consumption, along y-axis, for ith month
and the summation is over N pairs of data, x_i and y_i. The intercept is calculated from:

$$\text{intercept} = \frac{(\sum y_i)(\sum x_i^2) - (\sum x_i)(\sum x_i y_i)}{N \sum x_i^2 - (\sum x_i)^2} \tag{14.9}$$

The intercept represents the weather-independent consumption – the domestic hot water and the distribution and flue losses. It is sometimes called the **base load**. Any fuel for process heat or other weather-independent processes would also be included. The line can be determined from the monthly data in a heating season. Strictly, non-heating season data (i.e. summer consumptions) should not be included in the determination of the line as the summer consumption does not vary with weather and so gives a false impression of the line. However, problems can arise with the heating season having fractions of months.

An interesting statistical relationship which can be derived is the correlation coefficient, r. This quantifies the goodness of fit of the line to the data. For a perfect fit of the line to the data, r tends to one. If there is no fit, or relationship, as shown in Fig. 14.15, then r tends to zero. The value of r is obtained from

$$r = \frac{N \sum x_i y_i - (\sum x_i)(\sum y_i)}{\sqrt{\{N \sum x_i^2 - (\sum x_i)^2\} \{N \sum y_i^2 - (\sum y_i)^2\}}}$$

and r^2 is the **coefficient of determination,** a measure of how much data is explained by the straight-line relationship. For instance, if $r^2 = 0.6$, then 60% of the fuel consumption data is related to the weather (degree-days), but 40%

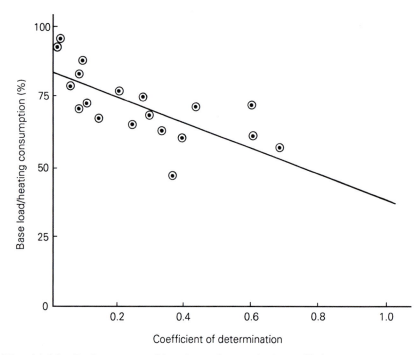

Fig. 14.16   Performance of heating schemes before efficiency measures.

is not. If the value of r, and consequently r^2, is low (and there are statistical confidence levels that can be worked out), then there is no reliable relationship between consumption and degree-days.

Surprisingly, in a study of 20 large heating installations (blocks of flats heated from a central boilerhouse), 16 of them did not have reliable relationships with the weather [15]. The r^2-values for the sites and the intercepts as a fraction of total heating season consumption are shown in Fig. 14.16. The significant value of r^2 for a statistically valid relationship is 0.57.

This poor relationship with weather suggests that the heating systems were poorly controlled. Also the high base loads implied large losses, again due in part to poor control. So a hypothesis can be inferred from this, that poorly controlled sites have high base loads and low r-values compared to similar sites that are well controlled.

There is some evidence for this hypothesis, because after the controls at the sites were improved and correctly set, as well as some minor energy conservation work, the base loads reduced and r-values improved (Fig. 14.17). Similar results have been found at a group heating scheme in north-east England and at a London school [15].

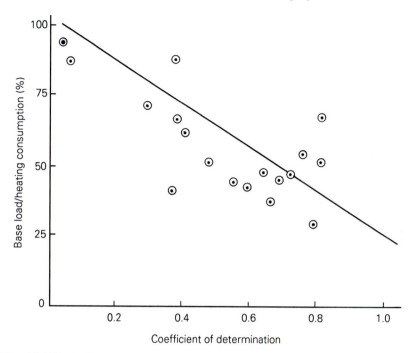

Fig. 14.17   Performance of heating schemes after efficiency measures.

Example 14.4

One of the large heating installations mentioned above was examined in detail. It consisted of five tower blocks of 214 dwellings supplied from a central boilerhouse with two 880 kW boilers and a 600 kW boiler. The calculated heat loss for the dwellings for an average inside air temperature of 19°C and a design outside air temperature of −1°C, assuming one air change per hour, was 979 kW. Assuming two air changes, the heat loss rose to 1117 kW. During the heating season, the heating was on continuously, although there were compensator and thermostat controls in the blocks. The system had the following consumptions.

Month	Consumption (therms)	Degree-days
June	5 737	81
July	4 754	25
August	5 901	24
September	5 737	61
October	16 720	97
November	21 966	241
December	18 688	273

Month	Consumption (therms)	Degree-days
January	27 868	367
February	28 360	337
March	18 196	252
April	22 786	221
May	15 737	134
Total	192 450	2113

Determine the performance line and analyse it in the light of the data. Take the heating season to be from 1 October to 31 May.

Solution

The performance line, to be determined by the method of least squares, is

$$y = sx + c$$

where y = monthly consumption (therms)
  x = monthly degree-days (K)
  s = slope (therms degree-day^{-1})
  c = intercept (therms)

For the least-squares analysis we have

$$\sum x_i = 1922 \quad \sum y_i = 170\,321 \quad \sum x_i^2 = 5.2 \times 10^5$$

$$\sum y_i^2 = 3.79 \times 10^9 \quad \sum x_i y_i = 4.35 \times 10^7$$

In equation (14.8) these values give

$$s = 44.4 \text{ therms K}^{-1}$$

and in equation (14.9) they give

$$c = 10\,620 \text{ therms}$$

The performance line is as shown in Fig. 14.14. The correlation coefficient is $r = 0.86$, indicating that 74% of the data is explained by the line, a reasonable fit.

The slope, s, should be related to the design heat loss of the dwellings, which was 979 kW for −1°C outside and 19°C inside. Under these design conditions for one day's continuous heating, with no heat gains considered, the balance point temperature of the building would be 19°C, and the degree-days would be 20 K. Hence the heat loss per degree-day would be

$$\frac{979 \text{ kW} \times 24 \text{ h}}{20 \text{ K} \times 29.3 \text{ kWh therm}^{-1}} = 40.1 \text{ therms K}^{-1}$$

With an air change rate of 2 h^{-1}, this would give 44.7 therms K^{-1}. This is very close to the performance line slope of 44.4, but it must be remembered that the air change rate is not known, and other factors such as boiler and system efficiency and controls will affect the slope and intercept [14].

The base load or intercept, c, should be close to the summer monthly consumption, but

$$\text{average base load} = \frac{\text{June + July + August + September}}{4}$$

$$= \frac{5737 + 4754 + 5901 + 5737}{4}$$

$$= 5532 \text{ therms}$$

which is well below the intercept at 10 620 therms. One reason for this may be that the dwellings were being overheated, as the boilers are capable of a combined output of 2360 kW, and the balance point temperature is above 15.5°C, the base temperature for the degree-days. The balance point temperature, and the degree-day base temperature to make the intercept equal the summer average base load, is found from the intercept of the performance line and the summer average base load:

$$5532 = 10\ 620 + 44.4x$$

$$x = -114.6 \text{ K}$$

To make the intercept coincide with the summer average base load at zero degree-days, 114.6 degree-days have to be added to each of the 8 points on the performance line. Hence the new degree-day total is

$$1922 + (8 \times 114.5) = 2838 \text{ K}$$

This produces a correction factor, φ, of

$$\frac{2838}{1922} = 1.48$$

From Fig. 14.11 this value of φ corresponds to a base temperature of 19.5°C. Considering that most of the dwellings will have heat gains to produce at least a 2–3 K room temperature rise, then it would seem that there is some overheating.

It is interesting to find what happens if the least-squares analysis is wrongly conducted on the whole 12 months' data. The performance line becomes

$$y = 70x + 3526$$

giving a higher slope nearer to an air change rate of 6 h^{-1} and an intercept below the summer average base load. The latter would indicate the need for a lower degree-day base temperature and a balance point temperature below 15.5°C. These facts show the errors due to the inclusion of the summer

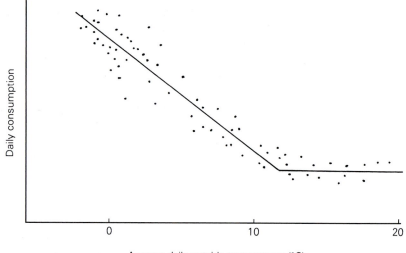

Fig. 14.18   An energy signature.

months' consumption. But as a lesson in interpreting statistics, the correlation coefficient is now 0.93, higher than the previous line, so giving a better fit to the data, even though some of the data should not have been included! Assuming that a building does have a good relationship with degree-days, then the performance line can be the basis for a future target. For instance, if the heating controls are set to maintain a lower inside temperature, then a target line below the current performance line can be set.

If the distribution losses are reduced by increased thermal insulation of pipework, then the intercept will be reduced, having a similar effect to setting down the temperature.

Further insulation of the building, however, will reduce the slope of the performance line as the building's heat loss is reduced.

If the heat loss of a building is known, perhaps from the design calculations for a new building, then a performance line can be derived from data collected by the BEMS and the slope compared with the heat loss.

Unfortunately, the ventilation and infiltration air change rate, N, will not be known and will probably vary with door and window opening. A pressurization test will indicate the air change rate [21], but this is difficult on large buildings and rarely done. Other factors affecting the performance line are the controls, solar radiation and wind (see later).

An alternative to the performance line that can easily be set up on a BEMS is the **energy signature**. As shown in Fig. 14.18, this consists of the daily consumption against the average outside daily temperature, t_{ao}. It yields the balance temperature of the building, which is the discontinuity in the curve. Summer consumptions can be included on this signature, giving further points above the balance temperature.

Unfortunately, the best-fit line for the signature by regression analysis is complicated by the discontinuity. The balance temperature needs to be known before the two parts of the line can be treated separately.

A method used in PRISM (PRIncetown Scorekeeping Method) determines the most appropriate base temperature for degree-days to give the best-fit straight line [17]. Summer consumptions are included in the data to determine the straight line, although the example above indicates that this might increase the correlation but give more error.

14.9   Fabric storage effects

All the methods that produce linear relationships between consumption and weather data are essentially steady state. This is satisfactory as long as dynamic effects, such as thermal storage in the building fabric (and variations in inside temperature due to controls), do not significantly affect the overall energy consumption. Over a short period of time, perhaps a day or so, the heat going into and out of the fabric will be a significant part of the energy consumption. Building fabric often has a time constant in terms of days (Chapter 9), and the heat stored and its take-up are affected by the outside temperature in the past, as well as the present. This has implications for the accuracy of degree-days over a few days and for degree-hours. The dynamic balance equation explains the effect mathematically. As in Chapter 9 the dynamic balance equation during the heating period is

$$pQ_{des} = \frac{H_{des}}{F_2} (t_{ai} - t_{ao}) + \sum m^*C_p \frac{dt_{ai}}{dT} \qquad (9.18)$$

where with the inclusion of the internal and external fabric of the building for storing heat, as well as the air, the total effective thermal mass of the building can be grouped into one summed term, $\sum m^*C_p$. The left-hand term is the heating system output, and the right-hand term is the steady-state heat loss and the storage of heat in the fabric and the air. In comparison with the fabric, the air stores very little heat. During the heating off period, the left-hand side is zero.

To determine the energy consumed, this equation must be integrated. The storage term (the differential term) is the only difference between equation (9.18) and the steady-state calculation for deriving the degree-day consumption. When the integration is over a considerable time (e.g. a week) then the weekly pattern is fairly stable and the storage effects do not impede the use of degree-days. But when individual days are considered then the storage of

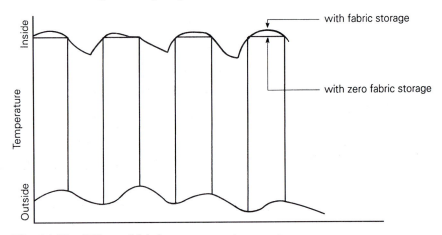

Fig. 14.19   Effect of fabric storage on temperature.

energy in the fabric can be a significant difference, even though the degree-days may be the same. This is especially true of a Monday morning after a cold weekend. The extra heat to warm up the fabric can be very significant, whereas for an equally cold day later in the week the energy consumption can be much less. An example will show this more clearly.

The influence of the thermal storage in a building's fabric is shown in Fig. 14.19, where the inside temperature graph of a building is compared to a theoretical building with no thermal storage (and perfect thermal control) that drops to the outside temperature when the heating is switched off.

Example 14.5
Compare the heat input to the office building in Fig. 9.12 at the start of the week to the input later in the week. Assume that the heating comes on after a weekend off, and that the heating is brought on by an optimizer before the start of occupancy at 9 am. Occupancy ends at 5 pm.

Compare this to an identical office with a larger effective thermal capacity and a time constant of 16 h. The heating plant is 25% oversized for both buildings.

Solution
From the example calculations on this building (Table 9.1) the heat stored in the fabric and air at design conditions was 1355 MJ, the fabric heat loss 48 kW and the air heat loss 16 kW. Hence the plant size is

$$(48 + 16) \times 1.25 = 80\,\text{kW}$$

By inspection the steady-state heat consumption over the occupancy time of one day, 8 h, operating at the design temperature difference of 20 K, is

$$80 \times 8 = 640 \text{ kWh}$$

$$= 2304 \text{ MJ}$$

So the heat stored is comparable to a severe day's consumption. Also as an indicator, the time for the heating system to get the heat into the fabric, if the steady-state heat loss were ignored, would be

$$\frac{1355 \text{ MJ}}{80 \text{ kJ s}^{-1}} = 16\,938\,\text{s}$$

$$= 4.7 \text{ h}$$

For more exact calculations, equation (9.27) gives the preheat time as

$$\frac{T_3 - T_2}{\tau} = \ln\left[\frac{pe^x}{\Phi + e^x(p - \Phi)}\right] \tag{9.27}$$

from this the preheat consumption is

$$pQ_{\text{des}}(T_3 - T_2) = pQ_{\text{des}}\tau \ln\left[\frac{pe^x}{\Phi + e^x(p - \Phi)}\right]$$

The consumption during the occupancy period should be the same for all days if the outside conditions are the same, provided that initially the heating has warmed the fabric sufficiently. With the short preheat times of some BEMSs, this will not be the case.

As the preheat consumption is the point of interest, the difference for a weekday preheat and the preheat at the end of a weekend is due to the different values of x in equation (9.27).

For the example office block with $\tau = 5.9$ h:

$$x_{\text{wkend}} = 64/5.9 = 10.85$$

$$e^{x_{\text{wkend}}} = 51\,534$$

and

$$x_{\text{week}} = 16/5.9 = 2.71$$

$$e^{x_{\text{week}}} = 15$$

So the ratio of weekend preheat to weekday preheat, with t_{ao} set to 0°C (i.e. $\Phi = 1.0$), is

$$\ln\left[\frac{1.25 \times 51\,534}{1 + 51\,534(1.25 - 1)}\right] \Big/ \ln\left[\frac{1.25 \times 15}{1 + 15(1.25 - 1)}\right] = 1.18$$

The weekend preheat uses 18% more energy than a weekday preheat. When the weather is milder, say, $t_{ao} = 10°C$ ($\Phi = 0.5$), the ratio is

$$\ln\left[\frac{1.25 \times 51\,534}{0.5 + 51\,534(1.25 - 0.5)}\right] \Big/ \ln\left[\frac{1.25 \times 15}{0.5 + 15(1.25 - 0.5)}\right] = 1.09$$

For the similar building with a time constant of 16 h, then

$$x_{wkend} = 64/16 = 4$$

$$e^{x_{wkend}} = 54.6$$

and

$$x_{week} = 16/16 = 1.0$$

$$e^{x_{week}} = 2.72$$

and with $\Phi = 1$ the ratio is

$$\ln\left[\frac{1.25 \times 54.6}{1 + 54.6(1.25 - 1)}\right] \Big/ \ln\left[\frac{1.25 \times 2.72}{1 + 2.72(1.25 - 1)}\right] = \frac{1.54}{0.703}$$

$$= 2.19 \qquad (14.10)$$

a considerable increase.

The numerator in equation (14.10) is 1.54, so the heating on Monday, including the preheating, is on for

$$(1.54 \times 16) + 8 = 32.6\,h$$

For Tuesday, and other weekdays under similar conditions, the heating is on for

$$(0.703 \times 16) + 8 = 19.25\,h$$

For the whole week, the ratio of Monday heating to that of the week is

$$\frac{32.6}{32.6 + (4 \times 19.25)} = 30\%$$

With 30% of the heating for the week being used on the Monday, and Sunday evening, a comparison of Monday's heating with that of Tuesday, for identical outside conditions, is not equal as a degree-day or degree-hour comparison would make it. For a number of outside conditions, the ratio is given in Table 14.1.

Monitoring with a BEMS should reveal the ratio of heating for the start of the week and other weekdays, and although outside conditions are unlikely to remain constant, it will give some insight into the fabric capacity and also the optimizer performance.

Table 14.1   Weekend-to-weekday preheat ratios

Outside temperature, t_{ao} (°C)	Weekend-to-weekday preheat ratio	
	$\tau = 5.9\,h$	$\tau = 16\,h$
0	1.18	2.19
10	1.09	1.71
15	1.08	1.65

It is often noticeable that the inside temperature of a building may creep up during the week as the fabric store of heat becomes 'full', as shown in Fig. 14.19. So degree-day and degree-hour calculations over days and hours may well give erroneous results due to this storage effect. For these short time periods, a more sophisicated model, such as the first-order model of equation (9.18), or a detailed dynamic computer simulation should be used.

14.10   Other influences on consumption

Degree-days are simply a measure of the outside air temperature, and although most of the above discussion of energy consumption has been related to degree-days, other weather influences such as solar radiation and the wind also affect consumption. These influences can partly be accounted for in the use of variable base degree-days, where the balance point temperature varies during the year. Controls and system efficiency also play a crucial role in a building's energy consumption for heating. All these influences are discussed in more detail in reference [18], but some of the salient details are considered here.

It has been shown [18], and confirmed [19], that there is a strong correlation between the monthly global solar radiation and the air temperature of the following month. The air temperature lags behind the solar radiation. This gives rise to a strong correlation between the degree-days for a heating season and the solar radiation. For nine months of the CIBSE Example Weather Year (1 October 1964 to 30 September 1965 [20]) for Kew, r^2 is 0.97 [17]. In general, the relationship between solar radiation and degree-days may be expressed as a straight-line equation:

$$S_m = S_0 - a\mathrm{DD}_{m+1}$$

where S_m = monthly global solar radiation for month m (W m^{-2})
$\quad\quad\quad S_0$ = intercept of straight line (W m^{-2})
$\quad\quad\quad a$ = slope of line (W m^{-2} K^{-1})
$\quad\mathrm{DD}_{m+1}$ = degree-days for month $m + 1$ (W m^{-2})

A **solar aperture** term can relate the useful contribution of the sun to the building's heating. Although the determination of a solar aperture for a

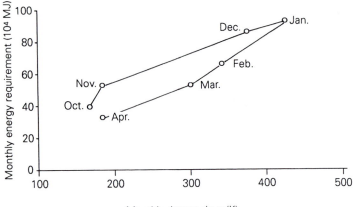

Fig. 14.20 Effect of large solar gain on the performance line.

building is complex [14], its dimensions are m^2 and it can be related to the building's glazed area.

Even though a number of papers report that the solar influence is difficult to detect from energy data [14], if a building were to use a lot of solar radiation then its performance line would exhibit the type of hysteresis shown in Fig. 14.20.

In April and November the degree-days were similar but the solar contributions were not, hence the difference in energy consumption. For Fig. 14.20 the building had an assumed solar aperture of 243 m^2 with 608 m^2 of glazing, of which 190 m^2 was facing south-east.

The effect of wind is more difficult to analyse. It affects the external surface heat transfer, as well as the infiltration through cracks and openings. Infiltration is influenced directly by the wind speed and also by the temperature of the outside air, due to the **stack effect**. The hot air rises in the building and generates a pressure gradient to pull cold air in at the lower openings and expel the warm air through higher openings [21]. Although the temperature difference can be related to degree-days, the wind velocity is not so well related [16], and the use of dynamic thermal models with comprehensive weather data would be more useful if the analysis were to be in greater depth than a simple degree-day assessment.

14.11 The effects of efficiency and controls

The heating system, and its controls to provide a constant required inside temperature, has a higher efficiency at the higher heat loads. These heat loads can be related to the outside air temperature and hence the degree-days. So

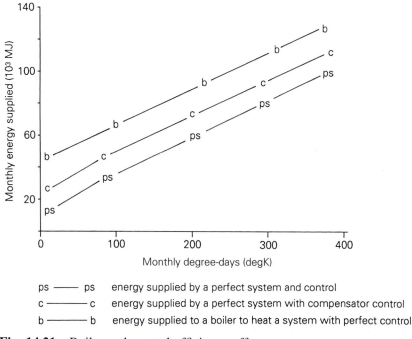

ps ———— ps   energy supplied by a perfect system and control

c ———————— c   energy supplied by a perfect system with compensator control

b ———————— b   energy supplied to a boiler to heat a system with perfect control

Fig. 14.21   Boiler and control efficiency effect.

the efficiency effect of the heating system and controls will reduce the slope of the performance line while raising its intercept (Fig. 14.21) compared to the useful heat required, determined from the heat loss calculation.

Derived from reference [16], Fig. 14.21 relates to a heating system with a compensator control and a single boiler. The inefficiency of the compensator relates to the rise in room temperature with rising outside air temperature and lowering load (Chapter 8).

14.12   Boiler signature

The boiler inefficiency can be determined from a steady-state boiler model whose parameters can be determined from a boiler energy signature (Fig. 14.22). The parameters are as follows:

(i) $\bar{Q}_{fuel}$ = average fuel supplied to the boiler in a certain time, including any boiler off time in the period (kJ)

(ii) $\bar{Q}_u$ = average useful heat supplied by the boiler to the heating system in a certain time, including any boiler off time in the period (kJ)

Such a boiler energy signature can be determined from a BEMS which monitors both the fuel input and the heat output. They are determined over a

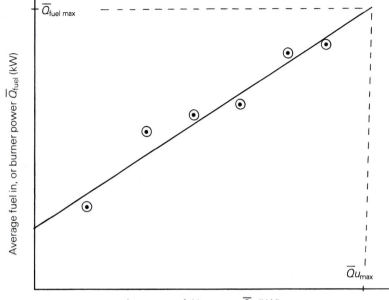

Fig. 14.22   Boiler energy signature.

period of time to give $\bar{Q}_{fuel}$ and $\bar{Q}_u$, which can be used to obtain the valuable parameter of boiler efficiency. If there are a number of measurements with an amount of experimental error, it will be necessary to determine the best straight line using the least-squares method.

But the boiler efficiency curve can also be determined from the energy signature and the load factor, φ. Load factor is more difficult to determine, but it can be obtained using a knowledge of the system design, including the flow rates, pipework and valves, and monitoring the principal flow and return temperatures.

If these factors are unknown, then the two extreme points of the signature may be determined. The full load, top point of the line (when $\varphi = 1.0$) is

$$\eta_{max} = \frac{\bar{Q}_{u\,max}}{\bar{Q}_{fuel\,max}}$$

where η_{max} is the maximum efficiency.

The intercept of the signature (when $\varphi = 0$) is given by

$$\bar{Q}_{fuel\,max}\frac{\beta}{1+\beta}$$

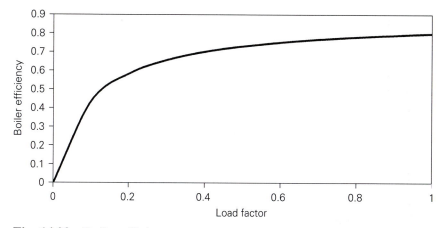

Fig. 14.23 Boiler efficiency curve.

where β is the ratio of heat loss through the boiler case and up the flue when the boiler is not fired, to $Q_{u\ max}$.

The boiler efficiency curve, in terms of the above parameters, is given by

$$\eta = \eta_{max}\left[1 - \frac{(1 - \varphi)\beta}{\varphi + \beta}\right]$$

A graph of this equation is shown in Fig. 14.23.

14.13 Normalized performance indicators

Once the BEMS has acquired a year's consumption, the performance of the building and plant can be assessed against performance yardsticks set by the Energy Efficiency Office and published in a series of twelve booklets entitled *Energy Efficiency in Buildings* [22].

To determine the building's performance against these yardsticks, its **normalized performance indicator** is calculated. For this, the heating consumption is adjusted for a standard degree-day year, with allowance for exposure. This is then added to the rest of the consumption and the resulting total is normalized for the area of the building, hours of occupancy and type of use of the building (e.g. whether a primary or secondary school).

This assessment may well feed back as adjustments to settings and controls on the BEMS and other efficiency measures with the production of energy targets. However, although the BEMS and the other measures may well be capable of achieving these savings, the human factor must be considered [23], as we discussed at the beginning of this book.

References

[1] CIBSE (1991) CIBSE Applications Manual, Energy audits and surveys, AM5, Chartered Institution of Building Services Engineers, London.

[2] Department of the Environment, Transport and the Regions (1993) *Energy Audits*, Fuel Efficiency Booklet 1, Department of Energy, London.

[3] Hughes, G. J. (1988) *Electricity and Buildings*, Peter Peregrinus, London.

[4] Fielden, C. J. and Ede, T. J. (1982) *Computer-based Energy Management in Buildings*, Pitman, London.

[5] Eyke, M. (1988) *Building Automation Systems*, BSP Professional Books, Oxford.

[6] Bentley, J. P. (1988) *Principles of Measurement Systems*, Longman Scientific and Technical, London.

[7] Hitchin, E. R. and Hyde, A. J. (1979) The estimation of heating energy use in buildings, Symposium on the Environment Inside Buildings, Institute of Mathematics and its Applications, Southend-on-Sea, May.

[8] Hyde, A. J. (1980) Degree-days for heating calculations. *Building Services and Environmental Engineering*, **2**(6).

[9] CIBSE (1986) CIBSE Guide, Volume B, Section B18, Owning and operating costs, Chartered Institution of Building Services Engineers, London.

[10] Hitchin, E. R. (1983) Estimating monthly degree-days. *Building Services Engineering Research and Technology*, **4**(4).

[11] ASHRAE (1997) ASHRAE Handbook Fundamentals, American Society of Heating, Refrigeration and Air Conditioning Engineers, Atlanta.

[12] CIBSE Building Energy Code, Part 2, Calculation of energy demands and targets for the design of new buildings and services, Section A, Heated and naturally ventilated buildings, Chartered Institution of Building Services Engineers, London.

[13] Billington, N. S. (1966) Estimation of annual fuel consumption. *JIHVE*, **34**.

[14] McNair, H. P. and Hitchin, E. R. (1980) The principal factors that influence gas consumptions in centrally heated houses. *Journal of the Institution of Gas Engineers*, December.

[15] Levermore, G. J. (1987) The performance of group heating schemes, Combined Heat and Power Association Conference, Torquay, July.

[16] Levermore, G. J. and Chong, W. B. (1989) Performance lines and energy signatures. *Building Services Engineering Research and Technology*, **10**(3).

[17] Fels, M. F. (1986) An introduction. *Energy and Buildings*, **9**(1–2).

[18] Owens, P. G. T. (1980) Figuring out energy needs. *Building Services* (CIBSE), **2**(7).

[19] Hay, N. and Levermore, G. J. (1988) Reduced weather data for heating calculations, Weather Data Symposium, Chartered Institution of Building Services Engineers, London, May.

[20] Holmes, M. J. and Hitchin, E. R. (1978) An example year for the calculation of energy demand in buildings. *Building Services Engineer*, **45**.

[21] Liddament, M. W. (1986) *Air Infiltration Calculation Techniques – an Applications Guide*, Air Infiltration and Ventilation Centre, Warwick.

[22] Department of Energy (1990) *Energy Efficiency in Buildings*; there are 12 booklets, entitled How to bring down energy costs in schools; Catering establishments; Shops; Further and higher buildings; Offices; Sports centres; Libraries, museums, art galleries and churches; Hotels; High street banks and agencies; Entertainment; Factories and warehouses; Courts, depots and emergency services buildings. The Department of the Environment, Transport and the Regions' Energy Efficiency Best Practice programme has many publications on energy efficiency available from BRECSU and ETSU (www.bre.co.uk/brecsu/ and www.etsu.com/eebpp/home.htm).

[23] Levermore, G. J. (1985) Motivation and Training for Energy Targeting, Energy Management Experience Conference Proceedings, London.

[24] Colantonio, E. S. (1989) *Microcomputers and Applications*, D. C. Heath, Lexington MA.

[25] Energy Efficiency Office (1993) *Degree Days*, Fuel Efficiency Booklet 7, Department of the Environment, Transport and the Regions, London.

[26] Hansen, E. G. (1985) *Hydronic System Design and Operation: Guide to Heating and Cooling with Water*, McGraw-Hill, New York.

[27] McLaughlin, R. K., McLean, R. C. and Bonthron, W. J. (1981) *Heating Services Design*, Butterworths, London.

[28] CIBSE (1999) CIBSE Building Energy Code 1, Energy demands and targets for heated and ventilated buildings, Chartered Institution of Building Services Engineers, London.

Index